Laboratory Manual for Organic Chemistry: A Microscale Approach

William R. Moore
West Virginia University

Anthony Winston
West Virginia University

The McGraw-Hill Companies, Inc.
New York St. Louis San Francisco Auckland Bogotá Caracas
Lisbon London Madrid Mexico City Milan Montreal New Delhi
San Juan Singapore Sydney Tokyo Toronto

Laboratory Manual for Organic Chemistry:
A Microscale Approach

4 5 6 7 8 9 0 QPD/QPD 9 0 9

ISBN 0-07-043052-7

The editor was Denise T. Schanck;
the production supervisor was Elizabeth J. Strange.
Quebecor-Semline was printer and binder.

Preface

This manual is a result of a six-year period of developing experiments for use in our undergraduate organic chemistry laboratories. The laboratory is important in emphasizing to students that they can be successful in a program, that not only reinforces the theory they learn in lecture, but also convinces them that they are capable of doing meaningful "hands-on" experimental work. The laboratory experience can be applied and extended to many different fields. Most students of chemistry will not become chemists, or even chemistry majors, but instead will enter another science-based area. The technique and methodology of problem solving that one learns in the laboratory, through the collection and interpretation of experimental data, will be very useful in coping with the problems faced in other fields as well as in chemistry.

We have tried to keep the manual concise to keep students focused on what they are actually doing. With a few exceptions duplication of discussions found in lecture texts have been minimized. The exceptions reflect our experience that in exercises on stereochemistry and spectroscopy, students are helped by having the discussion in hand while performing the exercises. Because students are novices in the laboratory, and some have a knack for "going wrong," fairly detailed experimental procedures have been provided, particularly in the earlier experiments. The experiments have been field-tested at the microscale level over a period of several years, and the majority of students should be successful in doing these experiments, and in acquiring increased confidence in their abilities.

With a few exceptions noted in the Instructor's Manual, each experiment is designed to be completed in one three-hour laboratory period. The average student should be able to complete the work on schedule, perhaps deferring an occasional melting point or infrared spectrum to be run during a subsequent experiment. In some experiments several alternative procedures are provided. In these cases it is not the intention that a student be required to perform every procedure. Some students may be assigned one part of the experiment, while other students are assigned other parts.

A central element of this manual is the use of the discovery, or investigative, approach by incorporating unknowns into many experiments. With this in mind, melting points, boiling points, and other physical data are not provided for the synthetic products. The only relevant physical constants are those determined by the students in the laboratory. These physical constants are available in the Instructor's Manual. In some experiments, "unknown" compounds are carried through the general procedures to generate products that must be characterized by their physical constants.

The terms "macroscale" and "microscale" have been widely applied to describe the quantities of materials used by each student. In this text, most experiments are on the millimole scale, which is 0.5 to 3 mmol, or 50 to 300 mg, of reagents. This scale is not greatly different from what practicing synthetic chemists use in exploratory research. At this level students will be able to isolate and characterize products conveniently and the quantities are small enough to reduce significantly the amounts of reagents and solvents consumed. An extremely important factor in reduced-scale operation is that the quantity of waste material generated is dramatically reduced over that encountered in performing experiments on macroscale. Waste disposal is a critical environmental consideration and the cost of dealing with waste products safely has risen to the point where it often costs as much to dispose of a chemical as it does to buy it in the first place. But because students still generate waste, we remind them repeatedly that all waste chemicals must be discarded in the appropriate waste containers provided in the laboratory.

Safety is an integral part of the experimental program. While the ultimate responsibility for safety lies with the institution and the instructors, we have included a core of safety rules and practices that should be observed in every laboratory program. Operating on a small scale has obvious advantages in terms in minimizing potential for serious accidents. While safety is emphasized, we avoid giving the impression that students are constantly facing life-threatening situations. Throughout, we have emphasized safe laboratory practices. If students learn to handle all chemicals with respect, they will be less apt to make mistakes when they are handling potentially hazardous materials. To ensure that they are alert, we have called attention to all chemicals that require special handling or precautions.

Today we must provide proper ventilation in laboratories. Reduced-scale operations may make it easier to avoid a contaminated laboratory atmosphere, but all operations involving volatile or potentially hazardous chemicals must be performed in a hood. This point is emphasized throughout the experimental discussions.

While certain "technique experiments" are presented on a macroscale level, all of the reaction chemistry is on a millimole scale. The equipment recommended is typical ground glass microscale glassware obtainable from several manufacturers. A relatively small amount of individual desk equipment is needed because some items are used only occasionally and can be handled quite well as common equipment, available at each work station.

Certain instruments are essential: balances weighing in milligrams, centrifuges, ovens, refractometers, and gas chromatographs. Infrared spectroscopy is used extensively, including fast FT-IR methods for characterizing products and unknowns. An experiment that combines the use of IR and NMR spectroscopy to determine molecular structures is included. However, we realize that NMR spctrometers are not yet available in all undergraduate programs, so the experiment is designed to allow students to work most of the problems with NMR spectra provided in the laboratory. These spectra recorded at 270 MHz, along with the corresponding FT-IR spectra, are available in the instructor's manual.

The nature and order of the experiments have been correlated with a typical lecture sequence. Stereochemistry has been included in a session that deals with molecular models in order to ensure that all students actually work with the models. Some instructors may have students do part, or all, of the experiment out-of-class, or in a separate non-laboratory session. In addition, a number of nontraditional experiments have been included, such as the study of dyes, the basis of color photography, and the synthesis of a liquid crystal. These experiments, along with an experiment on polymers, illustrate important synthetic chemistry leading to materials of commercial significance.

Report sheets for individual experiments have been included at the end of the manual, but are not intended to replace the laboratory notebook. The notebook should be detailed and complete, while the report sheets are to be used to summarize the results for the convenience of the instructor and the student.

William R. Moore

Anthony Winston

Acknowledgments

The authors thank the following persons for critical reviews of portions of the manuscript:

Ardeshir Azadnia
Michigan State University

Robert G. Bass
Virginia Commonwealth University

Albert W. Burgstahler
University of Kansas

Kent Clinger
David Lipscomb University

James R. Cox
University of Houston

Joseph M. Crockett
Bridgewater College

Jeffrey P. Fitzgerald
U.S. Naval Academy

Henry L. Gingrich
Princeton University

Floyd W. Kelly
Casper College

Robert D. Miess
Mayville State University

Anne B. Padias
University of Arizona

Douglas F. Taber
University of Delaware

James M. Takacs
University of Nebraska—Lincoln

We thank the staff of McGraw–Hill, Inc. for their encouragement and help in the organization and preparation of the manuscript and especially Christopher Fitzpatrick, Margaret Hollander, Shelley Kronzek, James M. Morriss, Nancy Rizza, Denise Schanck, and Jennifer Speer.

ABOUT THE AUTHORS

WILLIAM R. MOORE is a native of rural Minnesota where he received his elementary and secondary education. He earned a B.S. Degree in Chemistry from the University of California at Los Angeles in 1950 and a Ph.D. Degree in Organic Chemistry at the University of Minnesota in 1954. He then did postdoctoral research at the Massachusetts Institute of Technology on medium rings and elimination reactions and also designed and built gas chromatographs before they were commercially available while working with the late Professor Arthur C. Cope. In 1956 he joined the faculty at MIT and also married the former Phylis Grace Tocco, who earned her Ph.D. with Professor Cope in 1959. The author's research has included the first separation of the hydrogen isotopes and nuclear spin isomers by gas chromatography. His research in organic chemistry, supported by the Petroleum Research Fund of the American Chemical Society, the National Science Foundation and the National Institutes of Health, has dealt with reaction mechanisms, synthetic methods, carbenes and strained molecules. In 1972 he came to West Virginia University as Chairman of the Chemistry Department, a role he served until 1986. He is currently a Professor of Chemistry at WVU active in teaching and the undergraduate laboratory program. He has taught both organic and general chemistry and has directed undergraduate laboratories for nearly forty years. His interest in microscale methods, initiated by his colleague Dr. Winston, has been a focal point in recent years. He has also been an avid skier for forty years and he and his wife are dedicated gardeners.

ANTHONY WINSTON is a native of Washington D.C. He received a B.S. Degree in Chemistry from George Washington University in 1950 and an M.S. Degree (1952) and the Ph.D. Degree (1955) from Duke University. After several years as an industrial chemist, he came to West Virginia University in 1959. He was Chairman of the Chemistry Department from 1986 to 1990, and is now Professor Emeritus of Chemistry. He has over 30 years experience in the teaching of organic chemistry laboratories, including twelve years using microscale techniques. His research is in the areas of organic and polymer synthesis, with particular interest in the design of polymers for medical applications supported by the Cooley's Anemia Foundation of New York and the National Institutes of Health. He has published in the areas of organic chemistry, polymer chemistry, and microscale techniques. He is active in the American Chemical Society and has held offices in the local section and the Central Region. He is Director of the Microscale Institute at West Virginia University and has been an instructor in the summer microscale workshop for 10 years. A recent interest is the history and chemistry of the invention of photography. As a member of the Speakers Service of the American Chemical Society, he lectures to local sections on the chemistry of antique photographic processes. He is an amateur photographer, specializing in black and white prints. He has received many acceptances, medals, and other awards for prints submitted to international competitions approved by the Photographic Society of America.

Table of Contents

1. Safety in the Laboratory

A. Safety Rules and Regulations

There are a variety of real or potential hazards associated with laboratory work. To reduce the chance of personal injury and to avoid accidents, all laboratories have established safety rules and approved laboratory practices. Summarized below are relevant safety rules and practices which are enforced in most laboratories. Safety is the responsibility of all of the people working in the laboratory, students and instructors. Your instructors will provide you with information and guidance, but you are responsible for your own actions. The material which follows must not be read casually; you must study it carefully so that following proper safety procedures becomes automatic. Any additional safety rules and procedures established at your institution must always be obeyed.

Safety Rules

Approved eye protection must be worn at all times in the laboratory when experimental work is in progress.

Contact lenses should not be worn in the laboratory. If you have conventional spectacles, wear them under your goggles. If you must wear contact lenses, inform your instructor and wear properly vented goggles.

A laboratory apron or coat must be worn. Shoes are required (sandals should be avoided). Laboratory gloves will be available and should be worn when you are handling corrosive or irritant chemicals.

Food and beverages are not permitted in the laboratory.

Smoking in the laboratory is prohibited.

Unauthorized experiments and the use of unauthorized procedures are strictly forbidden.

Chemical spills or other accidents should be promptly reported to your instructor. Know the location and use of the eye wash, the safety shower, and the fire extinguisher in the laboratory.

Injuries, however slight, should be immediately reported to the instructor.

Do not use a burner in an organic laboratory without approval of your instructor — most solvents are highly flammable.

You are not permitted to work in the laboratory unless your instructor is present. Practical jokes, "horseplay," etc. are forbidden. Remember other students have a right to work in a safe environment.

Inform your instructor if you have a particular sensitivity to any chemical that might be encountered in the laboratory.

Laboratory Practice

Reagents: Use only the dispensing pipets that are supplied to measure out liquid reagents. Do not return reagents to the reagent bottles. Take only what you need. Do not waste reagents. Do not take reagent bottles to your desk. Leave the reagent bottles where they have been placed for common usage (side shelves, hoods, laboratory carts, etc.).

Waste chemicals must be placed only in the appropriate labeled containers provided in the laboratory. Do not discard waste chemicals in trash baskets or pour them down the drain.

Broken glassware must be placed in the container provided in the laboratory. Examine your equipment regularly for damage. Do not use cracked glassware.

Common equipment must be returned to the proper storage area or to your instructor at the end of the laboratory period.

Glassware should be cleaned at the wash-up sinks using the cleaning agents provided and the methods outlined by the instructor.

Wash your hands with soap and water periodically during the laboratory, and always at the end of the laboratory period. Wash your gloves before removing them and then wash your hands.

Clean-up — Keep your work area at your desk and hood clean and neat. Deposit waste materials in the correct containers. Before leaving the laboratory, clean all work areas with a sponge.

B. Accidents

1. Exposure to Chemicals can take several forms:

a. Chemicals in the Eyes: To avoid accidents that result in chemicals in the eyes, **safety goggles must be worn at all times in the laboratory.** There will be no exceptions. Failure to follow this rule can result in serious injury, even from substances that are not particularly dangerous otherwise. **If any material should splash into the eyes, wash immediately with copious amounts of water.** If an eyewash fountain is available, learn its location and use it if necessary.

b. Chemicals on the Skin. Chemicals are often absorbed readily by the skin and may result in an adverse reaction. Some people are more susceptible than others. Some may even experience an allergic reaction. In general, avoid contact with any chemicals found in the laboratory. If you do come in contact with a chemical, wash the skin thoroughly with soap and water. If you experience any reaction to a chemical, inform the instructor. If, in any experiment, there is a chemical that is especially dangerous, **this fact will be indicated in the procedure in boldface type.** Be sure to follow all of the indicated precautions.

c. Chemicals in the Lungs: Many chemicals are volatile and can generate vapors which will circulate in the air waiting for someone to breath them. Although most individuals can tolerate varying amounts of chemical fumes without adverse reaction, it is always important to minimize exposure. Since many vapors are toxic, several precautions must be taken. The most important precaution to take is to **conduct all chemical reactions and manipulations in the hood.** (Efficient hoods should be available at each laboratory station.) **Use the hoods for all processes where there is a possibility of toxic fumes.**

2. Burns;

Burns are a common source of injury. Burns can arise from gas burners, hot-plates, hot glassware, hot ring stands, etc. Gas burners are only seldom used, and then usually for sealing melting point capillaries or for drawing-out melting point tubes to make spotting pipets for thin layer chromatography. Burners must never be used when volatile flammable substances are in the immediate vicinity of the flame. Assume all hot-plates are hot. If there is any uncertainty, test carefully. The heating blocks, or sand baths, may also be hot. Be careful with them. Allow glassware, heating blocks, sand baths, and other equipment to cool before handling.

3. Cuts and Puncture Wounds:

Wounds can occur from broken equipment or by being jabbed with a hypodermic needle. Handle glassware carefully; not only is most of it expensive, but broken glassware can cause injuries. If some of the microscale glassware does not come apart easily after completion of a reaction, e.g. "frozen" ground glass joints, then see your instructor for assistance.

4. Handling of Hypodermic Syringes:

Standard hypodermic needles are very sharp and can cause injury. It is advised that all needles be blunted before use to reduce the possibility of injury. The sharp point of the needle is removed by cutting and grinding to give a flat blunt end. Such needles still pierce septa, but are far less apt to cause inadvertent "needle-sticks." Never use a hypodermic syringe as a means for measuring ordinary liquids. Use a micropipet for this. **Use the syringe only when specifically indicated in the experimental procedure**. And then follow the instructions carefully. When carrying reagents in a syringe from a reagent bottle to your work station, carry the syringe in a test tube, needle end down, to avoid the possibility of inadvertently "sticking" someone.

5. First Aid:

All colleges and universities should have established rules, regulations, and a protocol for handling an accident. The extent of first aid recommended or allowed will vary from institution to institution. Your instructor will be responsible for seeing that the local procedures are followed in the event of an accident. Your role will be to provide any assistance requested by your instructor.

C. The Chemicals We Use

The dangers associated with specific chemicals are described by the terms toxic, flammable, corrosive, and reactive. These terms, which are found on the labels provided by chemical manufacturers, give a quick general indication of the hazards associated with the substance. For every chemical used by any employee, federal law (OSHA) requires that a Material Safety Data Sheet (MSDS) be available describing the physical and chemical properties, including flammability, reactivity, accepted exposure limits, and potential health hazards. Thus, the information reported on an MSDS gives a starting point for assessing the relative hazards involved in using any given chemical.

In commercial manufacturing, or research laboratories with experienced personnel, very hazardous chemicals are used routinely. In the undergraduate laboratory, on the other hand, very hazardous compounds are not likely to be employed. Because of the fact that some chemicals which are extremely hazardous receive extensive publicity, e.g. methyl isocyanate (MIC), there is a tendency to view all chemicals with alarm. This view ignores a reality of life: all material substances are comprised of chemicals. The fact that a substance is naturally occurring does not guarantee that it is non-hazardous. For example, nitric oxide is a highly toxic chemical, but recent research has established that nitric oxide is an important compound in human biochemistry. Oxalic acid is toxic (orally in rats, LD_{50}: 480 mg/kg body weight), yet at the level found in spinach and rhubarb, it does not present a health hazard in a normal diet. Many people experience allergic reactions to common substances such as certain foods, cosmetics, detergents, cats, dogs, etc. Extreme sensitivity to a particular substance is best treated by avoidance of that substance.

In the laboratory, the only acceptable approach is to treat all chemicals with respect. By this we mean that all chemicals should be handled carefully. No rational person is going to eat or taste any chemical in the laboratory. But you must also strive to keep all chemicals off your skin. Some chemicals are irritants and some people are particularly sensitive to certain chemicals. Good ventilation is mandatory in a modern laboratory to eliminate the inhalation of fumes released by many substances and we repeatedly emphasize that **all work with volatile hazardous substances should be carried out in a hood.**

The microscale experiments described in this manual greatly reduce the potential hazards in the laboratory because of the small quantities of materials employed. Microscale procedures also reduce the amount of waste generated and dramatically reduce the problems and costs associated with waste chemical disposal.

D. Waste Disposal

Every institution is legally (and morally) responsible for establishing and enforcing correct procedures for the collection and disposal of all waste chemicals generated in teaching and research laboratories. Each student must understand and follow a cardinal rule:

All waste chemicals, including solvents, must be disposed of in appropriately labeled waste containers provided in the laboratory.

There should be separate containers for aqueous solutions, organic liquids and solvents, organic solids, inorganic solids, and contaminated paper products. Although it may seem repetitious, we will remind you in each experiment where wastes are generated that you have an obligation to correctly dispose of waste materials.

Designated laboratory personnel (employees) should remove waste containers regularly and dispose of these materials following the procedures and precautions outlined in *Prudent Practices for Disposal of Chemicals from Laboratories*, National Academic Press, Washington, D.C., 1983. In general, waste materials should be processed by established procedures before being consigned to an authorized disposal agency. The latter will dispose of waste materials according to federal, state, and local codes.

Students will often participate in the preliminary processing of waste materials by converting excess reagents to less potent end products, which can then be discarded safely.

2. Introduction

A. Glassware

1. Microscale Kit

A typical microscale kit consists of a two 5-mL conical vials, one 3-mL conical vial, a 10-mL round-bottom flask, water condenser, air condenser, drying tube, Craig tube, Hickman still, Claisen head, spin vane, hypodermic syringe, and needles. Assorted screw caps, O-rings, and septa are also included. Some of this equipment is shown on the following page.

2. Connections

Connections between the various items of equipment are made using the ground glass joints that are held together by screw caps. To make the connection, first place a screw cap on the inner ground glass joint and then push an O-ring completely over the joint. Place the inner joint into an outer joint and turn the screw cap so that the joints are firmly connected. By this technique, one can handle the assembly easily without fear that the joints will come apart. With an O-ring in place, grease is seldom needed for sealing the ground glass joints.

B. The Laboratory Notebook

A **bound** 8"x10" hard-cover laboratory notebook is necessary to report the procedures and the results of the experiments. In the notebook, keep an account of your laboratory experience by means of a formal write-up of each experiment. All entries should be **made in ink** (not pencil). Record your name on the outside front cover and also on the inside cover, together with your address and telephone number in order to help recover your notebook in case of loss. Number the pages of the notebook and leave a couple of pages blank in front to serve as a table of contents.

Experiments are to be written-up in the following manner:

Before the laboratory period: Write the title of the experiment on a fresh page, together with the date. Write the equations for all reactions that are taking place. Calculate and list all **pertinent** molecular weights of chemicals used. List the weights of materials to be used in the experiment and the corresponding number of moles, or millimoles. Answer prelab questions that are found at the end of the experiment.

During the laboratory period: Have your notebook with you in the laboratory while you are working. As you perform the experiment, write down everything that you do that is pertinent to the experiment: including the weights of materials that you actually use and the procedure in detail as you perform the experiment. It is important that you record what **you** do and what happens in **your** experiment. Do not write what is supposed to happen unless it really happens. **This is a record of your laboratory experience,** not someone else's. Write down the results as they occur **directly in your notebook,** not on scraps of paper. Calculate yields and other pertinent data and enter these directly in your notebook. Show the calculations, so that you can check the results later if necessary. **Do all this during the laboratory period, not later on at home**.

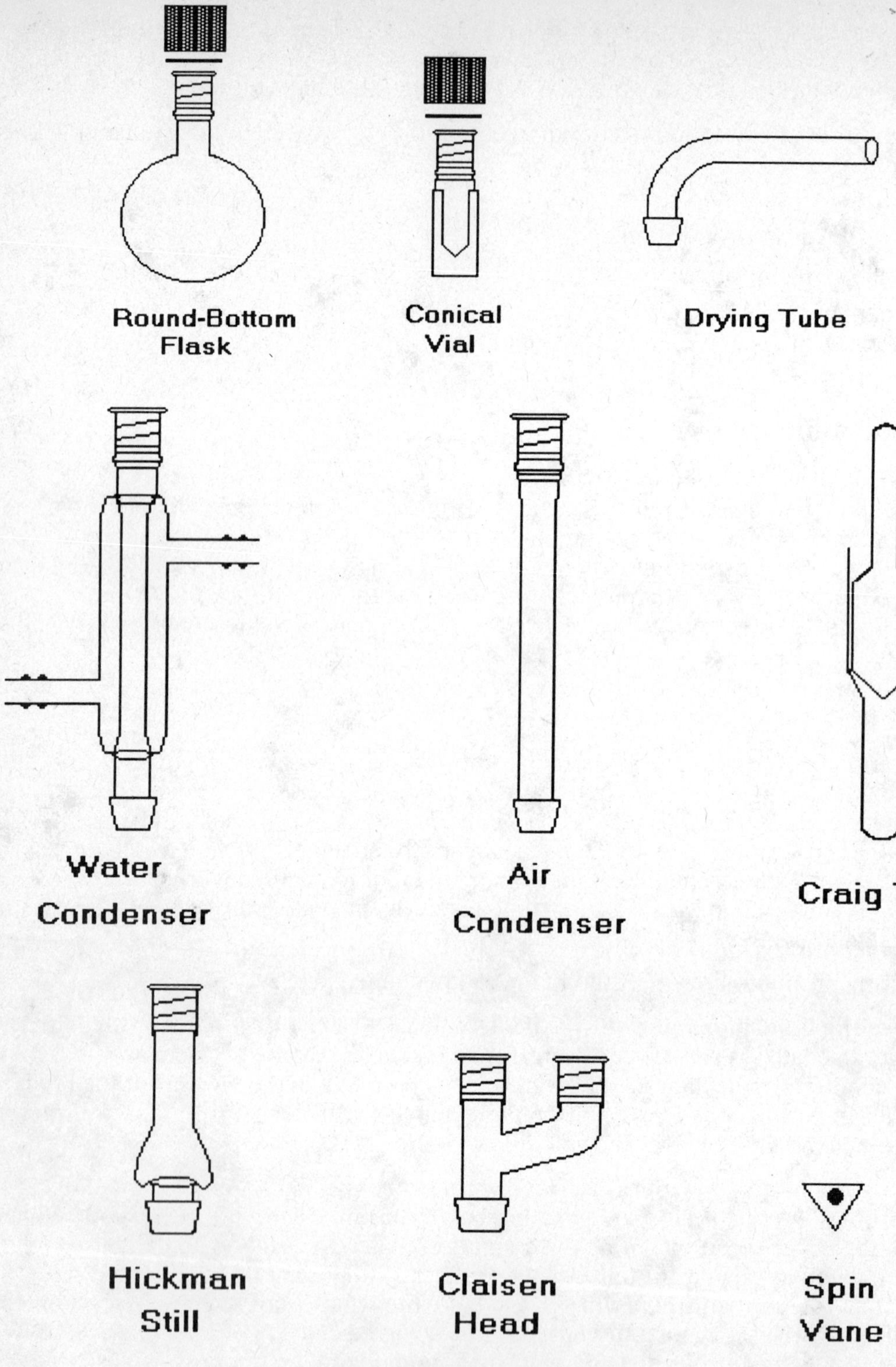

Microscale Glassware

After or during the laboratory period: Prepare graphs, if needed; do more complex calculations; add anything pertinent that is not already included. Write out any conclusions or explanations that may be necessary. Answer the postlab questions.

Remember the notebook is a daily record of your experiences. Prepare the write-ups carefully.

C. Reports Forms

Report forms for all experiments are included in the Appendix and serve as a guide to the data that will be needed. These forms may be used at the discretion of the instructor, who may require that they be turned in. They are not meant to take the place of the laboratory notebook. All data must still be recorded in the notebook at the time it is obtained.

D. Attendance

Regular attendance in the laboratory is required. If, for some reason, you cannot be present, usually because of illness, be sure to inform your instructor, in person or by telephone, as soon as possible.

E. Basic Laboratory Procedures

1. A Typical Reaction Set-up

Reactions are normally carried out in the conical vials with stirring by means of a magnetic spin vane and a magnetic stirrer. Often the reaction mixtures are heated to the boiling point of the solvent. In order to prevent the vapors from escaping into the laboratory, either an air-cooled or a water-cooled condenser is used to condense the vapors to liquid and to allow the liquid to flow back (reflux) into the vial. An air condenser is used when the temperature of the refluxing liquid is fairly high. In the case of low-boiling liquids, the water-cooled condenser is used. Deciding upon which condenser to use comes with experience, but if you find that the air condenser does not condense all the vapors, then you should switch to the water-cooled condenser. An air condenser is not recommended for liquids boiling below 100°C.

Often it is necessary to run reactions in such a way that water vapor from the air does not come into contact with the reaction mixture. In this case the drying tube is connected to the top of the condenser or to any other compatible fitting. The drying tube is filled with a drying agent, such as anhydrous calcium chloride, which is prevented from falling out of the tube by two cotton plugs, one near the joint and the other at the open end. Any air that passes through the drying tube is dried by the calcium chloride.

Sometimes it is necessary to add a reagent to a reaction mixture while the mixture is refluxing or distilling. In this event the Claisen head is a useful item. The Claisen head is first connected to the vial. Then a condenser, Hickman still, or drying tube is connected to one of the top fittings of the Claisen head (usually the right hand one in the drawing) and a screw cap and septum is connected to the other fitting. The reagent can then be added to the reaction mixture through the septum by means of a hypodermic syringe and needle.

If volatile liquids are being removed by evaporation, or if noxious fumes are being evolved in the reaction, then the procedure must be carried out in a well-ventilated hood. On larger-scale reactions, it is better to remove volatile solvents by distillation.

2. Cleaning Glassware

Glassware may be cleaned by rinsing with acetone or alcohol. Water may be used, but is effective only when the compounds are water soluble. Clean glassware immediately after the experiment is completed so that the glassware will be dry by the next laboratory session. If more rigorous drying is needed, the glassware may be placed in an oven set at about 80°C. Drying glassware with towels is not recommended because glassware will often break if a towel is forced into it.

3. Use of the Hickman Still, Distillation

In normal use, the Hickman still is connected to a vial to serve as a distillate collector. A liquid is placed in the vial together with a spin vane or a boiling chip. On heating, the liquid distills and is collected in the cup of the Hickman still. As the liquid collects in the cup, it is removed from the top with a micropipet. In the case of low-boiling materials, a water-cooled condenser may be attached to the top of the Hickman still to improve condensation. The Hickman still works well for the distillation of small amounts of material, from 0.3 to 2 mL. It is particularly useful for purifying liquid starting materials for reactions.

4. Use of the Craig Tube, Crystallization

The Craig tube is illustrated in Figure 3 of the chapter on crystallization.

Crystallizations on a small scale can be conveniently carried out in the Craig tube. A small amount of the solid is placed in the Craig tube along with about 1 or 2 mL of solvent. The tube is heated with stirring to dissolve the solid. Stirring is conveniently carried out by spinning a spatula in the tube while heating. If the solid does not dissolve completely, more solvent should be added and heating continued until solution is complete. The solution is then allowed to cool slowly to room temperature, or sometimes to 0°C in a bath of ice water. If crystallization does not occur spontaneously, crystallization can often be induced by scratching the walls of the tube or by "seeding" with a few crystals of the pure compound.

When crystallization is complete, the Teflon (or glass) top to the Craig tube is inserted, a section of thin wire about 6 inches long is wrapped around the stem, and the assembly is placed upside down in a 12-mL centrifuge tube (see Crystallization, Figure 3). The wire should extend above the lip so that the tube may be pulled out easily. This assembly is then placed in the centrifuge, an equal weight is placed in the opposite tube holder, and the centrifuge is run for about a minute. (The equal weight may be another Craig tube assembly, or it may be a centrifuge tube containing the proper amount of water.) The assembly is removed from the centrifuge, the Craig tube is removed from the centrifuge tube by means of the wire, the tube is opened, and the crystalline contents are removed. The crystals should appear almost dry. If not, the centrifuge was probably not run long enough. Check on this visually as soon as you remove the assembly from the centrifuge. If there is still liquid in the Craig tube, then centrifugation should be continued until all the liquid has been centrifuged out and the crystals appear dry.

If desired, the crystallization can be repeated by removing the Teflon plug, adding more solvent, heating to dissolve the compound, cooling to effect crystallization, and centrifuging. Alternatively, a few drops of cold solvent can be added to serve as a "wash liquid." The Teflon plug is re-inserted and the system is again centrifuged to remove the wash liquid.

The crystals can then be transferred to a small tared weighing dish and allowed to air dry. Sometimes the crystals can be oven dried. Drying to constant weight is encouraged to obtain accurate yield data. Store the crystals in a small screw-cap vial.

5. The Centrifuge

The centrifuge is useful in various ways in addition to the one involving filtration in the Craig tube. It is very important to **always** balance the centrifuge with a weight placed opposite to the holder in which the sample is placed. Often this is just another centrifuge tube containing a sufficient amount of water to achieve balance. Use the balance to adjust the weight. If the centrifuge is not balanced properly, there is great strain on the bearings and the centrifuge might actually "walk" off the counter.

6. The Balance

The balance weighs to ± 1 mg (0.001 g), but the readings are given in grams. When recording weights, **always** report three figures after the decimal. For example, 2.312 g. If the last digit is a zero, **always** report it as zero, such as 2.310 g. If the weight is less than one gram, **always** put a zero before the decimal, for example 0.310 g. **Always** include the units. In microscale, the basic weight unit is the milligram (mg) and the basic mole unit is the millimole (mmol). For example, a weight might be reported as 183 mg (1.50 mmol).

7. The Micropipet

Pasteur disposable pipets come in two sizes, 5 3/4 and 9 inches. In this manual we refer to them as "micropipets." The drops delivered by the micropipet will be referred to as "microdrops." Micropipets are useful in transferring liquids from one container to another. In this way they are used like any ordinary medicine dropper. However, the micropipets are fairly accurately made and the drop size is consistent from one pipet to another. The 9" pipet makes possible the removing of material from a deep bottle, from the Hickman still, or from other containers. The short pipet serves the same purpose, but is somewhat less fragile because of the shorter drawn-out section.

The 5 3/4" micropipets are useful in making micro-drying columns or chromatography columns. For this use it is necessary to place a small wad of cotton at the constriction of the micropipet. In order to do this, a small rod is needed to push down the cotton. Wooden applicators 6 x 1/12 inch are available and are suitable for this purpose. Bamboo skewers, available locally, are inexpensive and make ideal rods for inserting the cotton.

A filter pipet is a micropipet that has had a **very** small wad of cotton inserted in the tip. The cotton will filter solutions both going in and coming out of the pipet. To insert the cotton plug, a fine copper wire may be used, but the packing should not be too tight, or the flow will be restricted.

NOTE: When you first try to transfer a highly volatile solvent, such as methylene chloride or ether, with a micropipet, you must be very careful or else you will find that the solvent is spontaneously ejected out of the pipet, usually onto the desk-top, before you have time to transfer the material. This occurs because of the high vapor pressure of the solvent. To prevent this from happening, place a few drops of the pure solvent that is being used (either methylene chloride or ether) in a clean vial and using the rubber bulb, pull some of the solvent into the pipet and force it out again into the vial. Do this several times before you try to transfer the

sample. This process saturates the air in the pipet and bulb with the solvent vapors and thus prevents the build-up of pressure in the pipet. The transfer can then be made with the pipet without loss of material.

8. Extraction, Microscale

Extraction is a process whereby a compound is transferred from one solvent to another solvent that is immiscible with the first. For example, water and ether are immiscible, as are also water and methylene chloride. Consider the case of a compound dissolved in water: you want to transfer the compound from the water solution to an organic solvent, such as methylene chloride. The water solution is placed in a vial with about an equal volume of methylene chloride. The mixture is then stirred briskly for about a minute using a spin vane and the magnetic stirrer. The stirring is stopped and the two layers are allowed to separate. A micropipet is then used to transfer the lower layer, in this case methylene chloride, to another vial. This operation must be done carefully in order not to transfer any of the top layer. Some of the compound that was formerly in the water is now in the methylene chloride solution. However, there may still be some of the compound remaining in the water, and so a second extraction is performed. Normally three extractions will transfer about 95% of the compound. In the case of very water soluble compounds more extractions may be necessary. From a knowledge of the structure of the compound in question, it is often possible to decide whether or not three extractions will be sufficient.

9. Extraction, Macroscale

In macroscale, a conical separatory funnel is used to carry out liquid-liquid extractions (Figure 1).

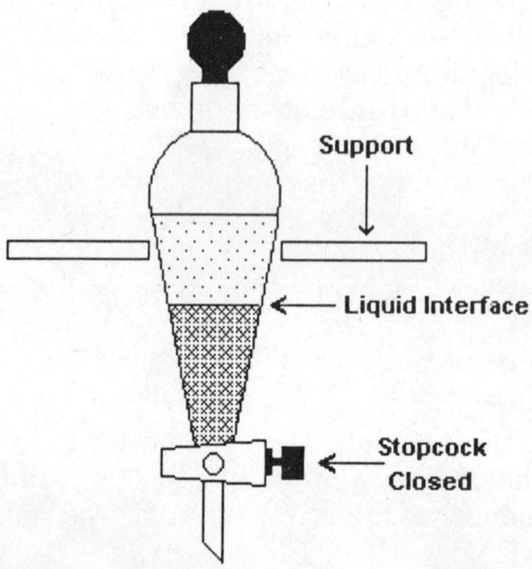

Figure 1. Separatory funnel.

To support the separatory funnel, a plate with a circular opening is placed on a ring clamped to the latticework or a ring stand in the hood. The separatory funnel is carefully lowered through the opening until it rests on the plate.

The stopcock is closed and two immiscible liquids are placed in the separatory funnel. One of these liquids is usually water and the other, an immiscible organic solvent. The object is to extract a substance from one solvent into the other solvent. Insert the stopper. Place one hand over the top of the funnel so that the stopper is in the palm of the hand. Keeping this hand positioned so that the stopper will not fall out, invert the funnel and **immediately release the pressure by opening the stopcock.** Close the stopcock and mix the contents by inverting the funnel at the rate of about one complete cycle per second. **Do not shake too hard** or you will emulsify the mixture. **Release the pressure frequently, using the stopcock while the funnel is inverted.** Two to three minutes of "mild shaking" is sufficient.

Return the funnel to the support ring, remove the stopper, and wait for the two liquids to separate. When the liquids have separated, draw off the bottom layer through the stopcock into a flask. After the bottom layer has been removed, the top layer is recovered by pouring it out of the top of the funnel into another container. This process may be repeated using fresh solvent in order to ensure a complete extraction of the substance. Further details of the use of the separatory funnel are presented in the experiment on extraction.

10. The Automatic Delivery Pipet

This description applies to the Wheaton pipet. Other makes of pipets may be used, with the only difference being the method for setting the quantity to be delivered.

The automatic delivery pipet is a device for measuring volumes very accurately. Do not use this pipet unless extreme accuracy is needed, such as measuring the reagents for a reaction. The pipets must be handled carefully, as they are expensive. First, place a plastic tip of the correct size onto the pipet. All liquids are contained in this tip. **Liquids are never drawn up into the pipet itself.** Using the calibrated screw cap on the top, set the volume desired. Then test the operation by pressing the top cap until the plunger hits the first stop. Release it. Does the pipet work easily? Depress the plunger again, to the first stop and then continue pressing until the plunger hits the second stop. Release. Does the pipet work easily? If it does, you are now ready to measure and transfer a liquid. Depress the plunger to the **first stop only.** Dip the tip into the liquid and release the plunger slowly and allow the tip to fill with liquid. Carefully move the pipet to the container and depress the plunger to the first stop and **then on to the second stop.** That last stop is used to blow the last drop of liquid out of the tip. When making the transfer, place the container very close to the reagent bottle to reduce the chance of liquid dripping out of the pipet before you can empty the pipet properly. Sometimes, for solvents with very high vapor pressures, such as ether and methylene chloride, the vapor will rapidly expel the liquid before you have a chance to transfer it. In this case, just work the plunger up and down a few times with the tip of the pipet still in the liquid. This will tend to equalize the pressure and help to keep the liquid in the pipet while the transfer is being made.

11. Evaporation of Solvents

If a nitrogen line has been installed in the laboratory then the evaporation of solvents may be conveniently carried out by directing a stream of nitrogen onto the surface of the liquid while warming gently. Connect a section of rubber tubing to the needle valve of the nitrogen line. At the other end of the tubing attach a micropipet. Turn on the nitrogen gradually, using the needle valve, and regulate the flow to a gentle stream. Test it with your hand. **The idea is to have a gentle stream of gas,** not one that will blow the solution out of the flask. Clamp the pipet in a downward position above the hot-plate. Carefully place the solution to be evaporated under the pipet making sure that the gas stream is not too fast. Using mild heating from the hot-plate, swirl the solution gently, or stir with the spin vane, and continue until the evaporation has been completed. Turn off the nitrogen. At the end of the laboratory period, **always** leave the nitrogen line in the off position.

If the laboratory is not equipped with a nitrogen line, the evaporation can be conducted using heat alone. In this situation place the vial on the stirrer hot-plate, insert a boiling chip, and heat gently to boiling. The boiling chip provides a surface upon which the bubbles of vapor can form. Without a boiling chip, the evaporating liquid often will "bump" (vaporize violently) and splash the liquid out of the vial, resulting in loss of most of the product. As an alternative to boiling chips, rapid stirring using the spin vane will prevent "bumping." Continue the evaporation to dryness, or until the proper amount of liquid has been removed.

Alternatively, evaporation can be carried out on the steam bath. Apply the steam gently, and there will be less chance of water collecting in the vial or other container.

12. Heating Blocks and Sand Baths

A heating block is a piece of aluminum (*ca* 2 cm thick) with a hole of a size to fit the vial or round bottom flask. The block is placed on the stirrer hot-plate, the vial is placed in the hole, and heat or stirring is applied. The heating block is an efficient heat transfer device which allows heat to come in from the sides, and thus speeds up the heating process. The heating block also serves as a stand for the vial and prevents the vial from being knocked over easily.

The sand bath is another device which can be used to promote heating of the vial during a reaction. The sand bath consists of a small dish containing sand and is used on the hot-plate to heat the vial or round bottom flask. The heating block, if available, is preferred because there is better heat transfer through the aluminum and there is no sand to spill.

13. Sublimation

Sublimation is the direct conversion of a solid to a gas and back to a solid, without going through a liquid phase. The procedure is best carried out under vacuum using a sublimator (Figure 2).

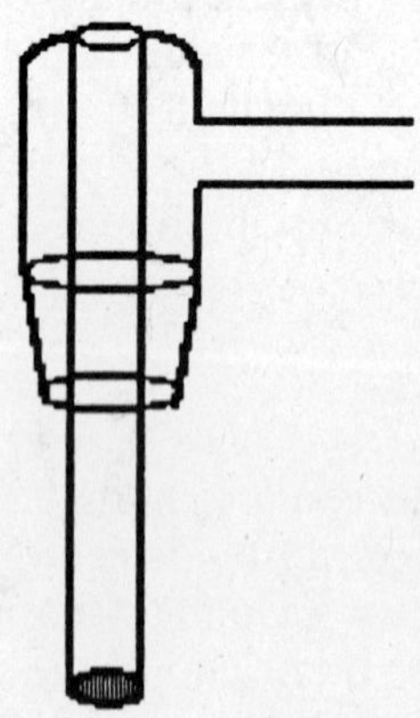

Figure 2. Sublimator

A sublimator consists of a cold-finger with a ground glass joint that fits a conical vial (see picture above of one type of sublimator). About 50 mg of the sample to be sublimed is placed in the conical vial and the sublimator is attached with the joint, O-ring, and screw cap. Ice water is carefully placed in the cold-finger without spilling any over the sides (see below), the side arm is attached to a vacuum line, vacuum is applied, and the vial is heated on the hot-plate. At a certain temperature the sample will begin to sublime (evaporate) and collect as a solid on the flat portion and sides of the cold-finger. The temperature should be maintained at this level until the sublimation is complete. The vacuum is then broken by removing the hose from the side-arm. The sublimator is removed from the vial with **great care** so that none of the material collected on the cold-finger is inadvertently knocked off. The sublimed product is then scraped off the cold-finger onto a watch-glass, or into a container, using a spatula.

For prolonged sublimations, the ice water in the cold-finger should be changed periodically using a micropipet. Changing the water in the cold-finger should be carried out **with great care** so as not to spill any onto the top joint of the sublimator. If water does spill, it may be sucked past the O-ring, through the joint, and into the vial. If this happens, the system must be shut down, the water removed from the vial, the sample dried, and the procedure restarted. As a precaution, make sure that the O-ring is in place and the cap is tight.

14. Calculations of Theoretical and Percent Yield

The determination of the **theoretical yield** and the **percent yield** of a reaction is a fundamental calculation of organic synthesis and one that is performed whenever a chemical reaction is carried out. **The theoretical yield is the "maximum" amount of the desired product that would have been obtained if the reaction had gone to 100% conversion and there had been no losses of material during the work-up of the reaction mixture.** This situation is almost never encountered because reactions generally do not go to 100% conversion and there are always losses of product on work-up. Thus, the amount of product isolated will almost always be less than the theoretical amount. **The percent yield is the amount of product that is actually obtained in the reaction divided by the theoretical amount of product obtainable based on the amounts of starting materials used, times 100.** A knowledge of the percent yield tells us the overall efficiency of the reaction as it was performed and allows us to predict the quantity of product that is likely to be isolated in subsequent reactions. A particularly low percent yield indicates either that the reaction is inefficient or that the procedure was not carried out properly. Perhaps a better method, or more care, should be employed. The theoretical yield and the percent yield will be routinely calculated during the laboratory program for all experiments involving organic syntheses.

In microscale, weight is generally given in milligrams (1 mg = 1/1000 of a gram). Molar quantities are generally expressed in millimoles (1 mmol = 1/1000 of a mole). Molecular weights have the units g/mol or mg/mmol.

The calculation of theoretical and percent yield is not difficult and involves four basic operations: **1**) Writing a balanced chemical equation for the reaction. **2**) Calculation of the molecular weights of the compounds involved in the reaction. **3**) Use of the stoichiometry of the reaction to determine the limiting reagent. **4**) Conversion (in microscale) of mg to mmol and mmol to mg. All of these operations should be familiar from your former training in chemistry. The procedure is illustrated by the following example:

Consider the reaction in which we start with 200 mg of benzoic acid, $C_6H_5CO_2H$, 200 mg of ethanol, CH_3CH_2OH, and 2 microdrops of sulfuric acid catalyst. On completion of the reaction, the product, ethyl benzoate, $C_6H_5CO_2CH_2CH_3$, is isolated, purified, and weighed. In this example, the weight of ethyl benzoate obtained is 210 mg. What is the percent yield of the ethyl benzoate isolated in this reaction?

1. Write a balanced equation for the reaction.
2. Below each compound write the molecular weight of the compound to three significant figures. Molecular weights may be found in handbooks or they may be calculated by adding the atomic weights of the individual atoms of the compound.
3. Below the molecular weights list the weight of each starting compound used: in this example 200 mg of each reactant. At this point leave a blank space under the product ethyl benzoate.
4. On the next line convert the mg of **starting materials** to mmol of starting materials. To do this divide the weight by the molecular weight. For example 200 mg / 122 mg/mmol = 1.64 mmol. Keeping track of the units will reduce errors in calculation.

$C_6H_5CO_2H$	+ $HOCH_2CH_3$	$\longrightarrow$ $C_6H_5CO_2CH_2CH_3$	+ H_2O
122 mg/mmol	46.0 mg/mmol	150 mg/mmol	
200 mg	200 mg	246 mg*	
<u>1.64 mmol</u>	4.35 mmol	1.64 mmol**	

* Added in step 6
** Added in Step 5

5. The next step is to determine the **"limiting reagent"** of the reaction. Since the equation states that one mole of each reactant gives one mole of each product, the limiting reagent will be the one that is in least amount in terms of moles. In the example, the **limiting reagent is benzoic acid** and the maximum amount of ethyl benzoate possible from this reaction can be no more than 1.64 mmol. Although there is more than 1.64 mmol of ethanol present, there is not enough benzoic acid to react with it all. Therefore, all subsequent calculations are based on the 1.64 mmol of benzoic acid. At this point <u>underline</u> 1.64 mmol under benzoic acid to identify the limiting reagent. Thus 1.64 mmol of ethyl benzoate is the **"theoretical yield"** of the reaction. **Write 1.64 mmol under ethyl benzoate.** Omit calculations of the amount of water because water is a by-product in this reaction.
6. Convert the 1.64 mmol theoretical yield to mg of theoretical yield by multiplying by the molecular weight of ethyl benzoate; 1.64 mmol x 150 mg/mmol = 246 mg. **Write 246 mg below ethyl benzote as shown in the reaction above.**

Steps 1, 2, and the determination of the limiting reagent should be done before coming to the laboratory. The theoretical yield may be calculated when the weight of the limiting reagent that you actually use in the reaction has been determined by measurement in the laboratory. The amount you take may be a few milligrams larger or smaller that the quantity listed in the experimental procedure. After the experiment is completed and the product is isolated and weighed, the percent yield of product can then be determined. In the example, the weight of ethyl benzoate isolated was 210 mg. The percent yield is thus:

$$\frac{210 \text{ mg}}{246 \text{ mg}} \times 100 = 85\% \text{ yield}$$

This calculation should be recorded in your notebook following your description of isolating and purifying the product. In practice, yields of less than 100% occur because of several factors: the reaction may not go to completion; side reactions may consume some of the reactants; losses occur in isolating and purifying the product, etc.

Finally: always look at your answer to see if it is reasonable. For example, if the percent yield is over 100%, you have certainly made a mistake somewhere. There are a variety of sources of error.

There may be an error in the way the reaction was written and the molar ratios of reagents and products are wrong.

The molecular weight data may be incorrect.

The calculations may be incorrect.

The weights of the reagents used in the reaction may be in error.

The product may be contaminated, often with water, and thus weighs more than it would if it were dry.

The calculation of yield is an integral part of synthetic organic chemistry and will be performed many times throughout the course. It is very important for you to learn these procedures so that you can carry them out automatically for each synthetic experiment.

3. Melting Points

A. Introduction

A pure compound has a definite melting point which can be used to help characterize and identify the substance. The melting point of pure water is used to define 0°C. If the sample of water is pure (free of any contaminants), we always find that water completely melts or freezes at the same constant temperature, 0°C, and any mixture of pure solid and liquid water (ice-water) will be at a temperature of 0°C.

Organic compounds behave in the same way. A pure crystalline compound always exhibits the same melting point, a characteristic physical constant, which can be used to help identify the substance. If a substance is not pure, the melting point is almost always **depressed.** Thus a sample of **pure A** might be observed to melt at 130.0-130.5°C, while a sample of **A** contaminated with some impurity might be found to melt at a lower temperature and over a broader range; for example, 124-127°C. The "sharpness" of the melting point, as well as the actual value of the melting point, indicates the degree of purity of a sample.

The lowering of the melting point of a substance by the presence of impurities is a phenomenon often treated in general chemistry: the depression of the freezing point of liquids by the presence of impurities. We learned that the freezing point depression, ΔT, is expressed as:

$$\Delta T = -(T_x - T_o) = K_F m_x,$$

where T_o is the freezing point of a pure solvent, T_x is the (lower) freezing point of a solution which has a molality, m_x, of solute X in the solvent (molality is the number of moles of solute per kilogram of solvent), and K_F is the molal freezing point depression and is a constant for the solvent being used. This relationship can be used to determine the molecular weight of an "unknown" solute in a standard solvent. The major point to understand is that:

A substance dissolved in a liquid lowers the freezing point of that liquid.

Since a substance dissolved in the liquid form of a compound **lowers** the freezing point of that compound, the same substance dissolved in the solid form of the compound will lower the melting point of that compound. Melting point lowering can be used to determine whether or not two materials having the same melting point are the same, or different, substances. For example, suppose sample **A** and sample **B** both melt at 80.0°C. Are these two samples the same substance, or are they different? To decide, we **thoroughly mix equal amounts of finely powdered** **A** and **B** and determine the melting point of the mixture. If the mixture melts at 80°C, there is a strong possibility that the compounds are identical. If the melting point is lowered to say 68-72°C, then clearly **A** and **B are different compounds.** This process of determining the melting point of mixtures is most commonly referred to as taking **"mixed melting points."**

For confirmation it is often advisable to vary the composition of the mixture by taking different amounts of **A** and **B,** and determining the melting point of each mixture. If the melting point remains 80°C for the mixture, our conclusion is that **A** and **B** are identical. Although there are some exceptions to the rule, the probability of having an exception is very low. To confirm the conclusion based on melting points, spectroscopic measurements are used.

B. General Procedure for Determining Melting Points

NOTE - In determining melting points, be very careful to avoid contaminating a sample with impurities. Make sure that watch glasses, micro-spatulas, etc., are clean.

Melting points are determined on very small samples, < 5 mg, using a commercial instrument called a Mel-Temp. The unit is heated electrically and the rate of heating is controlled by setting a variable voltage control. The sample is contained in a small capillary tube (1.5 mm OD x 90 mm long and sealed at one end) that fits into a port on the Mel-Temp. There are three ports so that three samples can be inserted at one time. A thermometer is used for determining the temperature at which the sample melts. The bulb of the thermometer is situated next to the sample in the heating block. The melting behavior of the sample is observed through a small lens in the front of the heating block. (If a different type of melting point instrument is used, see your instructor for operating procedures.)

A sample is first crushed to a fine powder on a watch glass using a micro-spatula. To load the melting point capillary, the open end of the capillary is pushed down onto a small pile of the powdered sample. When a few crystals have entered the tube, turn the tube over and tap it on the desk top until the crystals are forced down to the sealed end. A column of sample **2 to 3 mm in length is all that is needed.** Usually it is advisable to pick up small samples in two or three passes, because if too much material is forced into the capillary at one time, it may be difficult to pack down the sample. It is important that the sample is **tightly packed** to provide good heat conductivity during the melting point determination.

The melting-point capillary is placed in the melting point apparatus and, as the sample is heated, the sample is observed through a magnifying eyepiece, while at the same time, the temperature is monitored by means of the thermometer. The temperature of the unit is increased steadily, at a heating rate of 2 °C per minute. Adjust the voltage to achieve this rate. If a faster rate is used, the melting point will not be accurate because the thermometer generally lags behind the actual temperature of the sample and this lag increases with increasing heating rates.

A melting points determination really gives a temperature range. Record the temperature of the first appearance of liquid. Continue heating at the same rate and record the temperature at which the last crystals melt. Indicate these two temperatures as a melting point range, for example, mp, 110.0-111.5°C, indicating that the first sign of melting occurred at 110.0°C and the final melting was at 111.5°C. If the sample is really pure, the temperature range is narrow, perhaps less than 0.5°C. If the sample is not pure, the temperature range will be greater, sometimes as much as 10°C.

If a second melting point determination is to be made on the same compound, allow the unit to cool about 10-15 °C **below** the melting point of the sample. Insert a **fresh** sample and go through the procedure again. It is important to always use a **fresh** sample that has not been previously melted because some compounds decompose upon melting. Also, upon crystallizing from a melt, a sample may form a different crystalline modification with a melting point different from that of the original material (usually lower). For these reasons, **never remelt a sample that has already been melted once.**

When the melting point of a sample is completely unknown, it will save time to first determine an approximate melting point using a fast heating rate and then to repeat the melting point on a **fresh** sample while raising the temperature at 2 °C per minute. If the approximate melting point is known, time will also be saved by heating at a fast rate up to about 15 °C below the known melting point, and then to cut back the voltage and heat at the 2 °C rate until melting is observed.

Mixed Melting Points. To prepare a sample for a mixed melting point, make small equal-sized piles of each component on a watch glass and then grind the two substances together thoroughly with a micro-spatula. **Thorough mixing is essential.** Load a melting-point capillary and determine the melting point as before. Since the Mel-Temp can hold up to three melting point capillaries, the melting points of the pure samples and that of the mixture can be determined simultaneously.

In recording a melting point, write down everything you observe. Some compounds, even when pure, tend to "sinter" a few degrees below their melting points, suggesting that some realignment of the molecules in the crystal is occurring. In other cases, we may be seeing the beginning of melting of an impure sample without being sure we see any liquid. In any event, **record what you observe.**

C. Experimental Procedure

In this experiment you are to determine the melting points of benzoic acid (mp 122°C), 2-naphthol (mp 123°C), a 1:1 mixture of benzoic acid and 2-naphthol, and a solid unknown selected from the following table. The identity of the unknown is then to be determined from melting point and mixed melting point data.

Melting Point Unknowns

Compound	mp, °C	Compound	mp, °C
Fluoranthene	110	Urea	132
meta-Toluic Acid	112	*trans*-Cinnamic Acid	133
Acetanilide	114	Acetyl Salicylic Acid	140
Mandelic Acid	120	Salicylic Acid	159
Benzoic Acid	122	Benzanilide	163
Succinimide	126	Triphenylmethanol	163

First determine the melting point of the two pure compounds, benzoic acid and 2-naphthol. When this has been performed and reasonable agreement with the accepted values has been obtained, prepare a mixture of equal amounts of the two "knowns" as described above and determine the melting point of the mixture. How does the melting point of the mixture compare with the melting points of the two pure compounds?

Then proceed to the identification of the unknown sample. First determine the melting point of your unknown. Then refer to the table, **Melting Point Unknowns,** and select those compounds that might possibly be the same as your unknown. Prepare a 1:1 mixture of your unknown with each compound which **might** be the unknown. Generally there will be two or three possibilities. Then determine the melting points of these mixtures along with the melting point of a fresh sample of your unknown. From the "mixed melting point" data, you will be able to identify the unknown. **Record all of your data** in your notebook and report the results to your instructor.

If time permits after completing the identification of an unknown, prepare about 10:1 and 1:10 mixtures of benzoic acid and 2-naphthol and determine the melting points of these mixtures. How do these melting points compare with the pure compounds and with the 50:50 mixture?

Dispose of waste chemicals in labeled waste containers provided in the laboratory.

D. Prelab Questions

1. In the winter, many highway departments "salt" ice-covered roads. Why? What is the physical explanation of the phenomenon which is observed?
2. Suppose that compound **A,** mp 80°C, and compound **B,** mp 100°C, are completely insoluble in each other, both in the solid state and in the molten state. A 50:50 mixture of A and B is prepared and the melting behavior is determined in the usual way. What melting behavior should be observed? (Note: these conditions are not commonly encountered in the organic laboratory, unless one mixes nonpolar organic solids with inorganic salts.)
3. As outlined earlier, mixtures of two substances usually melt over a range, not "sharply." Why is a melting range observed, rather than a "sharp" melting point? (Hint: it may be easier to think in terms of freezing a molten mixture, where the observed freezing point drops as solidification occurs.)
4. It is fairly common to find that at some definite composition, two substances form an **eutectic** mixture. For example, 1-naphthol melts at 95.5°C and naphthalene melts at 80.5°C. When mixtures of varying composition of these two substances are prepared it is found that the lowest melting point is obtained with a mixture consisting of 60.5 mol % of naphthalene and 39.5 mol % of 1-naphthol. Such a mixture melts sharply at 61.0°C. As this eutectic mixture melts, the composition of the melted mixture and the composition of the solid mixture remain unchanged, and as a consequence the melting temperature is constant. In terms of its melting point, such a mixture behaves like a pure substance. Given a sample of this eutectic mixture, how could you prove that it is not a pure substance? What would you do and what would you observe?

E. Postlab Questions

1. In a melting point determination, a column of 2 to 3 mm of sample in a melting point capillary is recommended. What are the disadvantages of using an 0.5-mm column; a 10-mm column?
2. Why should you use the same melting point apparatus for all melting points?
3. What happens in using a melting point apparatus if the heating rate is too great?
4. What procedure worked best for you in filling a melting point capillary?

4. Crystallization

A. Introduction

Organic solids are purified by the method of **crystallization,** which is one of the most important techniques for purification, and one which will be used many times during the laboratory program. This technique takes advantage of the fact that most substances are much more soluble in a solvent when it is hot, than when it is cold. The compound is dissolved in the hot solvent to near saturation; the solution is filtered to remove insoluble impurities and then allowed to cool. The compound crystallizes from the solution in a higher state of purity because the impurities tend to remain in solution. The process can be repeated (recrystallization) until the substance is almost 100% pure.

The choice of solvent is an important consideration. The compound to be recrystallized should be fairly soluble in hot solvent, but much less soluble in cold solvent. A solvent which is "too good" will not allow recovery of much of the compound. On the other hand, if the solvent is "too poor," an excessively large volume of solvent would be needed.

A solvent should be fairly volatile, because after the compound is collected, it must be freed of adsorbed solvent. Most of the common solvents have boiling points ranging from 35°C to 120°C. If the boiling point is very low (35°C), there will not be a very large temperature difference between the **hot** and **cold** solvent because we usually do not cool solvents below the temperature of ice water (0°C).

Some solvents, such as benzene and carbon tetrachloride are toxic and should be avoided. Most organic solvents are highly flammable and must be handled with extreme care to avoid fire. Obviously, a low flammability is a desirable safety feature for a solvent.

Solvent polarity is of major importance in determining how soluble a compound is in a given solvent. Polar solvents are good solvents for polar compounds, while non-polar solvents tend to be good solvents for nonpolar compounds. Water is a very polar solvent, hydrocarbons such as hexane are nonpolar, while many solvents have intermediate properties. A few of the more useful solvents for crystallizations are listed below:

Water, H_2O, bp 100°C, mp 0°C, very polar, non-flammable.

Ethanol, CH_3CH_2OH, bp 78°C, polar, miscible with water in all proportions, flammable, widely used.

Acetone, CH_3COCH_3, bp 56°C, intermediate polarity, excellent solvent power, miscible with water in all proportions, flammable.

Methylene chloride (dichloromethane), CH_2Cl_2, bp 41°C, intermediate polarity, immiscible with water, denser than water, relatively non-flammable.

Hexane, $CH_3(CH_2)_4CH_3$, bp 69°C, nonpolar, immiscible with water, less dense than water, highly flammable.

Pentane, $CH_3(CH_2)_3CH_3$, bp 36°C; **Heptane,** $CH_3(CH_2)_5CH_3$, bp 98°C; **Cyclohexane,** cyclo-C_6H_{12}, bp 81°C; all similar to hexane.

Toluene, $C_6H_5CH_3$, bp 111°C, slightly polar, immiscible with water, less dense than water, flammable.

Diethyl Ether, $(CH_3CH_2)_2O$, bp 35°C, slightly polar, immiscible with water, less dense than water, highly flammable.

Ethyl Acetate, $CH_3CO_2CH_2CH_3$, bp 77°C, intermediate polarity, immiscible with water, less dense than water, highly flammable.

Solvent Pairs. Sometimes a compound might be too soluble in one solvent and relatively insoluble in another. In cases such as this, recrystallization is best performed using a mixture of two solvents, one in which the compound is very soluble and the other in which the compound is not. A mixture of the two solvents, **which must be miscible with each other,** may give the desired solvent properties for recrystallization. Usually the compound is dissolved in the better solvent and, while keeping the solution hot, the poorer solvent is added until crystallization is imminent. The solution is then allowed to cool to induce crystallization. Some common solvent pairs are listed below:

Water-Ethanol
Water-Acetone
Hexane-Diethyl ether
Toluene-Hexane
Methylene Chloride-Hexane
Ethyl Acetate-Hexane

In order to induce crystallization it is often necessary to add a few crystals of the compound to the solution. The crystals provide a surface on which crystallization can begin and then crystallization will generally follow rather rapidly. This is called "seeding" and it is a good idea to put a few crystals aside for just this purpose. In the event seed crystals are not available, another method is to scratch the walls of the flask with a spatula or stirring rod. This will usually induce crystallization.

In some cases a compound will separate as a liquid or oil. In that event, the mixture must be reheated until a clear solution is obtained. Then slower cooling should be tried, preferably while trying to initiate crystallization by seeding with the crystalline compound. If oiling-out continues to be a problem, a little more of the better solvent may have to be added. Of course the compound will not crystallize if the temperature of the solution is above the melting point of the compound.

The first experiment is a small macroscale recrystallization of 1 g of acetanilide from water. The second experiment is a microscale crystallization of 100 mg (0.1g) of naphthalene from ethanol using the Craig tube. The Craig tube is a very convenient item of equipment used to perform crystallizations on very small scale and will be used often in the laboratory program.

B. Experimental Procedure

1. Macroscale: Crystallization of Acetanilide on a 1 g Scale

Weighing. Weigh out **about** 1 gram (*ca*[*] 1 g) of impure acetanilide.

---

[*] *Circa*, abbreviation *ca*, Latin, about, used to indicate an approximate amount.

In this experiment it is not necessary to take **exactly** the quantity of the substance specified for recrystallization, but try to get within ± 10% or so of 1.0 gram, and **accurately** determine and **record** the weight of the sample taken. In other experiments it may be necessary to take a specified amount of a reagent. Often the quantity of one reagent will determine the **exact,** or nearly exact, amount of a second reagent. Fill a melting point capillary with impure acetanilide and set it aside for a subsequent melting point comparison with your purified acetanilide.

Dissolving. Transfer the sample to a clean dry 50-mL Erlenmeyer flask. Add no more than 20 mL of water and one boiling chip, and heat the mixture to boiling on the hot-plate. (A boiling chip prevents "bumping" in a boiling solvent.) Swirl frequently to dissolve the acetanilide. If necessary, add small increments of water (1-2 mL) until the acetanilide dissolves at the boiling point. If there is insufficient water, the acetanilide will melt and appear as a liquid on the bottom of the flask. Add water until all molten acetanilide disappears. There may be some solid material that will not dissolve at all. These impurities will be removed by filtration.

Handling Hot Flasks. An easy, but secure, way to handle small hot flasks without burning your fingers is to use finger protectors. Cut-off two 2-inch lengths of thin-walled rubber tubing and slit each lengthwise along one side. These sections can be opened and slipped over your thumb and forefinger to provide "insulated grippers."

Decolorizing with Charcoal. When the sample has dissolved, **remove** the flask from the hot-plate and add an additional small volume of water (10 to 20% of the amount required for solution). Be sure the solution is **slightly below the boiling point** and add about 100 mg of decolorizing carbon (charcoal).

> **Note: If the solution is not cooled somewhat when you add the charcoal, it will froth badly and material may splash out of the top of the flask!**

Swirl the flask to suspend the charcoal, then add about 200 mg of **Celite** filter aid, and again swirl the flask. The Celite is added to keep fine particles of charcoal from passing through the filter. The quantities of charcoal and Celite do not have to be exact; just use approximate amounts.

Filtration of the Hot Solution. Set up a **hot filtration assembly** as shown in Figure 1. Fit a **fluted filter** into a **stemless glass funnel** and place the funnel on top of a clean 50-mL Erlenmeyer flask. Pour 3 or 4 mL of water into the flask, place the assembly on a hot-plate, and bring the water in the flask to a boil. The water vapor will heat up the entire assembly and keep it hot during the subsequent filtration. If, while you are getting your sample ready to be filtered, the water all boils off, simply add a little more to the filter. The wire is important, because without it, the funnel will form a seal at the top of the flask and the vapor escaping through the funnel may prevent the liquid from running through the filter.

Return the flask containing the acetanilide solution to the hot-plate and again heat carefully. **Be alert, the charcoal tends to cause frothing.** When the mixture is boiling, filter it through the fluted filter paper in the hot filtration assembly. Filter as rapidly as possible, returning the flask to the hot-plate after each "pour" to keep it hot. Do not fill the filter paper in the funnel more than 3/4 full and pour the mixture into the **center** of the filter so that the unfiltered mixture does not go over the sides of the filter paper. If the **filtrate** is not clear of solids, it will have to be reheated and refiltered.

To remove any acetanilide remaining in the "original" flask, wash with 2-4 mL of water. To do this, add the water to the flask, heat it to boiling, and pour the hot wash onto the filter. After the wash has been filtered, let the water in the filter flask boil for a few minutes to allow the hot condensate to extract any acetanilide which may be retained on the filter. Then remove the funnel.

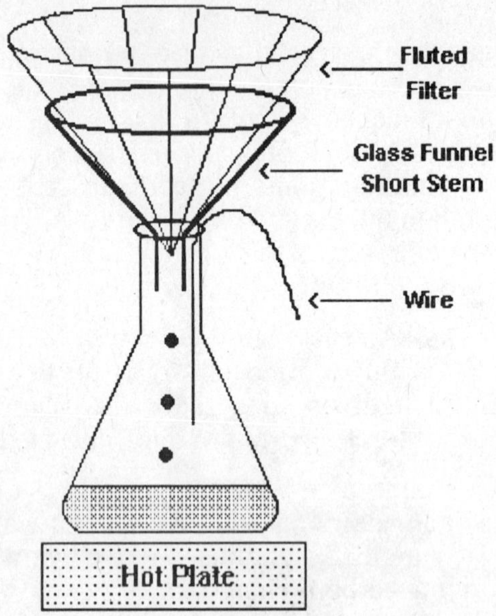

Figure 1. Hot filtration assembly.

Crystallization. If the volume of the solvent has been increased significantly beyond the amount needed to dissolve the acetanilide, the excess water must be boiled off, otherwise the solution may be too dilute and crystallization may not occur. Perform this operation if necessary, and then remove the flask from the hot-plate and cover the top with aluminum foil or a clean cork and allow the solution to cool undisturbed. Often crystallization will start spontaneously, but in the case of acetanilide the solution may become **supersaturated.**

If no crystals have appeared by the time the solution has reached room temperature, open the flask and add a few seed crystals of acetanilide, or scratch the walls of the flask with a spatula. If crystallization commences, allow it to proceed. When crystallization seems to be complete, cool the flask in an ice bath for at least 15 minutes before collecting the crystals. If crystals do not form using these techniques, the solution is probably too dilute and some of the water must be boiled away. In this event, reheat the solution, boil off some of the water, and recool.

Filtration to Collect the Product. "Vacuum" filtrations may be done either by using a house vacuum system, or by using a vacuum created by a water aspirator. Set up a **Hirsch funnel** in the manner shown in Figure 2. If a water aspirator is used, then the water trap on the right is necessary to prevent an inadvertent backflow of water from the aspirator into the system. If a house vacuum is used, then the trap is not necessary. **Clamp the filter flask to a ring stand or lattice to keep it from tipping over.** Connect a section of vacuum tubing to the filter flask and then to the source of vacuum. Place a circle of filter paper over the holes in the Hirsch funnel, wet it with a little water, and "seat it" by applying the vacuum. In the case of the water aspirator, the vacuum is applied by turning on the water full force and then closing the stopcock. To discontinue the vacuum, **always open the stopcock first and then turn off the water**, otherwise water may be sucked back into the filter flask.

When you are ready to collect the solid, swirl the crystallization flask to suspend the crystals, and pour the contents onto the filter in the Hirsch funnel while applying vacuum. A spatula may be needed to aid in the transfer of the crystals. To recover crystals still adhering to the flask, add 2-3 mL of ice-water to the flask, swirl, and pour this through the filter. It is best to **turn off** the vacuum **before** adding the wash liquid; let the wash liquid stand for a few seconds and **then** turn on the vacuum to suck the liquid through. Draw air through the crystals for a few minutes; then transfer them to a **clean, weighed** watch glass (note that you should record the actual weight of the watch glass, don't just "tare" it) and let the crystals air dry at room temperature, (probably until the next laboratory session). To speed up drying, slightly damp acetanilide can be dried more quickly in an oven set at 80°C. Determine the weight and melting point of your dry product. Take the melting point of the impure acetanilide at the same time and compare the value with the melting point of the purified sample.

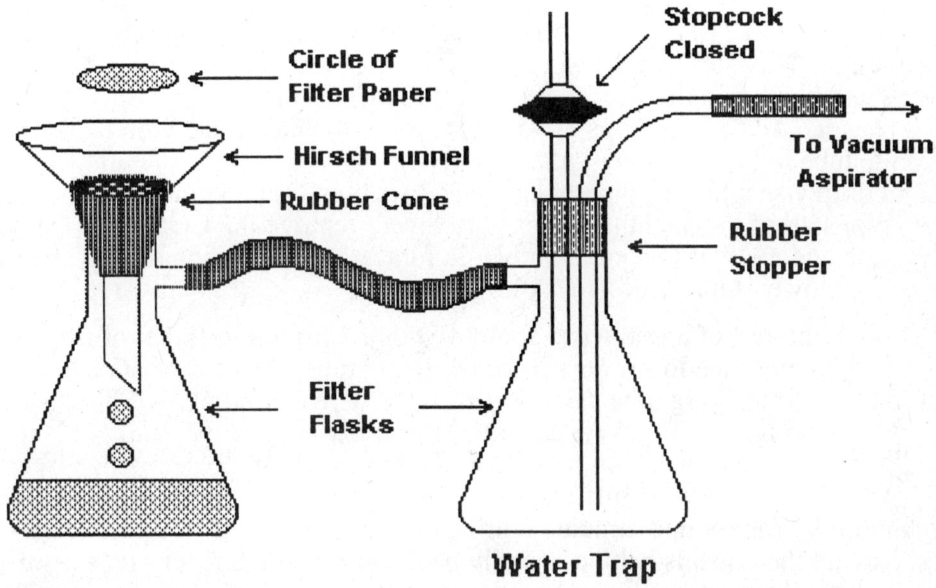

Figure 2. Hirsch funnel assembly.

2. Microscale: Crystallization of Naphthalene on a 0.1 g Scale. Use of the Craig Tube.

Review the discussion on the use of the Craig tube and the centrifuge given in the Introduction chapter of this manual.

A Craig tube assembly (Figure 3A) consists of two parts: a 75-mm glass tube about the size of a small test tube and a Teflon or glass filter plug, which fits into the lower part of the Craig tube. The tapered region provides a fairly snug fit, but not so snug as to prevent liquid from flowing past.

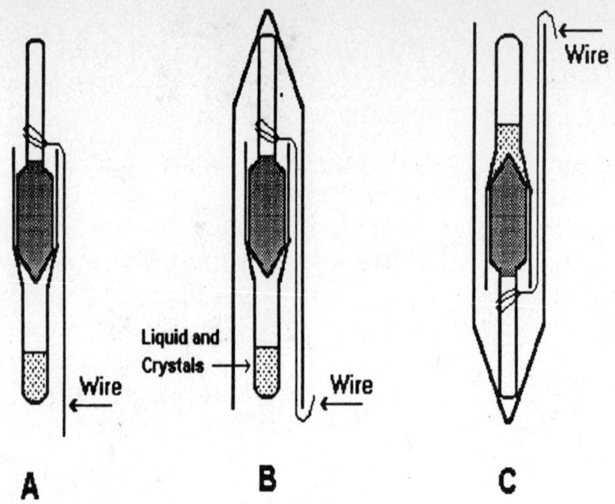

Figure 3. Craig tube assembly.

Fill a 100-mL beaker about half-full of water and heat to near boiling on a hot-plate. While the water is heating, accurately weigh out approximately 100 mg of naphthalene and place it in the bottom section of the Craig tube that is supported upright in a small beaker. Add a few drops of 95% ethanol to the naphthalene in the Craig tube and place the tube in the hot-water bath. While stirring with a microspatula, continue to add ethanol dropwise until the naphthalene just dissolves. (The volume should be **less** than 1 mL.) When all of the naphthalene has dissolved, remove the Craig tube from the hot water bath and allow it to cool as shown in Figure 4. The optional outer inverted beaker allows for slower, more uniform cooling.

Wrap the end of a fine wire (about 4 in long) around the stem of the filter plug and insert the plug into the lower portion of the Craig tube. Place a plastic centrifuge tube over the top of the Craig tube assembly with the wire extending over the lip of the centrifuge tube, see Figure 3B. Invert the entire assembly (Figure 3C) and place it the centrifuge. Make sure that the assembly is **counterbalanced** in the centrifuge by another similar assembly, or by a centrifuge tube filled with about 12 mL of water. The assembly is spun rapidly for about a minute. During centrifugation, the liquid leaks past the filter plug, leaving the crystals behind, usually packed against the filter. Fast centrifuging will remove virtually all "free" liquid and the crystals should look dry. Remove the assembly from the centrifuge, remove the filter plug using the wire, and scrape the crystals out using a spatula. Allow the crystals to dry for a few minutes on a watch glass.

After drying, the crystals of naphthalene are placed in a screw cap vial for storage. The cap must be tight, otherwise the naphthalene will sublime and be lost. Obtain the weight and melting point of the pure material. Determine the percent recovery of the naphthalene.

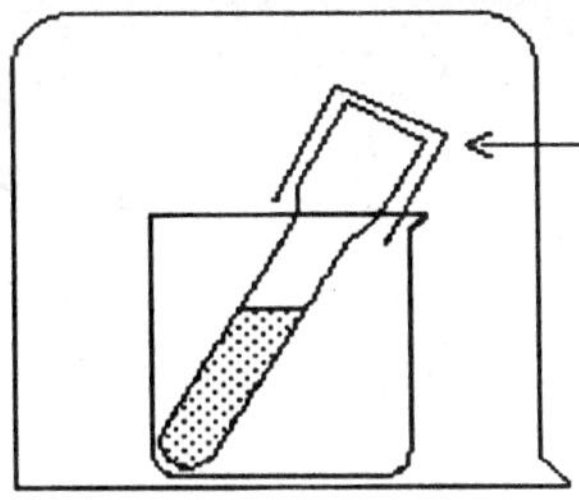

Figure 4. Craig tube crystallization arrangement.

Dispose of waste chemicals in labeled waste containers provided in the laboratory.

C. Prelab Questions

1. Explain whether or not each of the following pairs of solvents might be useful as a **solvent pair** for recrystallization.

   (a) water-ethanol
   (b) acetone-ethanol
   (c) hexane-pentane
   (d) ethanol-hexane
   (e) water-hexane
   (f) ethyl acetate-hexane

2. Nitrobenzene, bp 210°C, is a fairly polar liquid which is a good solvent for many compounds. What drawback does it have as a routine recrystallization solvent?
3. Why is charcoal sometimes used in recrystallizations? How does it function? Why should you avoid using a large amount? "Activated carbon" is used in air and water purifiers. How does it work?
4. Why must you keep the filter assembly hot while you remove the charcoal?
5. Why must you counterbalance a Craig tube assembly when it is centrifuged?

D. Postlab Questions

1. Suppose you dissolve a compound in too much solvent in a recrystallization. How would you find out that you had used too much solvent? Given that too much solvent has been used, what can you do about it?
2. Suppose a Craig tube assembly has been chilled in an ice bath. Why must the "filtration" step be performed quickly?
3. Acetanilide has a mp of 114°C, yet it apparently melts in boiling water. Explain.
4. Why is it necessary to dry the purified acetanilide before taking a melting point?
5. Explain why slow cooling **tends** to give "better" crystals in a recrystallization than the crystals obtained upon rapid cooling.
6. A compound is soluble in ether at room temperature but not very soluble at -80°C ("Dry-Ice" temperature). Would ether be a useful crystallization solvent for the compound? Explain.

5. Distillation

A. Introduction and Theoretical Discussion

1. Distillation of a Pure Liquid

The purpose of distillation is to separate the components of a mixture of liquids by taking advantage of differences in their boiling points.

A pure liquid (one compound) has a characteristic **vapor pressure** at any given temperature. The vapor pressure is the pressure of the vapor (i.e., a gas) **in equilibrium** with the liquid. If the liquid is introduced into a closed, evacuated vessel, molecules will tend to escape from the liquid, creating a pressure in the gas phase until the vapor pressure is reached. At this point a dynamic equilibrium is attained, with a balanced state of molecules escaping and returning to the liquid and no further change in pressure will be observed. If we try to increase the pressure by compressing the vapor, the vapor will condense to the liquid phase. The vapor pressure thus represents an upper limit to the pressure that the liquid can exert and this pressure will be constant at a constant temperature.

If a "permanent" gas such as nitrogen is present, the liquid will still exert its vapor pressure, but the total pressure observed will be the sum of the vapor pressure of the liquid and the pressure of the nitrogen; i.e. Dalton's Law of Partial Pressures applies:

$$P_{total} = P_N + P_A$$

where P_A is the vapor pressure of liquid A and P_N is the pressure of the nitrogen.

If a flask which is open to the air is partially filled with a liquid, some of the liquid will evaporate, displacing some of the air from the flask. The total pressure in the flask will be equal to the prevailing atmospheric pressure (the "barometric pressure"), but that pressure will be due to the sum of the vapor pressure of the liquid and the pressure of the remaining air.

The vapor pressure of a liquid always increases with increasing temperature, as shown in Figure 1A. If the data are plotted as log P *vs* 1/T (the logarithm of the vapor pressure *vs* the reciprocal of the Kelvin temperature, Figure 1B) a straight line is obtained. Thus, if the vapor pressures are known at two different temperatures, we can derive the straight line shown in Figure 1B and can estimate the vapor pressure of the liquid at any other temperature.

How do we interpret the fact that vapor pressure increases with increasing temperature? In the liquid phase, the molecules of a substance are strongly attracted to one another. To overcome the intermolecular attractions and drive the molecules into the gaseous state, where intermolecular attractions are very low (zero for ideal gases), energy must be obtained from the surroundings. As energy is added to the liquid to raise its temperature, the molecules move around faster due to their increased kinetic energy, and thus the vapor pressure (escaping tendency) of the molecules increases.

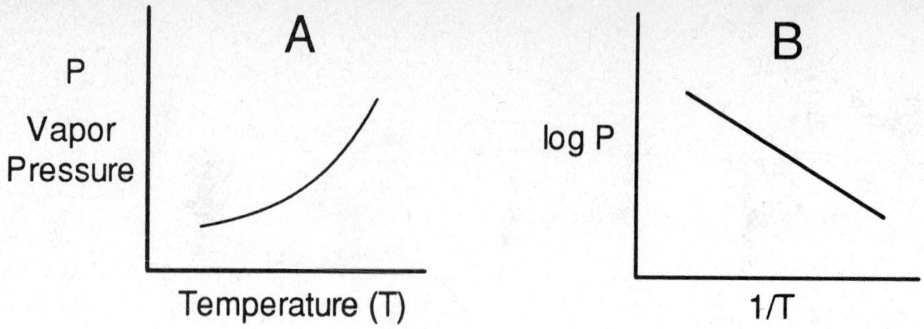

Figure 1. Vapor pressure - temperature plots for liquids, T in Kelvin.

When a liquid is heated in a vessel open to the atmosphere, the vapor pressure will increase until the vapor pressure reaches the atmospheric pressure. At this temperature, the liquid boils, and the temperature at which boiling occurs is called the **normal boiling point** provided that the barometric pressure is exactly one atmosphere (760 Torr or 760 mm Hg). Boiling converts the liquid to the vapor and the vapor dissipates into the atmosphere. The amount of energy required at a given temperature to convert a mole of a liquid to vapor is called the **molar heat of vaporization** of the liquid.

The phenomenon of boiling results because the vapor pressure is a property of all of the liquid, not simply of the surface. When the vapor pressure equals the applied pressure, molecules go directly from the liquid state to the gaseous state within the liquid, forming bubbles of vapor, which quickly rise through the liquid and escape from the surface.

If the bubbles do not form quickly enough, the liquid becomes superheated and violent "bumping" can occur due to very large bubbles forming abruptly - almost explosively in the body of the liquid. Bumping can be violent enough to throw the liquid out of the vessel. To avoid this problem, boiling stones are added to a liquid so that the irregular surfaces of the stones can effectively initiate bubble formation and reduce the chance of bumping.

A liquid can be heated in a flask joined to a condenser. A condenser is a device where the hot vapors encounter a cooled surface at which the molecules give up some of their energy and condense to the liquid state (Figure 2). The process can continue indefinitely: energy (heat) goes into the liquid in the boiling flask, some of the liquid evaporates, the hot vapors rise to the condenser where the molecules give up energy, condense to a liquid and drain back into the flask. A system operating in this fashion is said to be "at reflux" or we say that the liquid is "refluxing." Under these conditions the temperature of the vapor over the refluxing liquid remains constant.

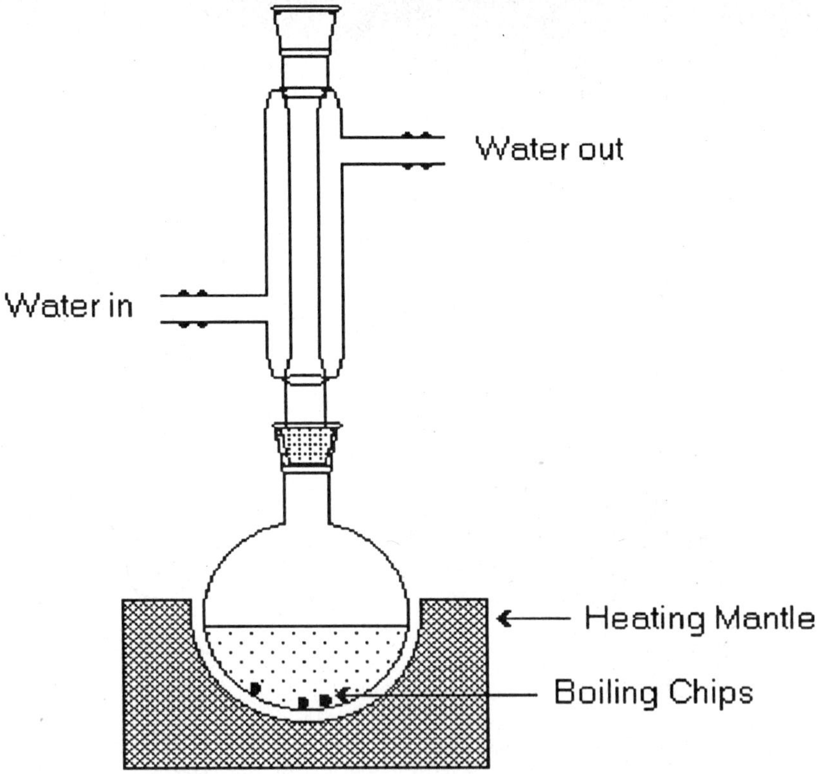

Figure 2. Apparatus for heating a liquid under reflux.

Suppose we arrange the apparatus such that the liquid formed by condensing the vapors is not returned to the boiling flask, but is drained off into another flask, or receiver. This apparatus is a "still" and the process is called distillation, see Figure 3.

In a still, the liquid is heated to boiling in the flask, or "pot." The vapors rise into the distilling head and then on into the condenser. Since the condenser is cooled by water passing through the glass jacket, the vapors condense in the condenser and the condensate flows down and collects in the receiver, in this case a graduated cylinder. A thermometer is positioned in the distilling "head" to measure the temperature of the vapors coming off at any particular time.

If a substance is pure, the material collected by distillation, called the distillate, will also be pure and the distillation will proceed at the normal boiling point. If the pressure is reduced below atmospheric pressure by attaching the system to a vacuum pump, the distillation can be carried out at reduced pressure. The boiling point will now be lowered and will correspond to the temperature at which the vapor pressure of the liquid is equal to the pressure maintained in the system. Any time a distillation is carried out in an "open" system, when the atmospheric pressure is less than one atmosphere (760 Torr), the distillation is at reduced pressure and the boiling points will be lower than the normal boiling points observed at one atmosphere pressure.

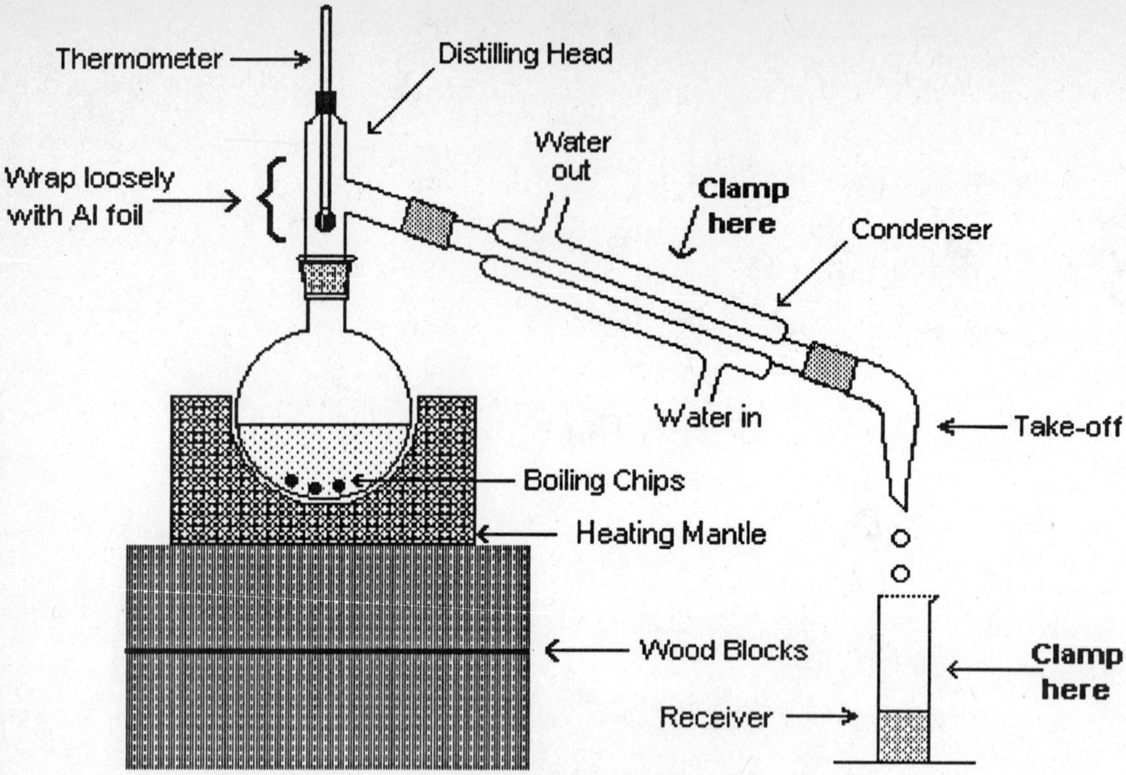

Figure 3. Apparatus for a simple distillation.

With a vacuum pump, the pressure can be reduced to very low values. Distillations are carried out routinely in research laboratories at pressures ranging down to 0.1 Torr or lower. Many substances tend to decompose if heated for any length of time at the normal boiling point at 760 Torr. In such cases, reduced pressures allow distillations to be carried out at much lower temperatures where decomposition is greatly reduced.

2. Distillation of a Liquid with a Dissolved Non-volatile Component

Consider the case where a nonvolatile solute is dissolved in a volatile liquid that is to be purified by distillation. At the same temperature, the vapor pressure of the solution will be lower than the vapor pressure of the pure liquid. Raoult's law states that:

$$P_A = P^\circ_A X_A$$

where P_A is the vapor pressure of the solution, P°_A is the vapor pressure of the pure liquid **A,** and X_A is the mole fraction of liquid **A** in the solution (the balance of the material is the solute). Since X_A is less than one (if $X_A = 1$, the substance is pure), the vapor pressure of **A** in the solution is lowered relative to the vapor pressure of pure **A**. Consequently the boiling point of the solution will be higher than that of pure **A**. When such a solution is distilled in the apparatus of Figure 3, although the boiling solution will be at a higher temperature than the boiling point of pure **A**, the vapors which condense on the bulb of the thermometer will yield pure liquid **A**, meaning that **the vapor-liquid equilibrium established at the bulb will occur at the boiling point of pure A**. Moreover, the distillate will consist of pure **A**.

As **A** is removed by distillation, the solution remaining in the boiling flask becomes increasingly concentrated with the solute. Thus the pot temperature will rise continuously during the distillation even though the head temperature will remain at the boiling point of pure **A**.

3. Distillation of a Mixture of two Volatile Components

Clearly we can obtain pure **A** by distilling it away from a nonvolatile solute (impurity). But, what happens when we have a mixture of **two volatile** liquids? The situation is complicated because **both components exert vapor pressures.** Let us consider the case of two liquids which form an ideal solution, i.e. Raoult's Law applies. For example, hexane (bp 69°C) and heptane (bp 98.5°C) form an ideal solution; consequently, by combining Dalton's Law of partial pressures and Raoult's Law, we find that the **total** vapor pressure of the solution will be:

$$P_{total} = P^{\circ}_{hex}X_{hex} + P^{\circ}_{hept}X_{hept}$$

Consider the behavior of an **equimolar mixture** of hexane and heptane; i.e., X_{hex} = 0.5 mf (mole fraction) and X_{hept} = 0.50 mf. From tabulated vapor pressure data for each pure compound, we can find that at 80°C: P°_{hex} = 1062 mm and P°_{hept} = 426 mm. Thus the **total** vapor pressure from this 1:1 mixture is:

$$P_{total} = 1062\ (0.50) + 426\ (0.50) = 531 + 213 = 744 \text{ mm} \quad (\text{at } 80^{\circ}\text{C})$$

We have taken a **liquid mixture** which consists of 50 mol% of each component. What is the composition of the **vapor** in equilibrium with this 1:1 mixture? Assuming that the vapor behaves as an ideal gas, the partial pressure of a gas is proportional to the number of moles of that gas (in a given volume).

$$P_A \propto \text{Moles of A} \qquad\qquad P_B \propto \text{Moles of B}$$

This fact leads to the conclusion that the **mole fraction (X′) of each component in the gas phase is equal to the ratio of its partial pressure to the total pressure.**

$$X'_{hex} = 531/744 = 0.714 \quad \text{and} \quad X'_{hep} = 213/744 = 0.286$$

The vapor is now richer in the more volatile component; the 50:50 liquid mixture gives a vapor consisting of about 71 mole percent hexane and 29 mole percent heptane.

Suppose this mixture is distilled at an atmospheric pressure which happens to be 744 mm. The initial boiling point will be 80°C and the initial distillate will consist of 71% hexane and 29% heptane. Note that the initial distillate is richer in the more volatile hexane. However, as we remove a distillate enriched in hexane, the liquid in the boiling flask becomes correspondingly enriched in heptane. This fact has two immediate consequences: the boiling point of the mixture increases and the vapor in equilibrium with the boiling liquid becomes relatively richer in heptane, the less volatile component. For example, at some stage, the "pot liquid" will consist of 25% hexane and 75% heptane. At 80°C this mixture would exert a total vapor pressure of:

$$P_{total} = 1062(0.25) + 426(0.75) = 266 + 320 = 586 \text{ mm Hg}$$

Since 586 mm is well below the atmospheric pressure (744 mm), to keep the solution boiling, the pot temperature must be increased, as it turns out, to about 88°C. And, we find that the vapor now consists of 45% hexane and 55% heptane. (The boiling point and the percentages can be calculated using tabulated vapor pressure data for the two compounds.)

What has been described is called a simple distillation. The initial distillate is enriched in the lower boiling component and the last distillate is enriched in the higher boiling component, but the distillation "cuts" are far from being pure.

Suppose we could take the initial distillate (71% hexane and 29% heptane) and redistill it. We would find that the initial boiling point is 74.5°C and that the initial distillate from this 71:29 mixture would consist of about 86% hexane and 14% heptane. Clearly we have gained a further enrichment. Unfortunately, we would also find that the amount of such an enriched mixture is quite small; each initial distillate will be only a small fraction of the original mixture. If somehow we could set up a practical way to carry out a series of successive distillations we could achieve increasingly improved separations.

This goal of separating components can be achieved, or at least approached, by performing a **fractional distillation** employing a packed **distilling column.** (The term fractional distillation originally referred to the process described above of successively redistilling the "fractions" or "cuts.") Many designs of fractionating columns are known, but they all work on the same principle; they produce the effect of a series of successive distillations along a vertical column. Figure 4 shows the kind of system to be used.

Each section of the column in which a vapor-liquid equilibrium is established is called a "theoretical plate." The term "plate" is derived from commercial distilling columns which do in fact have actual plates constructed in the column for the purpose of establishing vapor-liquid equilibria. The greater the number of theoretical plates in a column, the greater is the efficiency of the column for separating volatile components of a mixture. The "Height Equivalent of the Theoretical Plate" (the HETP) is the length of a distilling column that is sufficient to produce the equivalent of one equilibrium, or one theoretical plate.

In this experiment a glass tube is packed with a stainless steel sponge. This "packing" provides a high surface area to promote rapid vapor-liquid equilibrium. In operation, hot vapors continuously enter the column from the bottom and pass up through the column. All along the length of the column, vapor-liquid equilibria are established, and the result is that there is a steady flow of hot vapors up the column and a steady flow of hot liquid condensate down the column. The combination is called a countercurrent flow (vapors up, condensate down). For the column to function efficiently (in terms of separating components), there must be a temperature gradient along the column: it must be hotter at the bottom and cooler at the top. For example, if a perfect separation were to be achieved with our 50:50 hexane-heptane mixture, the temperature at the bottom of the column would be that of the boiling mixture, 80°C, but the temperature at the top of the column would be 69°C, the boiling point of pure hexane.

A sophisticated research instrument will have some means to ensure that the proper temperature gradient is established from the bottom of the column to the top. In our column the temperature gradient results solely from the changing composition of the hot vapors going up the column and the condensate flowing downward. Provided that we distill slowly, the system will work fairly well. But if we distill too rapidly, vapor-liquid equilibria will not be established along the column, resulting in poor separation of components. Moreover, there will be too much condensate running down the column, and it is apt to "flood" (the column becomes plugged with liquid), a condition which further thwarts separation.

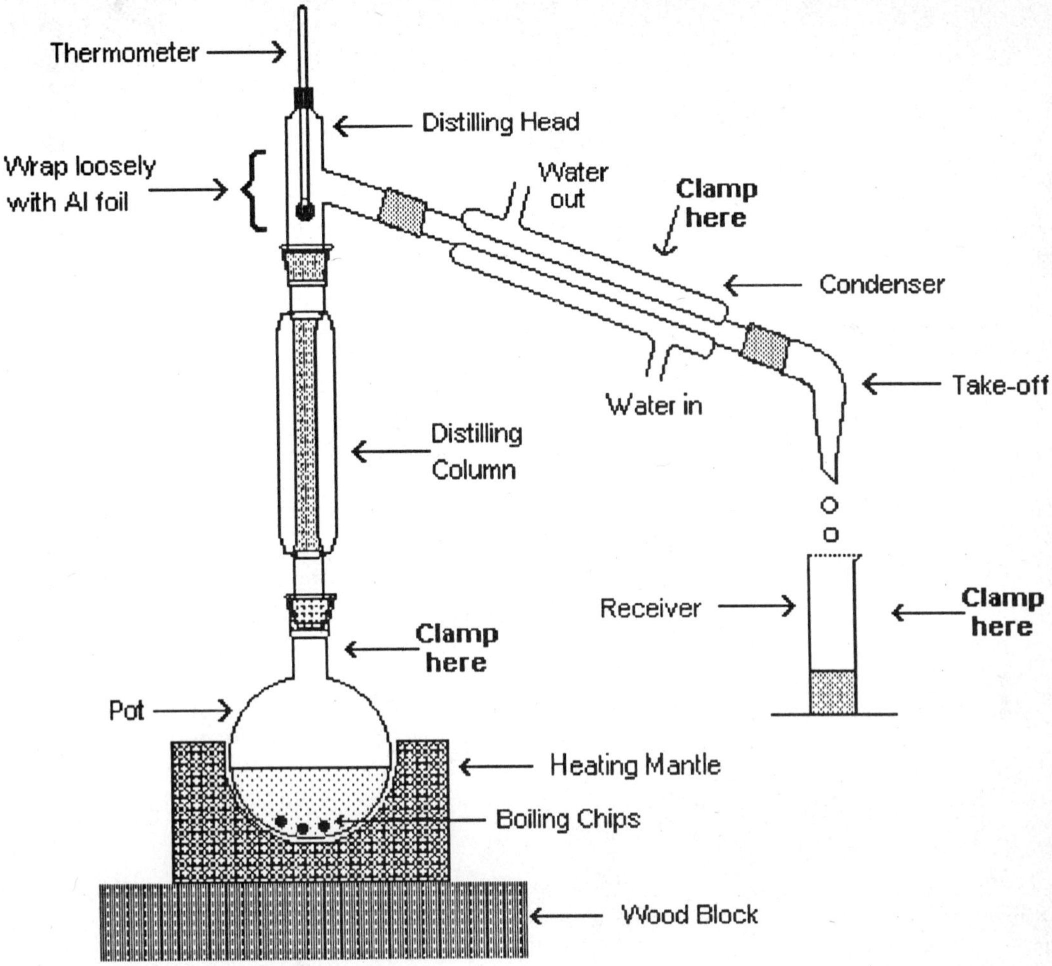

Figure 4. Apparatus for the fractional distillation of a liquid using a fractionating column.

Some compounds form highly non-ideal solutions; that is there is a great departure from Raoult's Law. In many cases such non-ideality does not permit a separation of the components by distillation. Instead a constant boiling mixture distills and further purification is thwarted. For example, ethanol-water mixtures distill to give a constant boiling distillate of 95% ethanol and 5% water. Such constant boiling, constant composition, mixtures are called **azeotropes.** Thus "alcohol" is often supplied as 95% ethanol (190 proof) because it is produced by an azeotropic distillation.

The following experiment demonstrates the process of distillation to separate the components of a mixture of compounds. The experiment not only demonstrates the use of a packed distilling column, but it also shows the value of the column in achieving a better separation of the components of the mixture than is possible in a simple distillation set-up where no distilling column is used. If the distillation is carried out slowly, a plot of the boiling point *vs.* volume distilled should show a marked difference between simple distillation and distillation using a packed fractionating column.

B. Experimental Procedure

1. Fractional and Simple Distillation

Many mixtures of liquids can be separated by distillation. If the boiling point differences of the components are large, a simple distillation should suffice to give good separation. However, when the boiling point differences are smaller; e.g., 30°C or less, an efficient distilling column will be required in order to get good separation of the components. In this experiment a mixture of compounds will be separated by simple or fractional distillation and the isolated fractions will be analyzed by gas-liquid chromatography and refractive indices.

Students will work in pairs, one performing a simple distillation and one a fractional distillation using a packed distilling column. The results of the distillations, particularly the distillation curves, should be compared and discussed in terms of the advantage of using the distilling column.

Equipment for Fractional Distillation: The fractional distillation set-up is shown in Figure 4 and consists of a round-bottom flask fitted with a distilling column filled with a packing material consisting of stainless steel sponge. The sponge is a simple, but very effective, material for maintaining good contact between the condensate and the ascending vapors. Such contact is essential for obtaining good separations of multi-component mixtures. The sponge must be packed snugly into the column, but not too tightly or the column will flood. There should not be any apparent voids in the packing, or the efficiency of separation of the components of the mixture will be diminished. A "distilling head" fitted with a thermometer is placed on top of the column. A condenser and a "take-off" adapter are attached to the distilling head. To collect the distillate a 50-mL graduated cylinder is positioned about 6 cm below the take-off adapter.

Equipment for Simple Distillation: The simple distillation set-up is shown in Figure 3. The only difference between fractional distillation and simple distillation set-ups is that a distilling column is not used in simple distillation.

A three-component mixture of acetate esters will be separated. Some physical properties of these esters are listed below.

Compound	MW	bp,°C	d^{20}, g/mL	n_D^{20}
methyl acetate	74.1	57.0	0.933	1.3614
ethyl acetate	88.1	77.1	0.900	1.3720
propyl acetate	102.1	101.6	0.888	1.3842

Note: It is important that in the following procedure you take exactly 30 mL of the mixture of acetates because subsequent directions are based on that volume.

Accurately measure 30 mL of the 1:1:1 mixture of methyl acetate, ethyl acetate, and propyl acetate and place the mixture in a 100-mL round-bottom flask to which you have added two or three boiling chips. Place the flask in the heating mantle and then clamp the flask to a vertical rod of a lattice or ring stand. Then attach the packed distilling column being sure to get it vertical. Attach the distilling head to the column. Then attach the condenser (you can temporarily hold it on with a rubber band) and support it with a clamp. When tightening the clamps, take care to avoid applying strain or breakage may occur. Tighten each clamp a little at a time, adjusting as needed to be sure the equipment is aligned properly. Finally the "take-off" can be held in place with two rubber bands. Carefully insert a thermometer through the adapter, **so that the bulb is just below**

the side arm of the distilling head. For a receiver, clamp a 50-mL graduated cylinder to a ring stand and center the top of the cylinder at least 6 cm below the tip of the take-off. This arrangement will leave room to slip a hand-held 4-dram vial under the take-off when a sample of distillate is to be collected. Attach the condenser hoses; the lower hose (toward the take-off) should go to the water inlet. Pass a slow, but steady, stream of water through the condenser.

Finally, loosely wrap **a piece of aluminum foil** around the distilling head in order to shield it from air currents that will cool the head and cause too much condensation and fluctuations in temperature. Make a small hole in the foil near the thermometer bulb. Observation of the liquid refluxing from the bulb will allow you to adjust the heating rate and to maintain better control of the temperature. Vapor and condensate should surround the bulb at all times during the distillation. You are now ready to start the distillation, **but first have your set-up checked by your instructor.**

Label five 4-dram vials (1 – 5) for collecting samples for later analysis. These vials will be used to collect about 0.5 mL samples of distillate. A 0.5 mL sample will give a 4 mm column of liquid in a 4-dram vial as measured to the bottom of the meniscus.

The distilling flask should be heated by applying a low voltage to the heating mantle. **It is crucial that you avoid heating too rapidly.** A fast distillation will lead to a poor separation of the components. Your only control is the heater voltage. Keep in mind that there is a "lag" in heating followed by an "overshoot." Initially the voltage should be set just high enough to cause gentle boiling. Then, as the distillation proceeds the voltage must be gradually increased. If rapid boiling occurs, immediately reduce the voltage. If absolutely necessary — very rapid boiling due to overheating — briefly slip the wooden block from under the heating mantle and drop the mantle to allow the liquid to cool to get the boiling rate under control. In fractional distillation, the vapors and condensing liquid will gradually rise up the column and eventually reach the distilling head and soon thereafter distillate will begin to collect in the receiver. In simple distillation, since there is no column, collection of liquid will begin almost immediately. In either case, try to adjust the heating to give a slow distillation rate — perhaps a drop every 4 or 5 seconds.

The heater voltage will have to be increased gradually; the greatest rate of change should occur during the change from a lower to a higher boiling component, at which point the distillation may appear to stop and the temperature may drop briefly until more heat is applied.

Using 4-dram vials, **collect 0.5 mL samples** of distillate at the following stages of the distillation (when the volume of distillate in the graduated cylinder reaches the specified level, collect the sample). The volumes given are based on starting with exactly 30 mL of the original mixture.

These vials must then be capped tightly to prevent evaporation because the samples will be kept for later analysis by GLC. If evaporation occurs, the sample may change composition through loss of the more volatile component, or the sample may be totally lost.

Vial	Volume collected in graduated cylinder
1	0 mL (the first few drops)
2	4 mL
3	13.5 mL
4	23 mL Stop distillation at this point.
5	Pot residue

Some of the highest boiling component will be "lost" due to "hold-up" in the column and flask. Even if the flask were to go dry, there still would be liquid and vapor in the column and vapor in the flask. Turn off the heater and allow the flask to cool. The vapors in the column will condense and drain into the flask, allowing you to collect the pot residue (vial #5). (Sometimes a higher-boiling liquid is added as a "chaser" to displace the last desired component from the flask and column.)

Record the volume and the corresponding distillation temperature in your notebook at frequent intervals throughout the distillation, including those points where samples are collected. Use the table on the report sheet as a guide (see Appendix).

If distillation is sufficiently slow and an efficient fractionating column is used, each compound, starting with the most volatile first, should come over at a fairly constant temperature that is reasonably close to the boiling point of the pure component. In practice we try to collect each fraction "cut" over a boiling range as narrow as possible. Each fraction should be predominantly one component. "Ideal" and "real" distillation curves are shown in Figure 5. The more efficient the distillation, the more closely will the curve approach the ideal example, leading to a cleaner separation of the components.

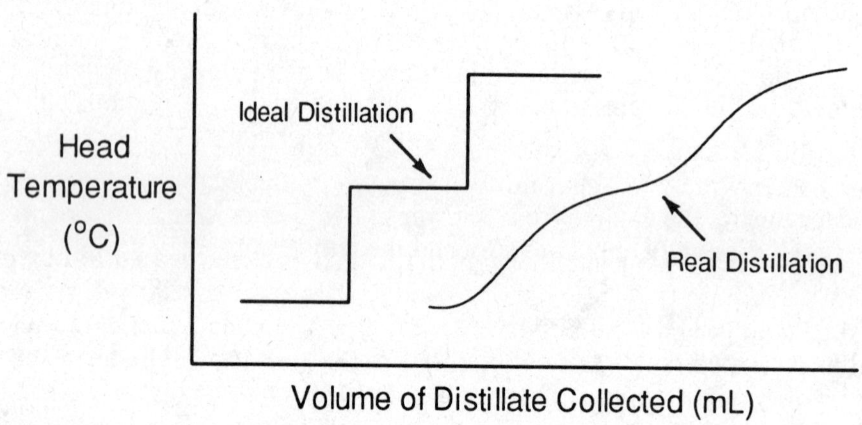

Figure 5. Ideal and real distillation curves.

On graph paper plot the observed "head" temperature (y-axis) versus the **cumulative volume** of distillate collected (x-axis), correcting for the cumulative volumes of the samples removed; e.g., at **Vial #4,** the total volume should be 23 + (4 x 0.5) = 25 mL. Mark the points where you withdrew samples with the sample number. If the report sheet of this experiment is used, plot your data on the grid reproduced on the reverse side.

2. Micro Boiling Point

We have learned how to determine the melting points of very small samples of solids. The boiling points of very small samples of liquids can also be determined. There are times when we have a only a few milligrams of a liquid, too little to measure the boiling point during a distillation. By use of the following procedure, microboiling points can be obtained. The equipment used for the determination of the micro boiling point is shown in Figure 6.

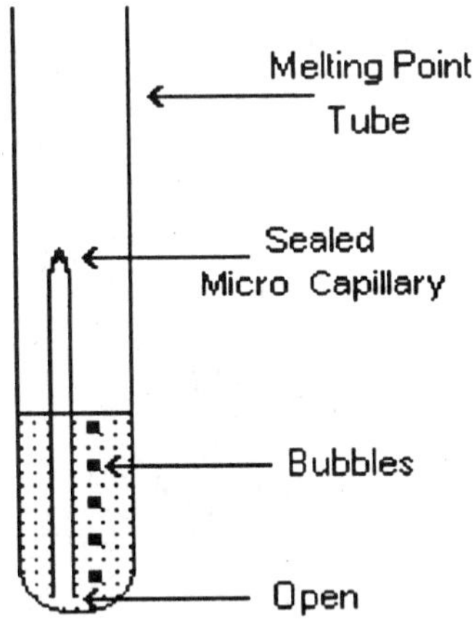

Figure 6. Apparatus for the determination of microboiling points.

A few microliters of a liquid are loaded into a melting point capillary. The loading is easily accomplished by heating the melting point capillary on a hot-plate while leaving one or two centimeters at the open end unheated. Heating drives out some of the air in the capillary. The open end of the capillary is then plunged below the surface of the liquid in question and as the tube cools, some of the liquid will be sucked into the capillary. When the tube has cooled to room temperature, tap the tube on the bench (as in loading a solid for a melting point), to move the liquid to the closed end. (Alternatively, one can carefully centrifuge the tube — not too fast — to force the liquid to the closed end.)

A section of very fine capillary tubing, which has been **sealed at one end,** is placed into the melting point tube with the open end of the capillary **directed downward,** (see Figure 6). These capillaries may be supplied, or they can be made by drawing out a melting point capillary in a small flame of a microburner. A one- or two-cm section of this capillary tubing is broken out and one end is sealed in a flame.

The melting point tube is placed in a Mel-Temp apparatus and the sample is heated. When the boiling point of the liquid is reached, fine bubbles will flow from the mouth of the fine capillary. The heat is then turned off, or better, reduced so that the Mel-Temp cools slowly. At some temperature a "last" bubble will escape and bubbling will cease. This temperature at which bubbling ceases is taken as the microboiling point. To obtain a second reading of the microboiling point, it is permissible to "recycle" the microboiling point sample — simply reheat and recool.

Take one or more of your samples from the distillation and determine the microboiling point. Compare the microboiling point to the boiling range of your sample as measured during the distillation.

Dispose of waste chemicals in labeled waste containers provided in the laboratory.

C. Prelab Questions

1. Calculate the vapor pressure at 80°C of a mixture of 35 mol % hexane and 65 mol % heptane.
2. Calculate the mole fraction of each compound in the vapor in equilibrium with the mixture in (1).
3. If a **simple** distillation were to be performed on this mixture (from Question 1), what would be the composition of the **initial distillate**?
4. Suppose the mixture were to be distilled using a **very** efficient fractional distillation column. What would the composition of the initial distillate be?
5. Why do we add boiling stones to the boiling flask to be used in a distillation?
6. In your own words, define "boiling point" and explain why a liquid boils.
7. Why do we observe "low" boiling points in our laboratories?
8. Why would it be dangerous to heat a liquid in a distilling apparatus which had no vent or opening to the laboratory?

D. Postlab Questions

1. Why does the boiling point rise continuously throughout the distillation?
2. At what stages of the distillation were the changes of boiling point the greatest?
3. What could you do to increase the amount of propyl acetate collected by distillation?
4. Consider the samples in vials 2, 3, and 4. Judging by the appearance of your boiling point curve, which sample appears to be most nearly pure (one component)? Which appears to be least pure?
5. Plot both simple and fractional distillation curves on the same graph and compare the results. Indicate which is your data and which is your partner's data. What do you conclude concerning the effectiveness of using a distilling column?

6. Gas Chromatography

A. Introduction

Chromatography is the general name for several extremely powerful methods for separating mixtures of many types of compounds. The word **chromatography** implies color, and dates back to the earliest separations of mixtures of colored materials, such as plant pigments, into the individual components. While most separations no longer involve colored compounds, the name has been retained for all of these related methods which are then subclassified according to the details of the separation method.

Some of the terms used to describe these methods are column chromatography, paper chromatography, gas chromatography, thin-layer chromatography, high-performance liquid chromatography, gel permeation chromatography, ion-exchange chromatography, size-exclusion chromatography, and electrophoretic chromatography.

All chromatographic methods are based upon a common principle. A **moving phase,** which may be either a liquid or a gas, moves through a **stationary phase** consisting of a finely-divided solid, which is often coated with a non-volatile, or non-soluble, liquid phase. In many forms of chromatography, the stationary phase is contained in a **column** constructed of metal or glass tubing. The moving phase is continuously passed through this column. A mixture of compounds is injected at the front of the column and swept along through the column by the moving phase. **Separation of the mixture is achieved because the components of the mixture travel through the column at different rates** and emerge from the column at different times. The times for the emergence of the components are monitored by a detector and the components may be collected individually if desired.

In **gas chromatography** the mobile or moving phase is a non-reactive gas, usually **helium.** If the stationary phase is an uncoated solid, the technique is termed **gas-solid chromatography** and it is used mainly for separating mixtures consisting of low molecular weight compounds which are normally gases at room temperature. If the stationary phase is a solid coated with a nonvolatile liquid, then the method is termed **gas-liquid chromatography.**

Gas-liquid chromatography (GLC), often referred to simply as gas chromatography (GC), is a powerful analytical method for both qualitative and quantitative analysis of mixtures of volatile components. By this method the components of the mixture can be identified and their concentrations measured, even at very low concentrations. In this experiment gas-liquid chromatography is used to analyze the various fractions obtained in the fractional distillation of the mixture of three esters. Sample sizes are of the order of 1–10 microliters (1–10 mg), but equipment is available commercially for either much smaller or much larger samples.

B. Gas-Liquid Chromatography

1. The Stationary Phase

In gas-liquid chromatography the **stationary phase** consists of a very fine granular **solid support,** which has been coated with a layer of a **nonvolatile liquid.** The solid support is comprised of an inert material (originally ground firebrick) which has been carefully "sieved" to a uniform particle size. These particles are very rough and thus have a large surface area. The solid support is coated with 5 to 20% by weight of a nonvolatile liquid, which forms a very thin film on each particle, perhaps 0.1 micron or less in thickness (1 micron = 10^{-6} m) (see Figure 1).

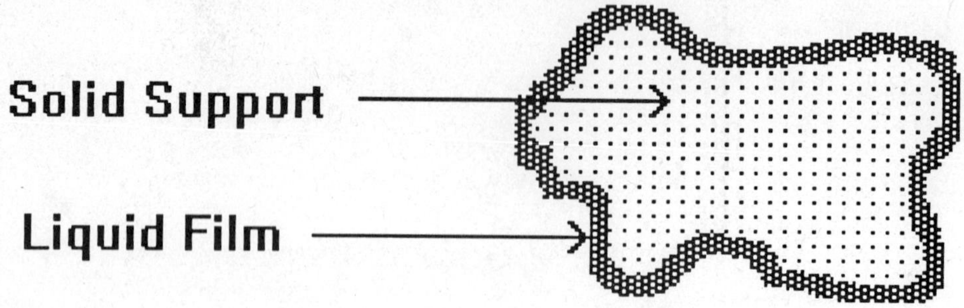

Figure 1. Magnified particle of solid support.

Many **nonvolatile** liquids are available for use as liquid phases, ranging from quite nonpolar substances to relatively polar materials. Thus the solvent properties of the liquid phase can be varied considerably. Even though the solid support is coated with a liquid, it appears to be a dry powder. When this powder is packed uniformly into a metal tube, typically 1/8- or 1/4-inch, one obtains, in effect, a highly-dispersed nonvolatile liquid anchored in the tubing. This **column** is coiled to fit into an instrument called a **gas chromatograph.**

2. The Gas Chromatograph

Figure 2 is a block diagram of a gas chromatograph. The mobile phase, **the carrier gas,** is helium and the gas is supplied at constant pressure to ensure a steady flow rate, usually about 60-100 mL/min. The helium first passes through an **injection port,** where samples are injected with a syringe through a septum into the helium stream. Since the injection port is heated, liquid samples are "flash vaporized." The components of the sample are then swept through the column by the carrier gas and then through the **detector.** The detector generates a signal when components from the sample pass through it and this signal is fed to a strip-chart recorder which continuously plots the magnitude of the signal as a function of time, as indicated by a distance measured on the chart paper. This chart is called a **chromatogram** (Figure 3).

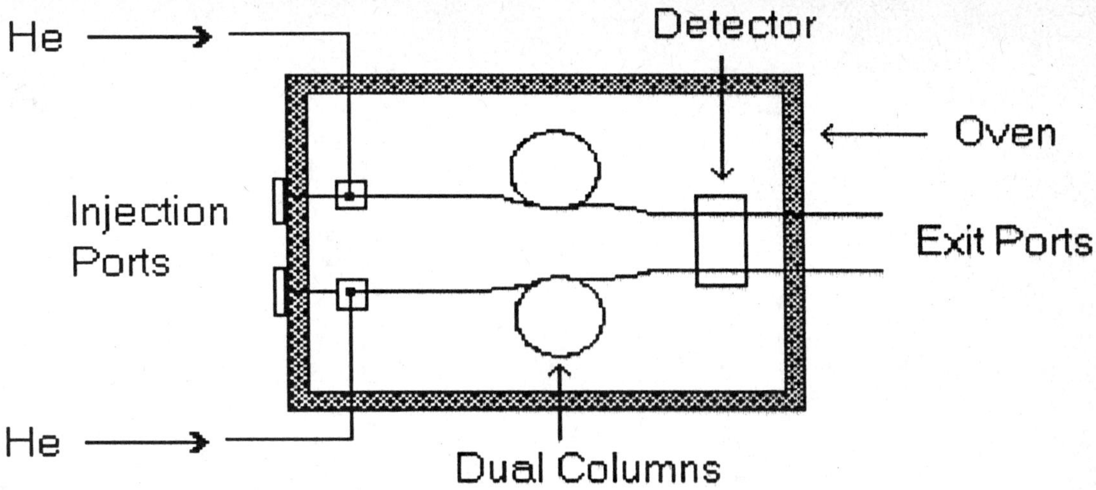

Figure 2. Gas chromatograph.

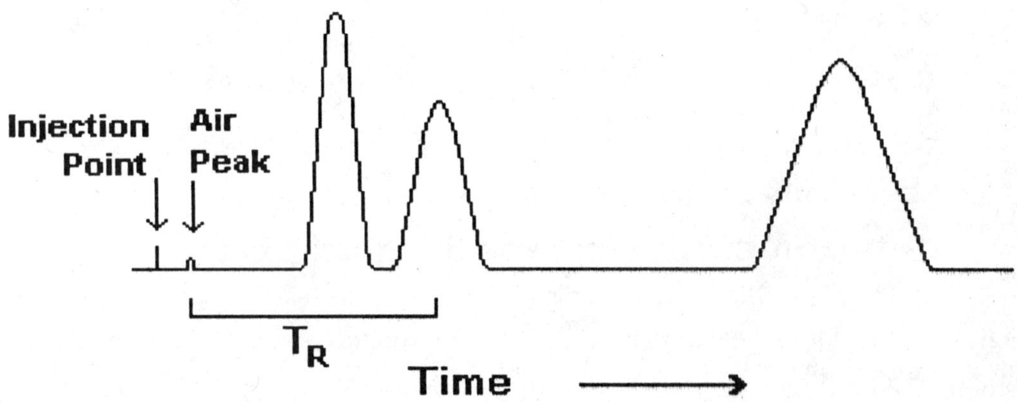

Figure 3. Appearance of a chromatogram.

Ideally, each component produces a symmetrical peak and, if the separation is good, the chromatogram will consist of a series of separate peaks. The time for a component to travel through the column, measured from the injection point to the peak maximum, is called the **retention time,** T_R. For a given column operating at a specific temperature and flow rate, **the retention time, T_R, is a characteristic property of each compound**; i.e., each compound will show the same retention time upon repeated injections under the same conditions. Thus, retention times can be used to identify components of a mixture.

Since the area under the peak is proportional to the **amount** of the substance passing through the detector, quantitative information can be obtained as well. We can find out what compounds are present in a mixture and how much of each is present.

3. Function of the Column

The basis for separations lies in the column. Why should different components travel through the column at different rates? To answer that question, let us consider a hypothetical section of the column (Figure 4). In this section we have the solid support coated with the nonvolatile liquid film and "free" space for the carrier gas. (No matter how tightly the column is packed, there still will be free space.) The carrier gas moves through this free space, passing around the stationary liquid phase.

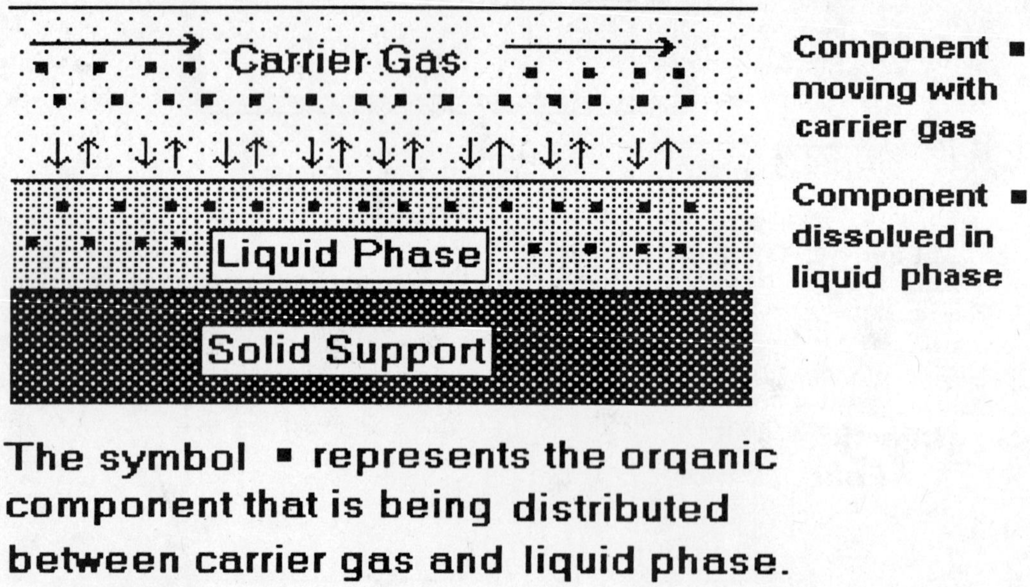

Figure 4. Section of column showing the components partitioned between carrier gas and liquid phase.

The carrier gas is moving through the column at a velocity represented by the long arrows. The sample component is represented by the small black squares and is being carried along by the carrier gas. What happens to a component of a mixture when it enters this section of the column? Once in this section (and in all sections of the column), the component will **partition** itself between the **gas phase** and the **stationary liquid** phase. Partitioning means that **some** of the component will **dissolve** in the liquid phase and **some** of it will be in the gaseous state being carried along with the carrier gas. (Some of the black squares of Figure 4 are now situated in the liquid phase as well in the gas phase.) According to Henry's Law, the partial pressure (and thus the concentration) of the component in the gas phase is directly proportional to its concentration in the liquid phase (and ***vice versa***). There is a **dynamic equilibrium** in which the molecules of the component rapidly transfer back and forth between the gas phase and the liquid phase. This partitioning of the component (small black squares) is represented by the vertical arrows of Figure 4. At any given instant, only a **fraction** of the component will be in the gas phase and it is only when the component is in the gas phase that it can move through the column. If this fraction is, for example 1/4, then this component will move along the column only 1/4 as fast as the carrier gas.

Each component of a mixture behaves similarly. A **low** affinity of a compound for the liquid phase means that at any particular instant, a **high** fraction of the component will be in the gas phase and thus the component will move relatively rapidly along the column. A **high** affinity of a compound for the liquid phase means that a **low** fraction of it will be in the gas phase and thus this component will move relatively slowly along the column.

In general, the more volatile a compound, the faster it will move through the column. However, compounds with similar boiling points may have very different retention times if they differ in other ways, such as polarity. If the stationary liquid phase is **polar,** then polar components will be more soluble in the polar phase and move relatively more slowly than nonpolar components. If the liquid phase is **non-polar,** then non-polar compounds will be more soluble in the non-polar phase and move more slowly.

4. The Chromatogram

The components of the mixture emerge from the column in the form of symmetrical peaks and the greater the retention times, the broader the peaks (Figure 3). With large differences in T_R the peaks are well separated. On the other, hand if the T_R values are close, the peaks move toward each other and may overlap (Figure 5). The width of a peak at half of the peak height, $W_{1/2}$, is directly proportional to the retention time. (This behavior is "ideal," but is closely followed provided that we don't "overload" the column with too much sample.)

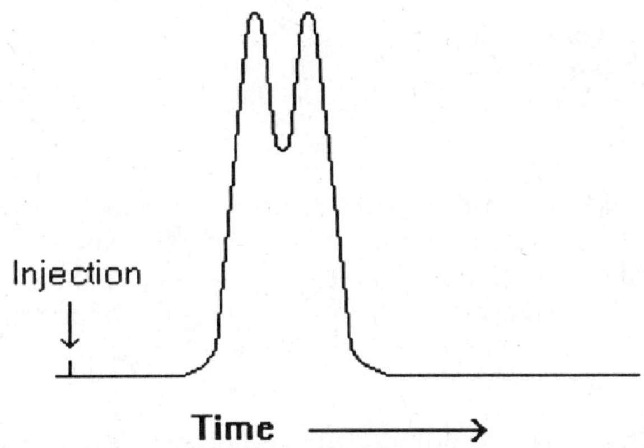

Figure 5. A chromatogram showing poor resolution.

The degree of separation of peaks is a measure of the **resolution** of the column (Figure 5). **High** resolution means sharp peaks and better separations, whereas **poor** resolution means broader, overlapping peaks, and poor separation. Many factors determine the degree of resolution of a column, but the most critical factors are set when the column is packed with stationary phase. Proper packing is essential for a column to give good resolution and the technique involved includes a high degree of art as well as science.

The operating temperature of a gas chromatography column is an important factor in controlling the separation of components of a mixture. As the temperature is increased, the fraction of sample in the gas phase increases, lowering T_R. The relationship is that log T_R is proportional to the reciprocal of the temperature given in degrees Kelvin (Figure 6). Normally the oven is maintained at constant temperature (isothermal operation), but sometimes it is advantageous to program a steady increase in temperature as separation proceeds. Programmed temperature gas chromatography is particularly useful in analyzing mixtures where the components have widely different boiling points.

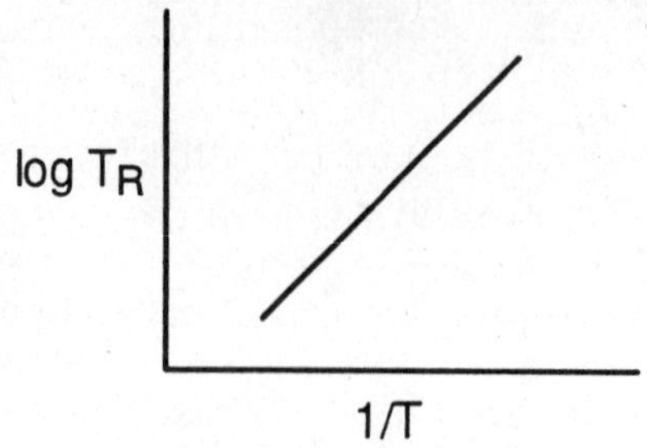

Figure 6. Relation between retention time (TR) and temperature. T in Kelvin.

Higher column temperatures give shorter retention times and result in faster analyses. But increasing the column temperature usually reduces the resolution of compounds with similar retention times and leads to overlapping peaks, as in Figure 5. The idea is to adjust the temperature to give fast analysis times, while still maintaining clean separation of the individual peaks.

A problem sometimes encountered is that stationary liquid phases are often somewhat volatile, and if the column temperature is too high, the liquid phase will "bleed-off" and destroy the effectiveness of the column. Thus, every column has a specified upper limit for its operating temperature.

5. Detectors

Several different kinds of **detectors** are employed in gas chromatographs. One of the simplest type is **the thermal conductivity detector.** Such a detector responds to a change in the thermal conductivity of the carrier gas due to the presence of another substance. A change in thermal conductivity of the gas generates a signal which is then translated into a response by a recorder. (Thermal conductivity refers to how fast heat is carried away from a hot body, in this case a hot wire).

With most detectors, including thermal conductivity detectors, the **area under a peak is proportional to the amount of component passing through the detector.** In very accurate work, we determine a **response factor** for each compound by determining the signal area produced by a fixed amount, e.g., one milligram, of each component. With thermal conductivity detectors, **equal weights of different organic compounds produce nearly equal peak areas** irrespective of the retention times of the compounds. As an example, for a mixture containing compounds A and B components, if the areas of peaks A and B of the chromatogram are in the ratio of 3:1, then the original sample was 75% A and 25% B.

C. Experimental Procedure

1. Running the Chromatograph to Obtain a Chromatogram

Most gas chromatographs have two columns, each consisting of a 1/4-inch or 1/8-inch copper or steel tube filled with the column packing and coiled to fit in the oven of the instrument. For a good pair of columns, one may contain a relatively nonpolar silicone oil liquid phase while the other has a more polar "Carbowax" (polyethylene oxide) liquid phase. Either column can be used, but the one selected will depend on the type of compounds to be separated. The instructor will provide operating procedures for your particular gas chromatograph.

Special microliter syringes are used to inject the small samples required for gas chromatography. **Handle the syringes with extreme care, as they are easily damaged.** The 25-microliter syringe has a glass barrel with a carefully fitted stainless steel plunger. The syringe should be cleaned several times with acetone before use to ensure that the sample does not become contaminated. The house vacuum can be used to evaporate the acetone. The instructor will demonstrate the proper handling of a syringe.

To load a syringe, place the needle below the surface of the liquid and pump the plunger up and down three or four times to remove air from the syringe. Then fill to the desired sample size. Wipe off the needle and back the plunger out a little (perhaps 1-2 microliter) to trap a slug of air in the end of the needle. The air provides an air peak on the chromatogram and also clears sample from the end of the needle to avoid premature volatilization of the sample as it is injected into the chromatograph.

To inject a sample, hold the syringe with both hands and carefully and gently push the needle through center of the rubber septum and continue to push it until the needle is all the way in. **If it hits something, do not force the needle further in because the needles are easily damaged.** Push the plunger in gently, but firmly, to eject the sample, **being especially careful not to bend the plunger.** Quickly withdraw the syringe from the injection system. A mark should be made on the chart at the time of injection. The instructor will demonstrate this procedure as well as how to use the other controls on the gas chromatograph and the recorder.

Record the chromatograms of the five distillation fractions, vials 1–5 from the previous experiment. In order to analyze the chromatograms properly, it is necessary that all peaks be on the chart paper. If one of the peaks goes off scale, either the attenuator is set too high, or too much sample has been injected into the chromatograph. To bring all peaks on scale, either inject less sample at the same chromatograph settings, or use the same sample size and reduce the sensitivity using the attenuator. On the other hand, very small peaks that do not fill up the chart paper will lower the accuracy of the quantitative analysis measurements. In this event, either inject a larger sample, or set the attenuator to higher sensitivity.

2. Analysis of the Chromatogram

Analyze the chromatogram according to the example shown in Figure 7.

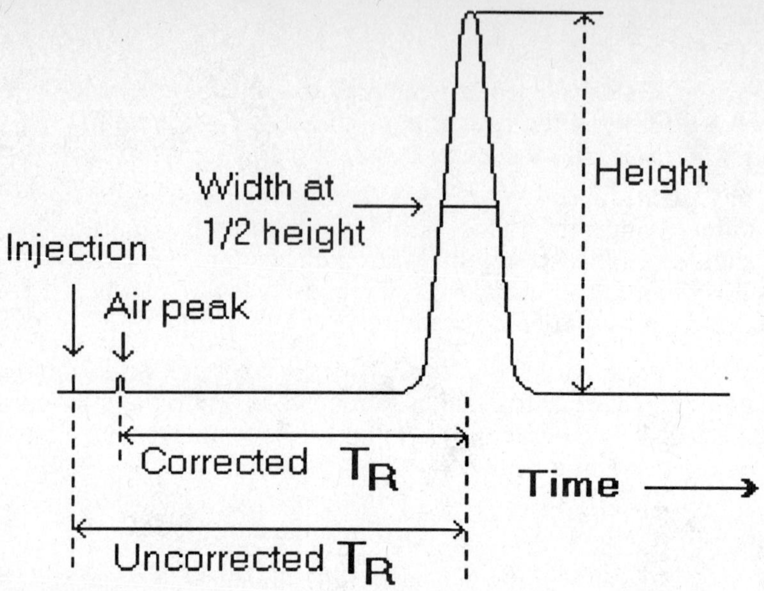

Figure 7. Analysis of the chromatogram.

Retention times measured from the time of injection are called "uncorrected." As noted earlier, usually a little air is injected along with the sample. This air moves through the column at virtually the same speed as the carrier gas. When the air emerges it generates an "air peak." Retention times are often measured from the air peak (taken as zero) and these are termed "corrected" (Figure 7).

Measure retention times with a ruler, or use the scale on the chart paper. If an air peak is present, measure retention times from the air peak; if not, measure the retention times from the moment of injection. Retention times depend upon the carrier gas flow rate, the column temperature, the amount and nature of the column liquid phase, and the length of the column. For a given column at a constant temperature, the relative retention times of different components (the ratios of the retention times) are constant, i.e., they are independent of flow rates, chart speeds, etc., and in fact do not vary much with moderate changes in temperature.

The area under a peak on a chromatogram is directly proportional to the amount of the component which produces the peak. On research instruments, peak areas are usually measured using electronic integrators. Without an integrator the area under each peak can still be estimated by one of the following methods.

1. Measure the peak width at one-half of the peak height, $W_{1/2}$, and multiply by the peak height, h (Figure 3).

$$\textbf{Area} \propto h \times W_{1/2}.$$

While this method can give good results, it requires reasonably wide peaks in order to measure $W_{1/2}$ accurately. This means that chromatograms have to be spread out by running at higher chart speeds than would otherwise be needed.

2. The area of a peak is proportional to the product of the peak height times the retention time (Figure 3).

$$\textbf{Area} \propto h \times T_R \quad \text{or} \quad \textbf{Area} = k \times h \times T_R$$

This relationship arises because the peak width increases linearly with retention time. If **all** areas are measured using the product h x T_R, you do not need to determine the proportionality factor k. In most cases, approximating each area by the product of the peak height times the retention time provides adequate accuracy and is the method of choice.

3. Refractive Index

The refractive index (R.I.) is a useful property for characterizing a liquid. The refractive index measures the ratio of the velocity of light in a vacuum to the velocity of light in the liquid (or solid). All light slows down as it passes through matter. Thus, we find refractive indices are all greater than one. Typical organic liquids have refractive indices in the range of about 1.3 – 1.7.

The instrument used to measure a refractive index is called a refractometer. There are many different types of refractometers and your instructor will explain the operation of the type of refractometer available. Only a drop or two of liquid is required.

Perform the operation carefully. It is particularly important to avoid scratching the prism. Also, be sure to clean the prism with a tissue soaked with ETHANOL when you have finished. DO NOT USE ACETONE, OR THE CEMENT HOLDING THE PRISM MAY BE DAMAGED.

Look through the eyepiece of the refractometer and adjust the prism so that a section of light resembling a spectrum appears. You should also see a dark band as well. Adjust the **dispersion control** so that the dividing line between light and dark is as sharp as possible. Then adjust the prism so that the sharp line of light intersects the cross-hairs precisely. At this point the view should appear as in Figure 8, left. Then switch to the scale to read the refractive index. The scale should appear as in Figure 8, right.

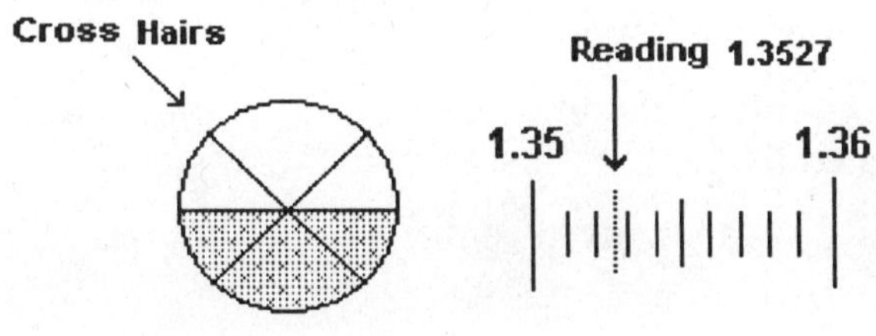

Figure 8. Determining the refractive index.

The dotted line in the Figure is the marker indicating the refractive index and it is now necessary to read the position of the marker. The numbers at the top will give the first two decimal places and immediately we see that the R.I. is between 1.35 and 1.36. Count the number of small divisions starting from the number on the left and proceeding up to the R.I. marker. In the example of Figure 8, the number of divisions is 2. Thus, the third decimal is 2. We now know that the R.I. lies between 1.352 and 1.353. Finally, estimate the distance from 1–10 that the R.I. marker is past the small division. In the example, this distance is estimated to be 7.

The reactive index is now known to 4 decimal places (5 significant figures) and is reported as n_D^{25} = **1.3527.** This is accurate to ± 2 in the 4th decimal place.

The "D" refers to the D-line of the sodium light. Although white light was used to measure the refractive index, the instrument is calibrated to give readings as if the orange sodium-D light had been used. The 25 refers to the **temperature** of the instrument in degrees Celsius. When you measure the refractive index, note the actual temperature of the instrument, usually the ambient temperature, and record this as indicated above. Recording the temperature is necessary because the R.I. changes with temperature. Actually the R.I. **decreases as the temperature increases** because the **density** of the liquid decreases.

For a mixture, the refractive index is generally closely related to the **weight fractions** of the components as shown in the following relation:

$$n = \Sigma n_i W_i$$

where n is the refractive index of the mixture and n_i is the refractive index of a component which has a weight fraction of W_i. Thus the R.I. can be used to measure the composition of a binary mixture when the R.I. of the two individual components are known.

Determine the refractive index of the contents of vial #3 and vial #5. Referring to the refractive indices of the pure components, estimate the composition of each fraction. What do these data suggest concerning the purity of the samples? Compare these results with the compositions of the samples as determined by gas chromatography. Explain your results.

Using the composition of each of these two fractions as determined by gas chromatography, together with the refractive indices of the pure components, calculate the R.I. expected for each fraction. Compare these data with the observed R.I. values.

Note: When all analyses have been completed, return the distillation fractions to the bottle provided in the laboratory.

D. Prelab Questions

1. Predict qualitatively how the retention times in gas chromatography will vary for a homologous series of compounds?
2. Does the fact that two different compounds have the same boiling point mean that they cannot be separated by gas chromatography?
3. Does the fact that two compounds have different boiling points mean that they definitely can be separated by gas chromatography?
4. Suppose that you find that two compounds have identical retention times. What can you do to (try) to get a separation?
5. The "air-peak" is assumed to represent the time it takes the carrier gas to move through the column. Why does air move as fast as the carrier gas while organic substrates move much more slowly?

E. Postlab Questions

1. Based on the analysis of your samples by gas chromatography, which of the three samples is most nearly pure? Which is least pure? Explain the results.

2. For otherwise identical conditions, how will the retention time of a substance vary with the **amount** of liquid phase used in the packing? How will **relative** retention times vary with the amount of liquid phase?

3. Suppose you do a GC analysis on a sample consisting of 15% A, 60% B and 25% C by weight. A and B are volatile and have different retention times but C is nonvolatile under the conditions used and therefore does not generate a peak on the chromatogram. Assuming that you do not know that there is nonvolatile material in the sample, what will you conclude from the GC analysis, i.e., what percentages of A and B would you calculate?

4. To get around the problems presented in problem 2, the **internal standard** technique is used. A volatile component which does not interfere with the GC analysis (i.e., it gives a **separate** peak) is added to the sample which is then analyzed. For example, 10.0 mg of D was added to 50.0 mg of the mixture of A, B and C (problem 2). The sample is then analyzed by GC (note that you need to inject only 1 or 2 μL) giving a chromatogram with peaks for only A, B, and D. The relative peak areas (i.e. relative to D) are found to be: A, 0.75; B, 3.00; D, 1.00. For 10.0 mg of D, what weights of A and B are indicated by this chromatogram; i.e. what weights of A and B are in the original 50 mg sample? Thus, what are the weight percentages of A and B in the original sample?

5. Now apply the concepts developed in problem 3 to analyze a different mixture containing A and B as well as unidentified nonvolatile materials. An 8.0 mg sample of D was added to 40.0 mg of an unknown. Analysis of this mixture gave the following peaks, listing the component with the retention time and the peak height (in parentheses), both given in chart paper units: A, 20.0 (90.0); B, 30.0 (70.0); D, 50.0 (24.0). Calculate the weight percentage of A and B in the original 40 mg sample. Show all work.

7. Thin-Layer Chromatography

A. Introduction

Thin-layer chromatography (TLC) evolved from paper chromatography and column chromatography.

In paper chromatography, a mixture of compounds is "spotted" on a section of filter paper along a pencil line called the origin. The filter paper is placed in a developing solvent with the origin above the liquid level. The solvent rises through the paper by capillary action, sweeping the components along with it, each component travelling at a different rate. When the solvent has risen some distance, the paper is removed, the line of the solvent front is marked with a pencil, and the paper is dried. If the components are colored, they can be seen as spots on the paper. Generally the components are colorless and some means of visualization is performed. Sometimes a reagent is applied to the chromatogram to give colored spots. Ultraviolet light may cause the compounds to fluoresce. Paper chromatography gives good results with certain hydrophilic compounds, such as amino acids, but most organic compounds do not work well on a paper substrate.

In column chromatography, a finely-divided solid adsorbent, usually alumina (Al_2O_3) or silica gel (SiO_2), is packed into a glass tube similar to a buret to form a column. A solvent is then run through the column to displace the air. A short layer of a concentrated solution of the mixture to be separated is then placed on the top of the column and allowed to run into the column of packing material. A solvent "the eluant," is then passed through the column and carries the components of the mixture along with it. The components move down the column, generally at different rates. The eluant is continually run through the column until the components emerge from bottom. Ideally, each component is eluted from the column separately in a series of fractions which can be collected in separate vessels.

In thin-layer chromatography (TLC), the same adsorbents (alumina, silica gel) that are used in column chromatography are supplied coated on thin sheets of plastic (or glass). The procedure is similar to paper chromatography in that the mixture to be chromatographed is spotted along a line drawn on the TLC plate. A developing solution rises by capillary action into the coating carrying the components with it.

TLC, as well as column and paper chromatography, is based on a partitioning of the compound between the moving phase (the developing solution) and adsorption on a polar stationary phase substrate (alumina, silica gel, or paper). Some compounds bind tightly to the adsorbent and others do not. This difference in adsorptivity means that different compounds spend a greater or lesser fraction of time in the moving liquid phase. The smaller the adsorptivity, the greater the time spent in the moving phase, and the greater will be the rate of travel through the substrate. Since different compounds have different absorptivities for the substrate, they move at different rates and are thus separated. After the chromatogram is completed, each of the separated components must be "visualized" by one or more of several techniques available.

B. Experimental Procedure

1. General Techniques for Thin-Layer Chromatography

TLC plates coated with silica gel and containing an ultraviolet fluorescent indicator, are used in this experiment. The silica coating must not be touched, because fingerprints can interfere with the analysis of the chromatogram. Always hold the plates by the edges. With a **pencil** draw a light line about 1 cm from the end of the plate, being careful not to break through the surface of the silica coating.

The developing chamber (Figure 1) consists of a 250-mL beaker having a height a little greater than the length of the TLC plate and covered with a watch glass or aluminum foil. Enough developing solution is added to the beaker to bring the level to about 0.5 cm. Place a circle of filter paper in the beaker so that one edge is in the developing solution and the rest of the paper is pressed against the wall of the beaker. Swirl the solution around so that the paper is saturated with the developing solution. The purpose of the paper is to help maintain an atmosphere that is saturated with the vapors of the developing solution. This prevents premature evaporation from the TLC plate which will cause distortion of the chromatogram.

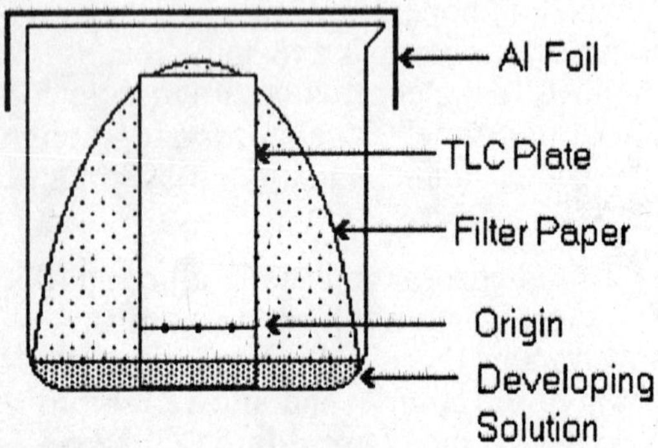

Figure 1. TLC chamber.

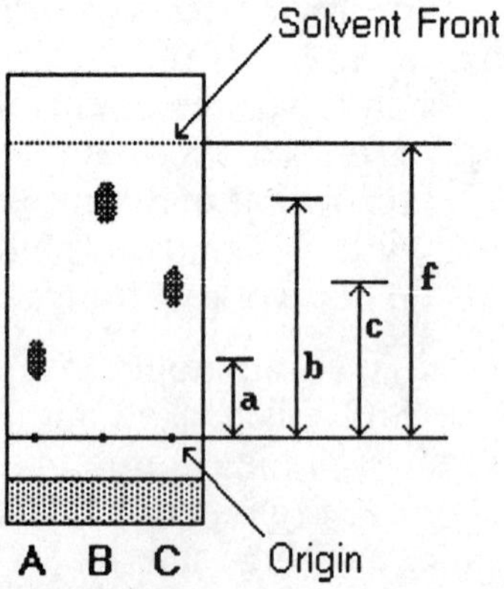

Figure 2. TLC chromatogram. $(R_f)_A$=a/f, $(R_f)_B$=b/f, $(R_f)_C$=c/f

The sample is dissolved in an appropriate solvent and spotted on the pencil line of the TLC plate using a capillary pipet. The pipet is made by drawing out a melting point tube in a flame to a very fine capillary. (Alternatively the capillary pipets may be supplied to you.) The pipet need only be dipped into the sample solution and the solution will run up into the pipet by capillary action. (Your instructor will demonstrate how to make and use a capillary pipet). The spot on the TLC plate should be **as small as possible,** perhaps 1 mm in diameter. A small identifying code number (**using pencil**) should be placed on the plate below the spot to identify the material spotted at that point. More than one sample, spaced about 1 cm apart, can be spotted on a plate. After spotting, allow the spot to dry, place the plate in the developing chamber, and cover the chamber with a watch glass or aluminum foil.

Make sure that the level of the developing solution in the chamber does not cover the spots. If this happens, your chromatogram is ruined, and you must start over. If the level of the solvent in the developing chamber is above the level of the spots, the solvent merely dissolves out the components of the spot and there will be no separation.

The solvent rises up the plate by capillary action, a process that takes several minutes. As the solvent moves through a spot, the components making up the spot move also, but at different rates. Thus the components are separated into individual spots on the plate (Figure 2).

When the solvent front nears the top of the plate, the plate is removed from the chamber and the solvent front is **quickly marked with a pencil before the solvent evaporates.** If the solvent front is not marked quickly, the solvent will evaporate and the position of the solvent front will be lost, and then it will be impossible to determine the R_f values for the compounds. After marking the front, allow the solvent to evaporate. The plate is now ready for "visualization" of the spots.

If the spots are colored they can usually be seen without aid. However, there may be spots that are colorless. Thus, special techniques must be employed to visualize the colorless spots. The TLC plates have a small amount of a fluorescent indicator bound to the silica. The short wavelength ultraviolet light (254 nm) of a UV-lamp causes the indicator to fluoresce, emitting an orange glow best seen in semi-darkness. Because many compounds quench the fluorescence, the components on the TLC plate appear as dark spots under the UV-lamp. Examine the plate with the UV-lamp to visualize the spots. Then with a pencil, draw a small circle around each spot that you have observed.

Using a millimeter ruler, determine the R_f value for each component. The R_f value is the distance a component (spot) has moved from the origin divided by the distance that the solvent front has moved from the origin (see Figure 2). With a given solvent system, the R_f value is a characteristic of a compound and can be used to identify the compound in an unknown mixture.

The UV lamp may also provide long wavelength UV light (366 nm) which does **not** trigger fluorescence by the indicator. This wave length may cause one or more of the spots to fluoresce, providing another way to make the spots visible. You should try both wave lengths and **record** the way the spots respond to UV light.

If a UV lamp is not available, an iodine chamber may be used to visualize the spots. Using this technique the dried plate is placed in a closed chamber containing crystals of solid iodine. The vapor of the iodine reacts with most compounds to give dark spots on the plate that are readily visible. When the spots are clearly visible, remove the plate and draw circles around the spots as they may disappear in time. Tape the plate into your notebook with Scotch tape with the tape completely covering the TLC plate. The tape will prevent the spots from disappearing.

TLC has the great advantage that very simple equipment is employed, yet remarkable separations can be achieved. The degree of separation depends upon the stationary phase (the silica on the TLC plate), the nature of the mobile phase (the developing solvent), and the care exercised in performing the procedure. It is essential that a sample be applied to a TLC plate as a **tiny, concentrated spot.** The spot **must be as small as possible** to reduce excessive spreading when the plate is developed. One must not use too much material, even with a very small spot, or the spot will spread too much and may overlap with other spots. Other conditions such as spreading, streaking, tailing, fronting, may also be observed. To achieve optimum results in TLC, it is often necessary to try a variety of solvents and solvent mixtures in order to achieve acceptable separations.

2. Separation of Acetaminophen, Acetysalicylic Acid, and Ascorbic acid.

Note: TLC plates from different manufacturers often give different results, even though the substrates may appear identical. Eastman Chromatography Plates with Fluorescent Indicator, Number 13181, work well for the experiments described here.

In this experiment, an unknown mixture of colorless compounds is to be analyzed by thin-layer chromatography and by examination of the chromatogram the components of the mixture are to be identified. The compounds in the mixture (all readily available at the drug store) are as follows: acetaminophen (Tylenol), acetyl salicylic acid (aspirin), and ascorbic acid (vitamin C). These compounds will be available in the laboratory as alcohol solutions or as solids and will serve as comparison samples to aid in the identification of the components of the unknown mixture. If the compounds are available as solids, obtain small samples and add a few drops of ethanol to each. It is not necessary that all of the compound dissolve.

Prepare several capillary pipets from melting point tubes as described above. Prepare a 4 x 8 cm TLC plate by drawing a light pencil line 1 cm from the end of the plate. Obtain a sample of the unknown. If the unknown is a tablet, crush it into a powder and add enough ethanol to cover the solid material. If the unknown is a solution, then obtain about 0.5 mL in a small test tube or vial. Using the capillary pipets, one for each sample, spot the three known standards and the unknown sample spaced out along the pencil line of the 4 x 8 cm TLC plate. The spots should be spaced about 1 cm apart and no closer than 0.5 cm to an edge. Mark each spot with a code number **in pencil** just below the spot.

To develop the chromatogram, a solution of 10% water in acetone will be used. Add the developing solution to the 250-mL beaker, making sure that the level of developing solution will be below the pencil line of the TLC plate when it is placed in the beaker. Add the sheet of filter paper, cover the beaker and swirl it to saturate the free space with solvent vapors. Stand the TLC plate in the beaker and cover the beaker with a watch glass or aluminum foil. Allow the plate to develop until the solvent nears the top of the plate. Remove the plate and quickly mark the solvent front with a pencil. Allow the plate to dry and observe the spots using the short wavelength tube of the UV lamp. Circle each spot with **a pencil** and determine the R_f values. Place the plate in an iodine developing chamber. When the spots are well developed, remove the plate; circle any additional spots and measure the R_f values. Cover the plate with transparent tape to preserve the iodine spots. Identify the components in the unknown sample.

3. Separation of the Pigments of Paprika

Place about 100 mg of paprika in a 4-dram vial or a conical vial and add about 0.5 ml of ethanol. Heat the mixture on the hot-plate and swirl for about 5 min. The ethanol solution is then suitable for TLC analysis directly.

Place two spots of the ethanol solution of the pigments along the pencil line of a 2.5 x 8 cm TLC plate. Allow the spots to dry and spot again at the same places with additional sample. Be careful when you do this and make sure to place each additional spot exactly on top of the original spot. Keep the spots as small as possible. Repeat the spotting about 8–10 times to build up sufficiently high concentrations of the pigment at the two spots. This repeat spotting is necessary, as otherwise the pigment spots of the developed plate will be faint and very difficult to see.

The developing solution in this case is pure methylene chloride. Develop the plates as before. To get a good separation, the developing solution should be allowed to travel almost to the top of the plate. Remove the plate and mark the solvent front quickly because methylene chloride evaporates rapidly. Observe the spots before the plate is quite dry because the spots tend to fade on drying. Mark as many spots as you can with a pencil. Number the spots and determine their Rf values. Record the colors of the spots. If a UV lamp is available, observe the spots under the short wavelength lamp and mark any additional spots that appear.

Dispose of waste chemicals in labeled waste containers provided in the laboratory.

C. Prelab Questions

1. Define the R_f value in TLC (use a sketch).
2. In TLC on alumina or silica, there is no stationary liquid phase as there is in GLC. What sort of property causes components to be "retained" by the substrate; i.e., what is the physical basis of the separations which are observed?
3. Consider two compounds of similar molecular weight: Compound **A** (nonpolar) and compound **B** (polar, with a carbonyl group and an OH group). Which one would be expected to have the **smaller** R_f value considering the fact that the alumina and silica gel are strongly polar? Explain.
4. Although we put a sample on a TLC plate as a **small** spot, we always find that the farther the component moves during development, the **larger** the spot gets. Suggest why this spot broadening occurs. What analogous behavior is found in gas chromatography?

D. Postlab Questions

1. Why does spotting too much sample on a TLC plate (overloading) lead to broad and/or irregular spots such as "streaking"?
2. In general, we assume that in a mixture, each component will show the same R_f value it shows when it is run alone. Sometimes the presence of a large amount of a second component **will affect** the R_f value of a component of interest. Suggest why this behavior is found.
3. **Thick-layer chromatography** is performed on a **preparative** scale in the same general fashion as TLC but the plates have much thicker layers of silica or alumina. Relatively large amount of a mixture are spotted in a virtual band along the origin (many spots are placed side by side). The plates are then developed in the usual fashion. Suppose a 75:25 mixture of A ($R_f = 0.80$) and B ($R_f = 0.40$) is run. Draw the developed chromatogram you should expect to see. How can A and B be **recovered** from the plate?
4. How could you use TLC as an aid in finding out how to separate a **large** quantity of material (5 g) by column chromatography?

8. Extraction: Separation of an Unknown Mixture

A. Introduction

Many organic substances can be separated by liquid-liquid extraction. Extraction is a process by which a compound is selectively transferred from one solvent to another. The two solvents must be dissimilar and mutually insoluble in each other, such as diethyl ether and water, or methylene chloride and water. In fact, a little ether dissolves in the water and a little water dissolves in the ether, but on mixing there is clearly an aqueous layer and an ether layer. Since ether is **less dense** (d = 0.70 g/mL) than water (d = 1.00 g/mL), the ether layer floats on top. In the case of methylene chloride (d = 1.325 g/mL) and water, the densities are such that the methylene chloride will form the bottom layer with the water on top.

Extraction theory will be explained through an example. Let us dissolve 4.0 g of compound **A** in 100 mL of ether and shake this solution with an equal volume of water. In the process of shaking, the layers are mixed thoroughly and the compound becomes distributed or "**partitioned**" between the two solvents. At equilibrium, the ratio of the concentration of **A** in the ether layer to the concentration of **A** in the water layer is a constant and this constant is called the partition ratio and is indicated by the Greek letter α. The symbol α is defined as the ratio of the concentration of the substance **A** dissolved in the ether to the concentration of **A** dissolved in the water.

$$\frac{[\mathbf{A}]_{\text{ether}}}{[\mathbf{A}]_{\text{water}}} = \alpha$$

If α is large (>>1) most of **A** will be in the ether layer. If α is small (<<1), most of **A** will be in the water layer. The partition ratio is a function of the nature of the compound **A** (the solute) and the properties of the two immiscible solvents.

In our example let us assume that α =3. Thus:

$$\frac{[\text{A}]_{\text{ether}}}{[\text{A}]_{\text{water}}} = 3.0$$

Since the system contains 4.0 g of **A**, distributed between equal volumes of the two phases, it is clear that 1.0 g of **A** must now be in the water layer leaving, 3.0 g of **A** in the ether layer for a total of 4.0 g. Thus:

$$\text{After 1st Extraction} \quad \frac{[\text{A}]_{\text{ether}}}{[\text{A}]_{\text{water}}} = 3.0 = \frac{(3.0 \text{ g}/100 \text{ mL})_{\text{ether}}}{(1.0 \text{ g}/100 \text{ mL})_{\text{water}}}$$

Now let us separate the layers and set aside the water layer containing **A.** Then let us extract the ether layer again with a **fresh** 100-mL portion of water. Again, the remaining 3.0 g of compound **A** will distribute between the two layers until equilibrium is reached, placing 3/4 of **A** in the ether layer and 1/4 in the water layer:

After 2nd Extraction $\dfrac{[A]_{ether}}{[A]_{water]}} = 3.0 = \dfrac{(2.25\ g/100\ mL)_{ether}}{(0.75\ g/100\ mL)_{water}}$

Again separate the layers, set the water layer aside, and extract the ether solution with a fresh 100-mL portion of water, giving:

After 3rd Extraction $\dfrac{[A]_{ether}}{[A]_{water]}} = 3.0 = \dfrac{(1.69\ g/100\ mL)_{ether}}{(0.56\ g/100\ mL)_{water}}$

At the end of three extractions with water, 1.69 g of **A** remains dissolved in the original 100 mL of ether and 2.31 g of **A** is dissolved in the 300 mL of water extracts.

It might seem to be a waste of time to make three successive extractions with 100-mL portions of water. What would have happened if we had used 300 mL of water in one portion at the outset? In this case the 4.0 g of **A** would be distributed as follows: If we let x equal the weight of **A** in the 300 mL of water extract, there must be 4 - x g of A dissolved in the 100 ml of remaining ether. Thus:

$$\frac{[A]_{ether}}{[A]_{water}} = 3.0 = \frac{(4 - x)/100}{(x)/300}$$

Solving for x gives x = 2 and 4 - x = 2. Thus the distribution is expressed as follows:

$$\frac{[A]_{ether}}{[A]_{water]}} = 3.0 = \frac{(2.0\ g/100\ mL)_{ether}}{(2.0\ g/300\ mL)_{water}}$$

Note that a single extraction with 300 mL of water removed 2.00 g of **A,** whereas three successive extractions with 100-mL portions of water removed 2.31 g of **A.** The point is important:

A series of successive extractions using smaller portions of solvent will always be more efficient than a single extraction with the same total volume of the extracting solvent.

In the example, we started with 4.0 g of compound **A** dissolved in 100 mL of ether and extracted some of the **A** with three portions of water. Suppose, instead, that the 4.0 g of **A** had been dissolved in water and we had used ether as the extracting solvent. In this case, using equal volumes of the two solvents, after each extraction 1/4 of **A** remains in the water. Thus after three successive extractions the amount of **A** remaining would be:

$$1/4 \times 1/4 \times 1/4 = 0.016 \text{ or } 1.6\%$$

Thus in three extractions 98.4% of **A** would have been extracted from the water by the 300 mL of ether. Show (by calculation) that had the 300 mL of ether been used in a single portion, only 90.0% of A would have been extracted from the water by the ether.

Separation of Mixtures Based on Acid-Base Properties: Clearly if we have a mixture of two compounds, **A** and **B,** which are to be separated by extractive methods, the process will not be successful unless the compounds have very different partition ratios.

In practice we find that many organic compounds are quite soluble in organic solvents such as ether, but are quite insoluble in water. If one of the components is an organic acid or an organic base, we can take advantage of the acid-base properties of the compounds to be separated by converting the acid or base into a water-soluble salt, which is insoluble in organic solvents. By so doing there is a dramatic change in the partition ratio of that component to favor water and this makes possible a clean separation of the components.

In this experiment a colorless organic acid will be separated from a yellow neutral compound. The course of the extraction can be observed by following the yellow color of the neutral compound.

The compounds are dissolved in diethyl ether, $CH_3CH_2OCH_2CH_3$, which is commonly called simply "ether." Ether is an excellent solvent for many organic compounds, but it does not dissolve salts. Since ether is immiscible with water and less dense than water (d = 0.714 g/mL), on admixture with water, ether will form the top layer. Ether is very volatile (bp = 35°C), which is an advantage because it is easily evaporated.

DANGER: Ether is <u>extremely flammable</u>. When it is in use, there must be no flames in the laboratory at any time. This is particularly important. Ether is more dense than air and will flow along the desk top and be ignited by a flame several meters distant. Never heat ether on a hot-plate because ether can be ignited by a hot-plate even though no flames are present. Always use a steam bath to heat ether solutions. Work in a hood at all times to avoid inhaling the vapors.

Both the acids and the neutral compounds are very soluble in ether, but are insoluble in water. However, if the mixture is extracted with aqueous sodium hydroxide, the organic acid is converted to its sodium salt which, being very water-soluble, is extracted into the water leaving the neutral compound in the ether. The reaction of the acid is shown below.

$$RCO_2H + Na^+ + OH^- \longrightarrow RCO_2^- + Na^+ + H_2O$$

Subsequently, the organic acid can be recovered from the aqueous sodium hydroxide by acidifying with hydrochloric acid. The strong acid reacts with the soluble sodium salt ($RCO_2^-\ Na^+$) to precipitate the water-insoluble RCO_2H.

$$C_6H_5CO_2^- + Na^+ + H^+ + Cl^- \longrightarrow C_6H_5CO_2H + Na^+ + Cl^-$$

The neutral compound is **not extracted** from a ether solution by aqueous sodium hydroxide and thus the organic acid and the neutral compound are cleanly separated by this technique.

Separation of an Unknown Mixture: The separation of an organic acid from a neutral organic compound by extraction, as described below, works well using the following compounds. The neutral compounds are all yellow; the organic acids are colorless. The composition of the unknown mixture will be determined through the use of melting points and mixed melting points of the final separated products.

Neutral Compounds	Melting Point °C
Fluorenone	84
Benzil	95
Dibenzalacetone	112

Organic Acids	Melting Point °C
Benzoic Acid	122
m-Toluic Acid	111
o-Toluic Acid	107

B. Procedure, Macroscale

Make sure that there are no flames in the laboratory before conducting this procedure. Do all manipulations in a hood.

In macroscale, a separatory funnel is used to carry out liquid-liquid extractions (Figure 1). To support the separatory funnel, a plate with a circular opening is placed on a ring clamped to the latticework or a ring stand in the hood. A conical separatory funnel is carefully lowered through the opening until it rests on the plate.

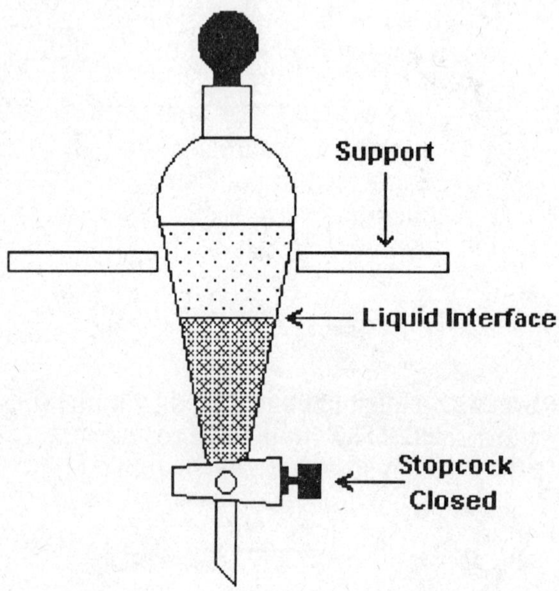

Figure 1. Separatory funnel.

1 Dissolve *ca* 2 g of the unknown mixture in 30 mL of ether contained in an Erlenmeyer flask.

If you are to separate a known mixture rather than an unknown (for example, benzoic acid and fluorenone), weigh out *ca* 1 g of each component and **record** the weights to 0.001 g (1 mg). Combine the samples in an Erlenmeyer flask, add 30 mL of ether, and swirl the flask to dissolve the compounds.

2. Transfer the ether solution to the separatory funnel **making certain that the stopcock is closed.** Add 15 mL of 5% aqueous sodium hydroxide and insert the stopper. Place one hand over the top of the funnel so that the stopper is in the palm of the hand. Keeping this hand positioned so that the stopper will not fall out, invert the funnel and **immediately release the pressure by opening the stopcock.** Close the stopcock and mix the contents by inverting the funnel at the rate of about one complete cycle per second. **Do not shake too hard** or you will emulsify the mixture. **Release the pressure frequently using the stopcock while the funnel is inverted.** Two to three minutes of gentle shaking is sufficient.

3. Place the funnel back in position on the support, **remove the stopper,** and allow the layers to separate. Gentle swirling might help the layers to form. Then **carefully** draw off the lower layer of aqueous sodium hydroxide into an Erlenmeyer flask. Before all of the lower layer has run out, it is good practice to close the stopcock and swirl the funnel gently to dislodge any droplets of the lower layer that stick to the walls of the funnel. On drawing off the last of the lower layer, be sure to close the stopcock before any of the ether layer enters it.

4 Add another 15 mL of the 5% sodium hydroxide to the separatory funnel, then insert the stopper. Invert the funnel, open the stopcock to relieve the pressure, close the stopcock, and shake gently as before. **Be sure to open the stopcock often to relieve the pressure.**

5. Draw off the aqueous layer into the same flask that contains the first aqueous extract.

6. Wash the ether with water. To do this, add *ca* 10 mL of saturated NaCl solution, shake briefly as before, draw off the bottom water layer and discard it.

7. Pour the ether layer out of the **top** of the funnel into an Erlenmeyer flask. (The **upper** layer should never be drawn off through the stopcock because it will pick up some of the lower layer remaining in the bore of the stopcock.)

Isolation of Organic Acid from the Water. To the combined water layers add about 8 mL of 6 **M** hydrochloric acid. A white precipitate of the organic acid will form. Stir the suspension and then check the pH with litmus or other indicator paper. The solution should be strongly acidic. If not, add more of the hydrochloric acid until it is. Cool the beaker in an ice-water bath, allowing it to stand with occasional stirring for at least 15 minutes. Collect the white solid on a Buchner or Hirsch funnel with suction. **Be sure to clamp the filter flask so that it does not tip over.** Disconnect the suction and wash the precipitate with a small portion (10 mL total) of ice water as follows: Pour a small portion of ice water over the solid. Allow the water to stand for 10 or 15 seconds, then reapply the suction. Repeat the process several times and then remove as much water as possible by suction. Transfer the solid to a **weighed** watch glass or weighing boat and allow it to "air-dry." The acid can also be dried rapidly in an oven set at 80°C. (Note: You must record the weight of the watch glass; you cannot simply "tare" it.) When dry, determine the dry weight and melting point. Calculate the percent recovery of the acid from the starting weight.

Identify the acid from its melting point and confirm the identification by taking a mixed melting point with an authentic sample.

Drying the Ether Solution. The ether solution from **7** is dried by adding a small amount of anhydrous $MgSO_4$ to the flask. The exact amount is not critical, but if there are no visible droplets of water, then 0.5 to 1 g should be sufficient. Swirl the flask for a minute or two to suspend the $MgSO_4$. Remove the $MgSO_4$ by gravity filtration using a glass funnel and filter paper, and collect the filtrate in a **dry** Erlenmeyer flask. Rinse the $MgSO_4$ on the filter with a few milliliters of ether and collect the filtrate in the same flask.

Isolation of the Neutral Compound from the Ether. After drying, the ether is removed by simple evaporation **on the steam bath in the hood.** Add a boilng stone and place the Erlenmeyer flask on the steam bath that has been regulated to provide a gentle flow of steam. **Do not turn the steam on to full force.** This is undesirable, because if too much steam is used, there is a great chance that water from the condensing steam will get in the flask, and then the product will come out wet.

As the ether evaporates, the neutral compound separates as a yellow solid, or it may separate as an oil that solidifies on cooling. Cool the flask and remove the compound from the flask with the aid of a spatula. Allow the solid to stand in the air until it is thoroughly dry, which should take only a few minutes. Determine the yield and the percent recovery.

> Evaporation of this quantity of ether into the atmosphere does not result in environmental problems. Ether is non-toxic, except in large doses, and was used for many years as an anesthetic in surgery. Ether has no effect on the ozone layer and in sunlight undergoes biodegradation to harmless materials.

Alternatively the bulk of the ether can be distilled and collected using the simple distillation equipment used in the section on distillation. A rapid flow of cooling water in the condenser may be necessary to condense the low-boiling ether.

Using the Craig tube, recrystallize *ca* 50 mg of the neutral product from 95% ethanol. Add a few drops of ethanol to the solid in the Craig tube and warm **gently** to boiling. Continue to add drops of ethanol and heat until the material is completely dissolved. Allow the solution to cool and crystallize, and then isolate the product by centrifugation. After drying, record the melting point and identify the product. Confirm the identification by taking a mixed melting point of the product with a sample of authentic material.

NOTE: This separation scheme is summarized in the flow chart (Figure 2).

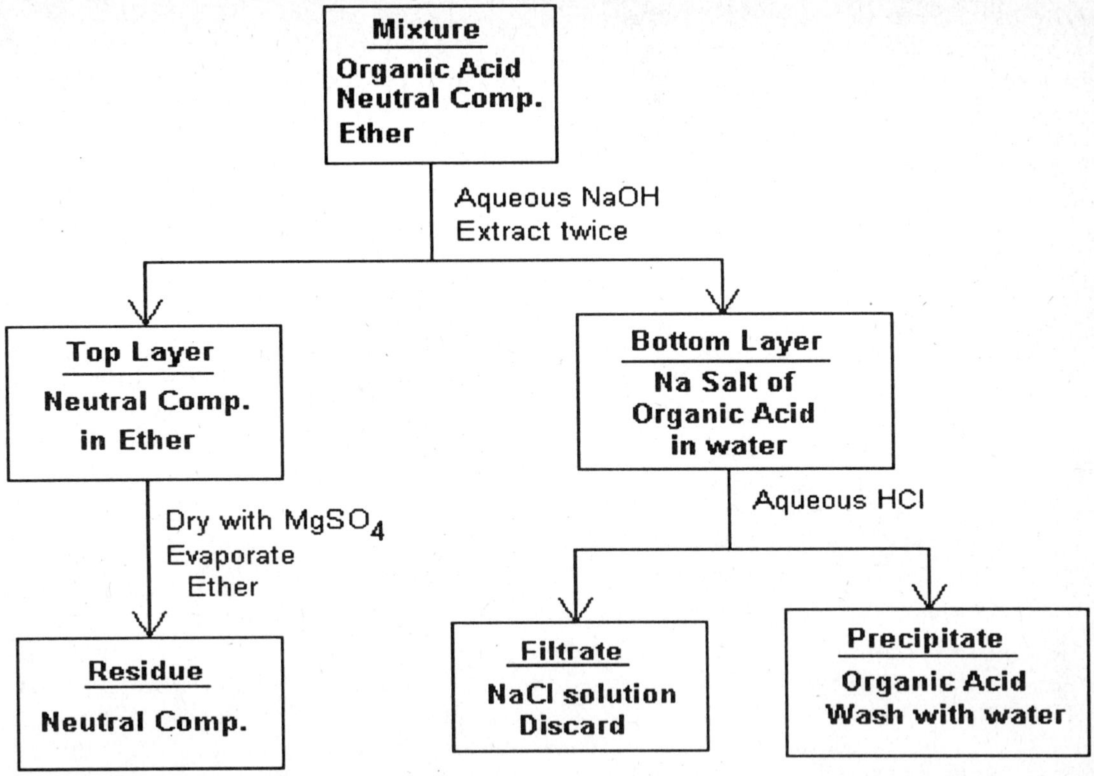

Figure 2. Chart for the separation of an organic acid and a neutral compound by extraction.

C. Procedure, Microscale

At this point, the discussion on extraction in the introduction chapter should be reviewed.

The following two procedures provide experience in conducting extractions in which the organic layer is the top layer (ether-water) or the bottom layer (methylene chloride-water). Which layer is the top and which is the bottom is governed only by the relative densities of the two liquids. Water is more dense than ether, and so water will form the bottom layer. Water is less dense than methylene chloride, so in this system water will be the top layer. Be very careful and know which layer is which at all times. Also, never discard any layer until the experiment is complete, and the products are isolated. This guards against inadvertently throwing away the wrong layer that later you wished you had saved.

1. Extraction from Diethyl Ether

Weigh out *ca* 200 mg of the unknown mixture (or 100 mg of benzoic acid and 100 mg of fluorenone) into a 5-mL conical vial (**vial A**) containing a spin vane. Add *ca* 1.5 mL ether (use vial marks to determine quantities) to the vial and stir until the solids are dissolved. Then add *ca* 1 mL of 5% sodium hydroxide solution and stir vigorously for about a minute, but not so vigorously so as to splash any of the solution out of the flask. The idea is just to get good mixing of the two immiscible liquids. Stop the stirring and allow the vial to stand until the liquids have separated and a clear line can be distinguished between the layers. Swirling gently will help. Using a micropipet, carefully transfer as much as possible of the bottom water layer to a second 5 mL conical vial **(vial B)**.

Note: When removing the bottom layer with a micropipet, squeeze the bulb to expel the air, dip the end down to the bottom of the vial, and squeeze out a few more bubbles of air into the solution to expel any of the top layer that may have gotten in the pipet during the immersion.

Add another 1 mL of 5% sodium hydroxide to **vial A** and stir as before. Remove the bottom layer as before and add it to the first extract contained in **vial B.** Add 0.5 mL of saturated NaCl to **vial A** and stir. Remove the bottom layer and discard. Add 0.5 mL of ether to **vial B.** Swirl briefly and transfer the bottom layer to a small beaker. (This washing with ether removes traces of fluorenone that might have been carried over with the water during the extractions.) Discard the ether washings.

2. Extraction from Methylene Chloride

Weigh out *ca* 200 mg of the unknown mixture (or 100 mg of benzoic acid and 100 mg of fluorenone) into a 5-mL conical vial **(vial A)** containing a spin vane. Add methylene chloride up to the 2 mL mark on the vial and stir until the solids are dissolved. Then add 1 mL (use marks on the vial) of 5% sodium hydroxide solution and stir vigorously for about a minute, but not so vigorously so as to splash any of the solution out of the flask. The idea is just to get good mixing of the two immiscible liquids. Stop the stirring and allow the vial to stand until the liquids have separated and a clear line can be distinguished between the layers. Swirling gently will help. Using a micropipet, carefully transfer as much as possible of the **top water layer** to a second 5 mL conical vial **(vial B)** without removing any of the bottom layer. Tipping the vial slightly will help in the removal. Add another 1 mL of 5% sodium hydroxide to **vial A** and stir as before. Remove the top layer as before and add it to the first extract contained in **vial B.** Add 1 mL of water to **vial A** and stir. Remove the top layer and discard.

In order to reduce contamination of the sodium salt of the acid, the aqueous solution contained in **vial B**, should be washed with a small amount of methylene chloride. To do this, using your tweezers, transfer the spin vane from the **vial A** to **vial B**. Add about 1/2 ml of methylene chloride to **vial B** and stir briefly. Remove the spin vane with tweezers and then using a micropipet, remove the bottom layer of methylene chloride and discard it. Transfer the aqueous sodium salt solution in **vial B** to a small beaker. The sodium salt of the acid and the neutral compound are now separated. The next step is to recover pure organic acid and the neutral compound from the two layers.

3. Isolation of the Organic Acid

To the sodium salt solution add 6 M HCl dropwise until precipitation is complete, as indicated by a strongly acidic solution (use pH or other indicator paper). Avoid a large excess of HCl, because the acid may be slightly soluble in water. Collect the product on the Hirsch funnel and wash it with a few drops of ice-cold water. Remove the solid from the funnel and paper, and allow it to air-dry (or oven-dry it at 80°C). Determine the yield, mp, and percent recovery. From the melting point, determine the identity of the acid. Confirm the identification through a mixed melting point with an authentic sample.

4. Isolation of the Neutral Compound

The neutral compound is to be isolated by evaporation of the methylene chloride, or ether, but before this can be done, the solution must be dried. In one of the short micropipets, insert a small wad of cotton and pack it in where the stem narrows, but not too tightly. Add about 1 mm of sand and then about 5 cm of anhydrous sodium sulfate. Tap the tube on the bench to pack it in. Clamp the tube to the stand just above a 3 mL conical vial or other container, see Figure 3. Then, using a micropipet, transfer the methylene chloride (or ether) solution to the sodium sulfate column and collect the effluent in the container.

NOTE: When you first try to pick up a highly volatile solvent, such as methylene chloride or ether, with a micropipet, you must be very careful or else you will find that the solvent is spontaneously ejected from the pipet, usually onto the desk-top, before you have time to transfer the material to the drying column. This occurs because of the high vapor pressure of the solvent. To prevent this from happening, place a few drops of the pure solvent that is being used (either methylene chloride or ether) in a clean vial and using the rubber bulb, pull some of the solvent into the pipet and force it out again into the vial. Do this several times before you try to transfer the extract. This process saturates the air in the pipet and bulb with the solvent vapors and thus prevents the buildup of pressure in the pipet. The transfer can then be made with the pipet without loss of material.

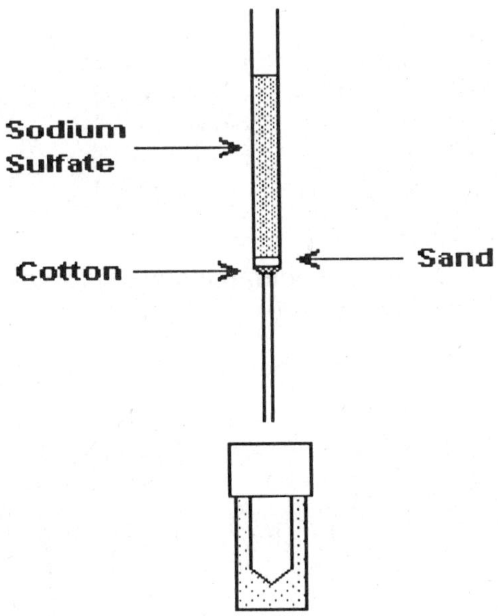

Figure 3. Sodium sulfate drying column.

When all of the solution has been run through the drying column, then run *ca* 1 mL of the pure solvent through the column to chase out the remainder of the solution containing the product. The solvent is then evaporated by a gentle stream of nitrogen flowing through a micropipet. During evaporation, the solution will become quite cold, and this retards evaporation. Gentle heating will speed up the evaporation. Alternatively, evaporation may be conducted on the steam bath. In this case, however, be careful not to let the steam enter the flask and get the product wet. After the solvent has evaporated, there may be a small amount of water remaining. (The product may appear to be damp.) If this occurs, just continue the evaporation until the crystals appear dry. Transfer the crystals to a weighing dish, allow them to dry further if necessary, and determine the yield, melting point, and percent recovery. The product may be crystallized from ethanol using the Craig tube. Determine the yield, mp, and percent recovery. From the melting point determine the identity of the acid. Confirm the identification by a mixed melting point with an authentic sample.

Dispose of waste chemicals in labeled waste containers provided in the laboratory.

D. Prelab Questions

1. *para*-Toluidine, mp 45°C, is very soluble in ether but only slightly soluble in water. It is an organic **base** which reacts with aqueous hydrochloric acid to form a water-soluble salt which is insoluble in ether.

$$CH_3\text{-}C_6H_4\text{-}NH_2 + H^+ + Cl^- \longrightarrow CH_3\text{-}C_6H_4\text{-}NH_3^+ + Cl^-$$

   Given a mixture of *para*-toluidine and fluorenone, devise a procedure based upon solvent-solvent extraction to separate these compounds. Give your answer in the form of a flow chart.

2. Suppose we extract an aqueous solution which is very dense (because it contains heavy salts) with methylene chloride and, when the layers separate, we are not sure which is the aqueous layer and which is the methylene chloride layer. What **simple** experimental procedure could be used to settle this problem?

3. (a) A 5.00-g sample of compound **A** was dissolved in 50 mL of water. This solution was then shaken with 50 mL of ether. At this point the ether was found to contain 2.00 g of compound **A.** Calculate the partition ratio: $[A]_{ether}/[A]_{water} = \alpha$.

   (b) The aqueous solution was then extracted **four** more times with 50-mL portions of ether (a **total** of five extractions). How much of the compound remained in the aqueous layer? How much of compound **A** was in the five combined ether layers? What percentage of A was extracted from the water?

   (c) How much **A** would have been extracted into ether if the original aqueous solution had been extracted **once** with a 500-mL portion of ether? What percentage of **A** would have been extracted by the ether?

4. Why is it necessary to wash the benzoic acid thoroughly with water; i.e., what is being removed?

E. Postlab Questions

1. You were told to avoid shaking the separatory funnel too hard when carrying out the extraction "....or you will emulsify the mixture." What is an emulsion? Why must an emulsion be avoided here?

2. Account for any loss of material you experienced in isolating the benzoic acid and the fluorenone.

3. Would ethanol be a suitable solvent to use in place of methylene chloride in this experiment? Explain.

4. In this experiment, how could you tell that the fluorenone was not being extracted by the aqueous sodium hydroxide?

9. An Introduction to Stereochemistry: Use of Molecular Models

Some of the terms and concepts applicable to stereochemistry, and which are covered in most lecture texts, are discussed below. This summary will be followed by exercises designed to help you understand structural stereochemistry. Refer to the section on stereochemistry in your lecture text while studying this chapter and while working the exercises using your molecular models.

A. Isomerism

Isomers are molecules which have the same molecular formula, but differ in the arrangement of the atoms. All isomers fall into two general categories: **structural isomers** and **stereoisomers.**

Structural (Constitutional) Isomers are molecules which differ in the connectivities of the atoms; i.e., the atoms are joined in different sequences. For example:

CH_3-O-CH_3 and CH_3-CH_2-OH; CH_3-$CHCl_2$ and CH_2Cl-CH_2Cl;

CCl_2=CH_2 and CHCl=CHCl; CH_2=CH-CH_2-CH_3 and CH_3-CH=CH-CH_3

Stereoisomers are molecules that have their atoms joined with the same connectivities (identical sequences), but differ in the spatial arrangements of the atoms. For example:

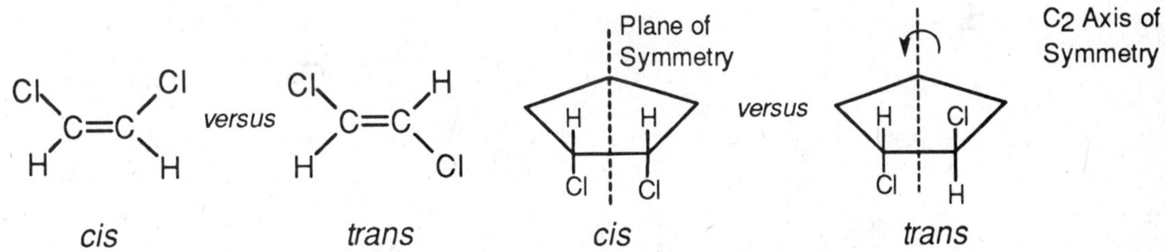

All stereoisomers fall into two classes: **enantiomers** and **diastereomers.**

Enantiomers are pairs of molecules that are mirror images of one another, but are not capable of being superposed.

Diastereomers are stereoisomers which are **not** enantiomers. The molecules must possess the same connectivities (a requirement for classification under stereoisomers), **but are not mirror images of one another.** The two 1,2-dichloroethenes above are diastereomers; they have the same connectivities, but are not mirror images.

Cis-trans **Isomers:** This type of stereoisomerism applies only to compounds that differ in the arrangement of atoms about double bonds or rings. *Cis-trans* isomers of this type are also called geometric isomers, a term often encountered. The examples above illustrate *cis* (on the same side) *trans* (on opposite sides) isomerism.

Conformational Isomerism: Groups such as the methyl group can rotate about the single bond that connects the group to the molecule. Such rotation is often called "free rotation." The term free rotation means that molecules can assume many different shapes, or **conformations,** as a consequence of rotation about single bonds. Every different shape that the molecule can assume as a result of these rotations is called a **conformer** of that molecule and the process is called **conformational isomerism.** Actually "free rotation" is a misnomer because, in fact, there are usually energy barriers to rotation about single bonds, due to various repulsive and attractive forces that increase and decrease as the rotation proceeds. However, unless these barriers are relatively high, free rotation will occur. In general, one can only freeze out conformations at very low temperatures. While a given molecule may be able to assume many different conformations, a few conformations may be preferred because they are lower in energy than other conformers. For that reason, the molecules of any given substance which can undergo free rotation consist of a collection of different conformers which readily interconvert with each other.

B. Symmetry

While we are all aware of a quality we call symmetry, we may not have considered the fact that there are different kinds of symmetry. We will look at some aspects of symmetry which are widely encountered in discussing the three-dimensional characteristics of organic molecules.

1. Symmetry about an Axis:

A simple symmetry axis is an imaginary line passing through the center of a molecule with the quality that a 360° rotation of the molecule about this axis will generate other orientations of the molecule which are indistinguishable from the original orientation. For example, the square planar platinum complex and the cyclobutane derivative shown below have C_2 axes because at 180° and 360° rotations the orientations are indistinguishable from the original orientations. The "2" of C_2 shows that there are two equivalent orientations or the "2" can be taken to mean that at 360°/2 = 180° the orientation is equivalent to the original orientation.

Trigonal planar molecules such BF_3 and the all *cis*-1,2,3-trichloropropane, have the C_3 axes shown below; in both cases a rotation of 360°/3 = 120°C gives an orientation indistinguishable from the original orientation. Note that the platinum complex has two additional C_2 axes and BF_3 has three C_2 axes in addition to the C_3 axis. The dibromodichlorobutane has one C_2 axis.

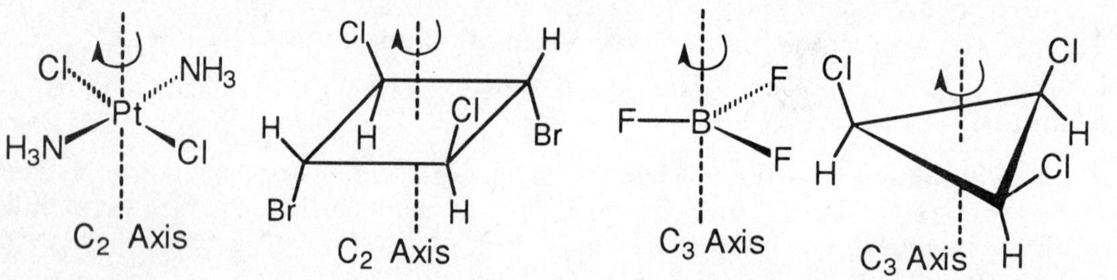

The existence of a simple symmetry axis, C_n, means that there are equivalent atoms which are interchanged by a 360°/n rotation. Thus in BF_3 all three F atoms are equivalent.

2. Symmetry about a Plane:

When we place an object in front of a mirror we see the reflection of the object in the mirror. We call this reflection the mirror image. The object and the mirror image may or may not be identical (superposable). For example, your two hands may be said to illustrate nonsuperposable mirror images. They can clearly be placed in an object mirror image relationship, but they cannot be superposed one on the other. Thus, your two hands are not identical. On the other hand, a cube and its mirror image are superposable and so the two are identical.

A molecule is said to have a **plane of symmetry** if an imaginary plane can be passed through the center of the molecule dividing it into two halves which are related to each other as object and mirror image. For example, the *cis*-1,2-dichlorocyclopentane shown above has a perpendicular plane of symmetry that passes through the bond connecting the C_1 and C_2 carbon atoms and the C_4 carbon atom. The all *cis*-1,2,3-trichlorocyclopropane has three symmetry planes and the 1,3-dibromo-2,4-dichlorocyclobutane has two symmetry planes.

3. Symmetry about a Point or Center:

A molecule is said to have a center of symmetry if the following condition is met: Consider all lines drawn through the center point of a molecule. Starting at the center, go out along each line any equal distances in opposite directions. If the molecule has a center of symmetry, identical atoms or environments will be encountered in the opposite directions. Three such lines are shown in drawing **A** below. Note that each line encounters the same atom going in either direction from the point of symmetry. Other lines may not encounter atoms, but the identical environments will be encountered in opposite directions. Drawing **B** is of a real molecule with a center of symmetry.

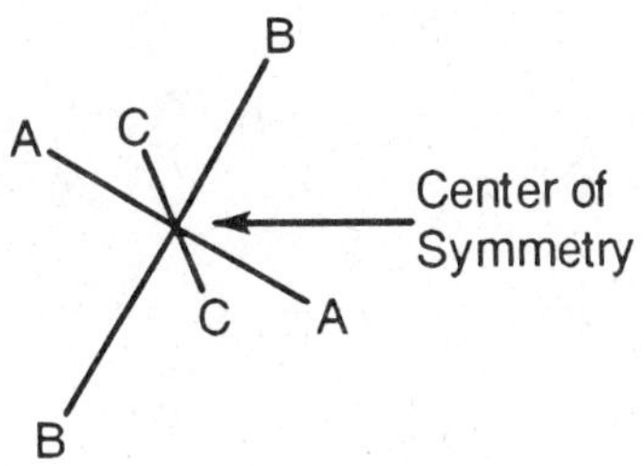

A

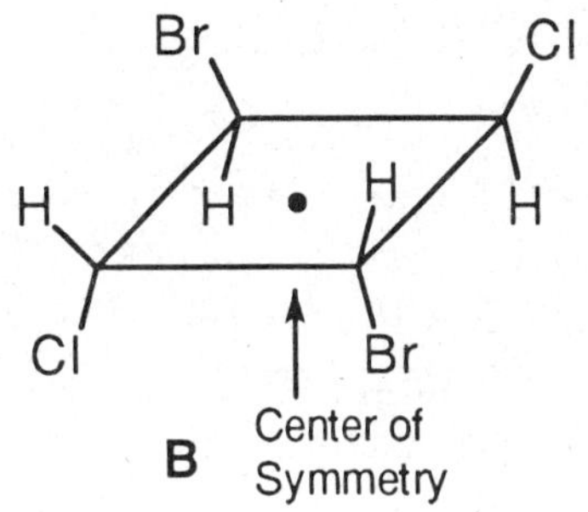

B

C. Chirality

Rule 1: A molecule, is said to be chiral if the object and its mirror image are nonsuperposable.

Rule 1 is the "necessary and sufficient test" for chirality. On the other hand, if the object and its mirror image are superposable, then the two are identical and the object is said to be achiral (not chiral). Chiral molecules related as object and **nonsuperposable** mirror image are **enantiomers.** In terms of symmetry, rule 2 applies.

Rule 2: If a molecule possesses a plane of symmetry, or a center of symmetry, the molecule is superposable on its mirror image and is achiral.

Effect of Chirality on the Physical Properties of a Molecule: Most physical properties of enantiomers, such as the melting point, boiling point, refractive index, IR and NMR spectra are identical. One physical property is different and that is the effect of the molecule on plane polarized light, usually called the **"optical rotation."** Enantiomers rotate the plane of plane polarized light equally, but in opposite directions; i.e. one enantiomer may rotate the light 20° (+20°) clockwise, while the other enantiomer rotates the light 20° counterclockwise (-20°). The actual degree of rotation depends on many factors such as the structure of the molecule, the wavelength of the light, etc.

Racemic Mixture (Racemate): A 50:50 mixture of enantiomers has an optical rotation of exactly zero because the rotation of one enantiomer exactly cancels the rotation of the other. This mixture is called a **racemic mixture,** or **racemate.**

Achiral molecules cause no rotation of the plane of plane polarized light and are said to be **optically inactive.** However, a substance which is experimentally found to be optically inactive either may be achiral, or it may be a racemic mixture. If it is a racemic mixture the substance is capable (at least in principle) of being resolved into its optically active enantiomeric components. However, resolution may be a simple or a very difficult experimental task, depending on the structure of the molecule. Resolution of an achiral compound is impossible because there is only one structure; i.e. the object and its mirror image are identical.

Chiral Center and Configuration: A carbon atom with four different atoms or groups attached to it is called a **chiral center** (the terms stereogenic center and asymmetric center are also used). The three-dimensional sense of attachment of the four different groups on a chiral center is called the **configuration.** Clearly enantiomers have opposite configurations at the chiral centers.

Although it is easy to differentiate between enantiomers by examining molecular models, how do we specify which is which? The absolute configuration can be assigned to a chiral center using the Cahn-Ingold-Prelog sequence rules which form the basis of the **R,S** system of nomenclature. A brief outline will be given here, but you should refer to your lecture text for a detailed treatment.

The R,S-System for Assigning Absolute Configuration to Chiral Centers: The four groups attached to a chiral center are assigned priorities, in decreasing order of priority: 1>2>3>4. The priorities are established by atom by atom comparison. In comparing atoms, **a higher atomic number means a higher priority.** One starts with the atoms directly attached to the chiral center and compares them to see if an order of priorities can be established. Consider the simple case of CHFClBr, in which all of the "groups" are atoms. The group priorities based on atomic number are Br(1) > Cl(2) > F(3) > H(4). A model of the molecule is then viewed with the group with the **lowest priority (4)** held away from the viewer as shown below.

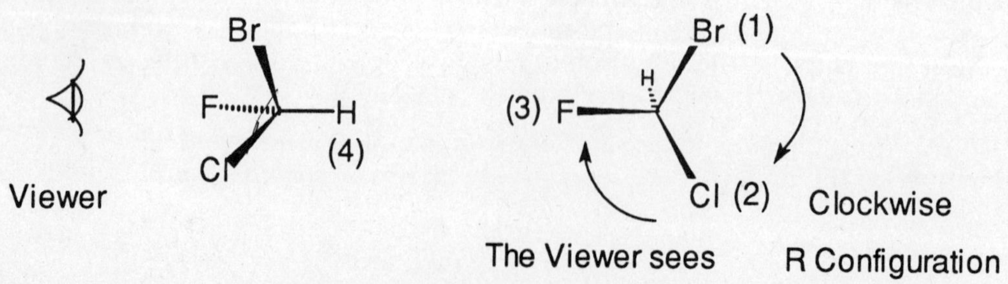

If groups 1,2,3 are arranged clockwise, the configurations is defined as **R.** If the 1,2,3 pattern is **counterclockwise,** the configuration is S.

In general, the groups will be polyatomic. Double and triple bonds are considered to have "phantom atoms" (or "imaginary" or "equivalent" atoms) which are primed below as a reminder that these phantom atoms carry no substituents. Phantom atoms are equivalent to "real atoms" in a direct comparison, but lose when the priority comparison moves to the next attached atoms because phantom atoms have no attached atoms.

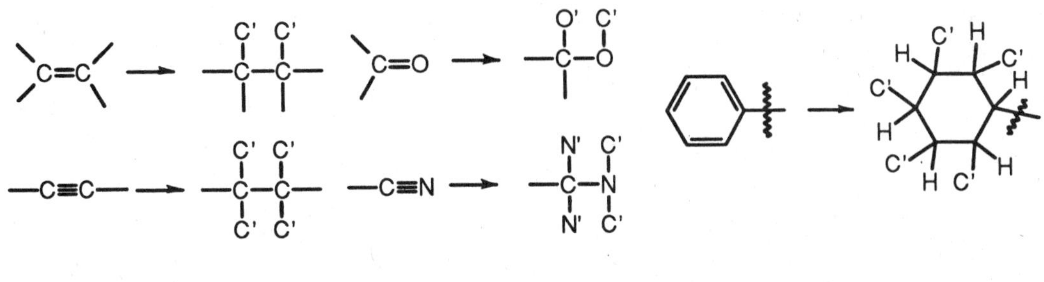

Start at the chiral center and move out one atom at a time in each group, making priority comparisons at each step until an order of priorities is established. If two or more of the atoms attached to the chiral center are the same (typically carbon atoms) compare the next attached atoms (three on carbon), one by one, with the corresponding next attached atoms in the other groups. At all times, the basic rule applies: in a direct comparison, **one atom of higher atomic number has priority over any number of atoms of lower atomic number.** If priorities cannot be established, move one atom down the chain or branch of highest priority for each group and then compare the next attached atoms. Repeat the process until a priority decision is reached. If the highest priority chains or branches of two groups are identical, go back to the first branch point where they differ and move along the next highest branches in the manner outlined above. A number of groups in decreasing order or priority are given below:

$-Br > -SCH_3 > -F > -OCH_3 > -OH > -CH_2Br > -CO_2H > -COCH_3 > -CH_2OH > -C_6H_5 > -C{\equiv}CH > -C(CH_3)_3 > -CH{=}CH_2 > -CH(CH_3)_2 > -CH_2CH_2Br > -CH_2CH(CH_3)_2 > -CH_2CH_2CH_3 > -CH_2CH_3 > -CH_3 > -H$

***Cis* and *Trans* Diastereomers:** The terms *cis* and *trans* are used to describe the stereochemical relationship of 1,2-substituents on disubstituted carbon-carbon double bonds or substituents on different positions of a cyclic system as illustrated in the first figure in this chapter.

The E,Z-System for Tri- and Tetra-Substituted Double Bonds: The terms *cis* and *trans* do not work well in assigning the configuration about tri- and tetra- substituted double bonds. For these systems the **E,Z**-system of designation is used. Priorities are assigned to substituents on the double bond using the same sequence rules devised for assigning absolute configurations R and S. Consider the following example:

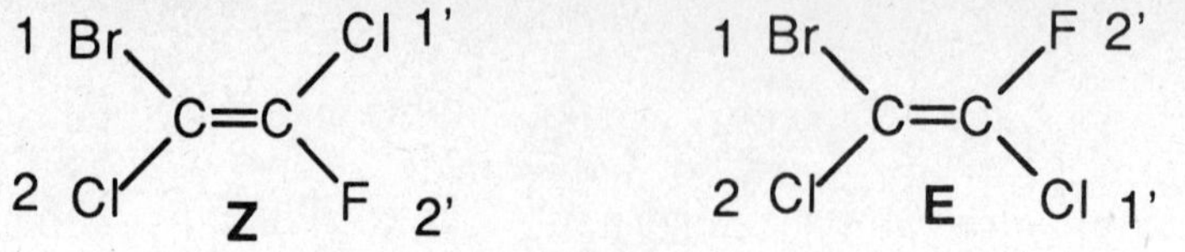

Each end of the double bond is considered separately and priorities are assigned: 1,2 at one end and (independently) 1',2' at the other end. If the higher priorities at each end (1 and 1') are on the same side, the configuration is called **Z.** If 1 and 1' are on opposite sides of the double bond, the configuration is called **E.**

Multiple Chiral Centers: When there are **n** different chiral centers in a molecule, there are theoretically $\mathbf{2^n}$ possible stereoisomers. For two chiral centers, four stereoisomers are possible. Consider the following example (Fischer projections, used below, are defined later.):

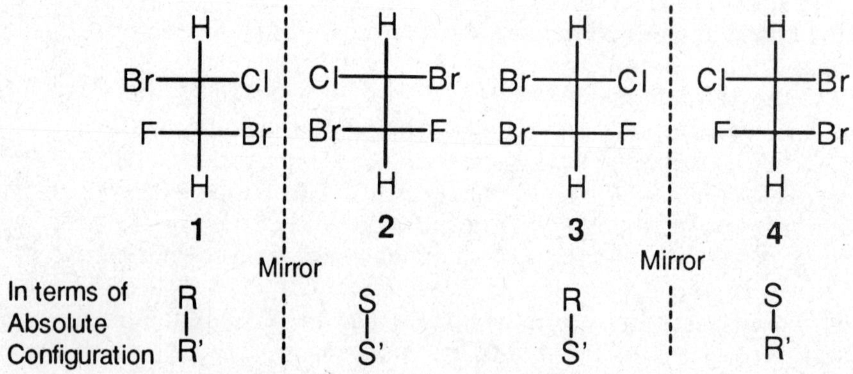

In this example, **1,2** and **3,4** are pairs of enantiomers. But **1,3** and **1,4** and **2,3** and **2,4** are pairs of stereoisomers which are all related as diastereomers, i.e., stereoisomers which are not enantiomers.

When the chiral centers carry the same four groups, the stereochemical possibilities are more restricted. In the example below, there are two chiral centers bearing the same groups:

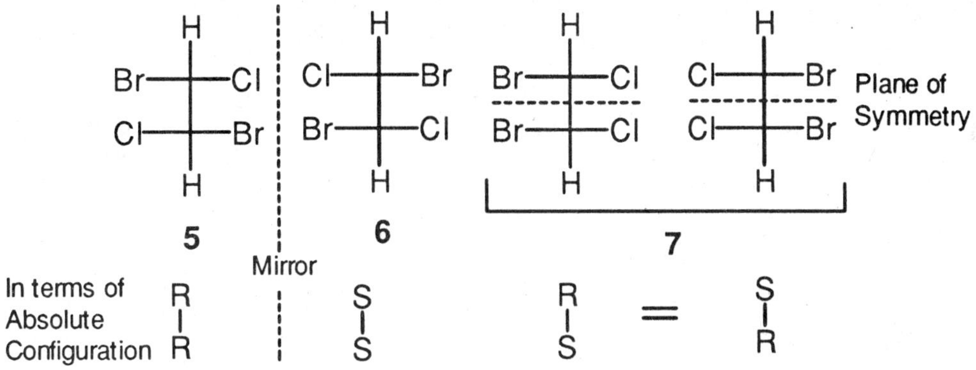

Molecules **5** and **6** are chiral and represent a pair of enantiomers. But **7** is achiral and it is said to be a ***meso*** molecule. Even though *meso* molecules have chiral centers, they are achiral as a whole because one half of the molecule (which is chiral) is the mirror image of the other half (but not superposable). In other words a *meso* molecule possesses a plane of symmetry, which means that the molecule and its mirror image are superposable (see rule 2 above). In this example **5,7** and **6,7** are pairs of diastereomers. Note that in this case there are only three stereoisomers.

D. Drawing Stereochemical Structures

Wedge Drawings: Heavy lines indicate the bond is above the plane of the paper, dashed lines are below the plane, and conventional lines are in the plane of the paper.

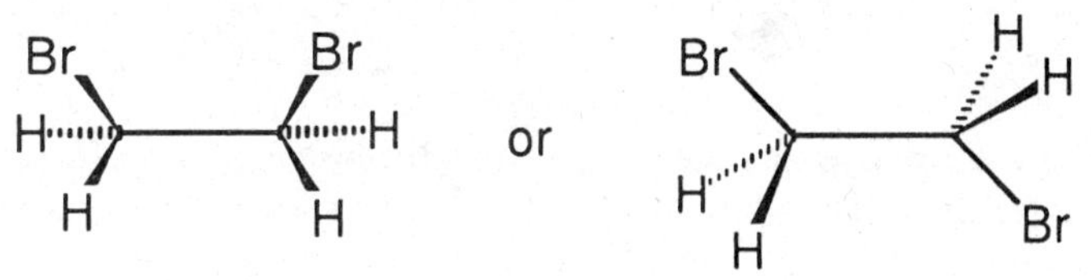

Newman Projections: The viewer looks along a C-C bond and draws a set of three lines radiating from a point to represent the front carbon atom with its attached groups. Then a circle is drawn to represent the rear carbon atom with radiating lines to show the groups attached to it.

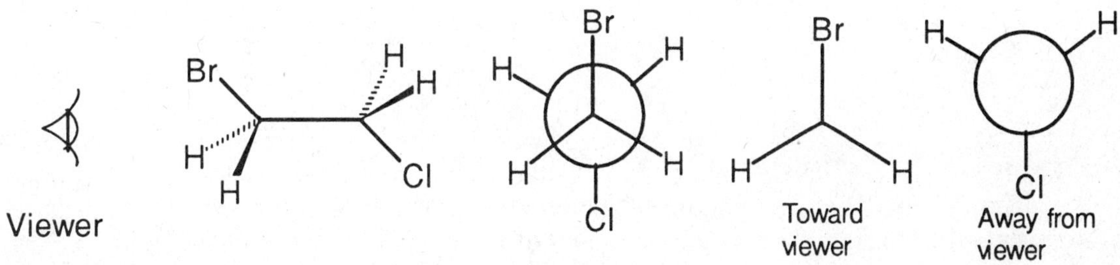

Newman projections are useful in showing the rotation of groups about single bonds. Note the terms used to designate the different conformations. The infinite number of conformations intermediate between the **staggered** and **eclipsed** conformations are termed **skew** conformations; of the staggered conformations, the **anti** and **gauche** forms are particularly important. Eclipsed conformations must be drawn slightly askew (to show both front and back sets of bonds), but they imply complete "lining up."

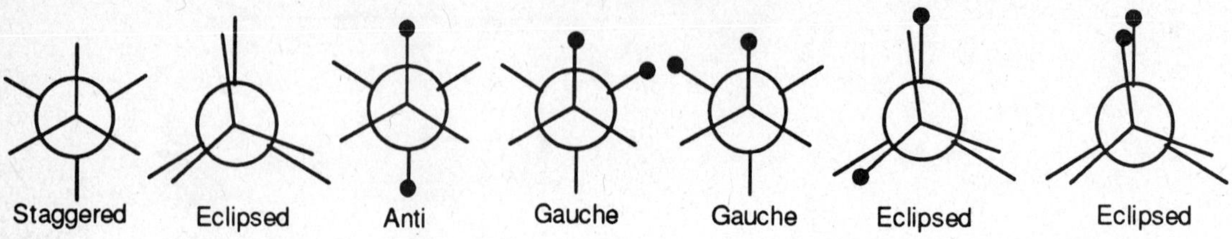

Conformations of Cyclohexane: The cyclohexane structure may be drawn in the chair form, the boat form, the twist-boat form, or the half-chair form. All these forms interconvert through rotation about single bonds. Note that in the chair-chair interconversion all axial bonds become equatorial and all equatorial bonds become axial.

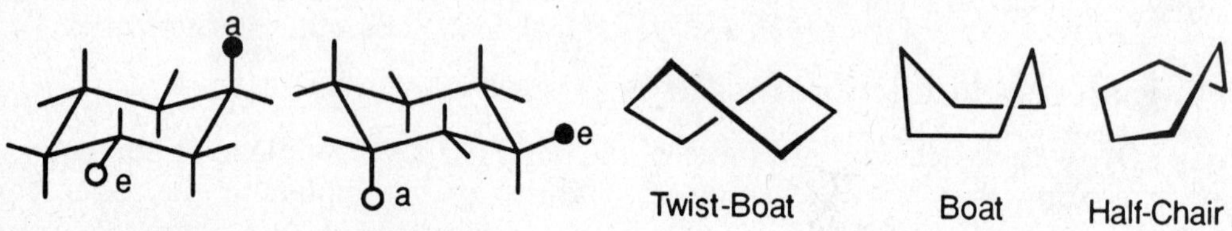

Chair - Chair Interconversion
a - Axial e - Equatorial

Fischer Projections: These drawings are two-dimensional representations of molecules which indicate the three-dimensional structures and they define the configurations at chiral centers. Fisher projections consist of horizontal and vertical lines only. **All horizontal lines project above the plane of the paper and all vertical lines project below the plane of the paper.**

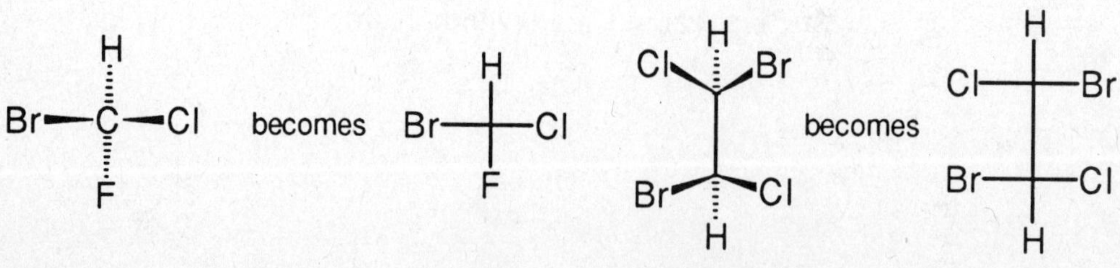

A Fischer projection can be rotated 180° in the plane of the paper without changing the implied configuration. But a rotation of 90° inverts the configuration; i.e., the rotated formula represents the configuration opposite to that of the original Fischer projection.

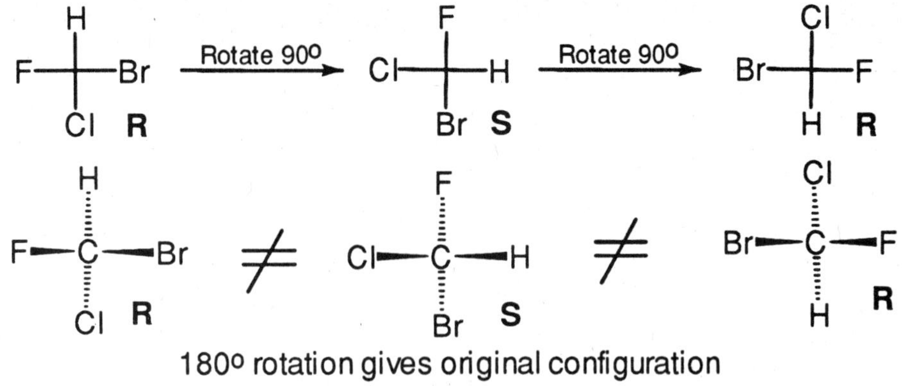

E. Exercises

1. (a) Prepare models of the four isomeric butenes, C_4H_8. Note that restricted rotation about a carbon-carbon double bond is responsible for *cis-*, *trans-* stereoisomers. Name each isomer and indicate pairs which are structural isomers and pairs which are stereoisomers. Are any of these isomers chiral? Identify any of the symmetry elements discussed above. Are the stereoisomers related as enantiomers or diastereomers?

   (b) Assemble models of the three $C_2H_2Cl_2$ isomers. Which are structural isomers and which are stereoisomers? Are any chiral? Identify any symmetry elements. In each case explain whether the molecule will or will not have a dipole moment.

   (c) Prepare models of the 2-chloro-2-butene isomers. Designate which is **E** and which is **Z.** Is either molecule chiral? Identify any symmetry elements. Why is the **E,Z** nomenclature rather than *cis-*, *trans-* designations necessary?

   (d) Prepare a model of 2-pentanone. Are there any *cis- trans-* isomers in this case? Is the molecule chiral?

   (e) Prepare a model of 3-penten-2-one. Are stereoisomers possible here? Explain. Are any of these chiral? This system is said to be **conjugated.** What is the meaning of conjugation?

2. Assemble models of ethyne (acetylene) and 2-butyne. How many (and which) atoms are colinear in an alkyne? Are there **E,Z** isomers based on an alkyne linkage?

3. Use different colored spheres to represent different atoms; e.g., white = H, red = F, blue = Cl, yellow = Br.

   a) Prepare a model of CH_3F. Identify the three symmetry planes and the C_3 axis.

   b) Prepare a model of CH_2F_2. Identify the two symmetry planes and the C_2 axis.

   c) Prepare a model of CH_2FCl. Then make a second model which is the mirror image of the first model. How are the models related? Is the molecule chiral or achiral? Identify any symmetry plane(s). Are there any symmetry axes?

   d) Prepare a model of CHFClBr. Then make the mirror image of this model. Are these models superposable? Are they chiral? Do they possess any symmetry planes? Do they have any symmetry axes?

Assign the **absolute configuration** to each model that you have made using the **R,S** system. Now on the model you designated as **S,** interchange any two substituents. What is the absolute configuration now? On the same model, interchange any two substituents. What is the absolute configuration now? Repeat interchanging substituents until you are sure that you understand the principle. Determine the result of an odd number and an even number of interchanges.

4. Prepare all possible stereoisomers of 2,3-dibromobutane. How many are there? How many are chiral? Identify enantiomers and diastereomers. Draw Fischer projections to represent these stereoisomers. Assign the absolute configuration (**R** or **S**) to each chiral center.

5. Prepare all possible stereoisomers of 2-bromo-3-chlorobutane. How many are there? How many are chiral (be careful)? Draw a Fischer projection for each stereoisomer. Identify enantiomers and diastereomers. Assign the absolute configuration at each chiral center. You should have one *meso* stereoisomer (you may have made an identical or duplicate model) Rotate the *meso* model internally until you find the conformation with a symmetry plane; continue internal rotation to find the conformation with a center of symmetry. Note that a *meso* compound does not have to have adjacent chiral centers. For example, in 2,6-dibromoheptane, in which the chiral centers are separated by three methylene groups, $-CH_2CH_2CH_2-$, exactly the same stereochemical relationship is found, *viz.*, there are three stereoisomers: a *dl*-pair and a *meso* compound.

6. (a) Assemble a model of ethane and make the **staggered** and **eclipsed** conformation. Which one has a plane of symmetry? Which one has a center of symmetry?

   (b) Prepare a model of 1,2-dibromoethane and make the **anti** and (Br,Br)-**eclipsed** conformations. Is either conformer chiral? Which conformer has a center of symmetry? Which has a plane of symmetry? Identify symmetry axes. Would either conformer have a dipole moment?

   Make a second model of $BrCH_2CH_2Br$ and prepare the two **different gauche** conformers. Are there any symmetry axes? Are these conformers superposable? Are they chiral? How are they related? Now do a 120° internal rotation (going through the Br, Br eclipsed conformation) about the C-C bond of one of the conformers and compare the new conformer with the second "nonrotated" conformer. What do you find? Now do a similar 120° internal rotation on the second conformer. What do you find?

   An important principle is illustrated by this example. Transient conformations may be chiral, but they can "average out" by interconverting. In this case the conformers interconvert rapidly. Thus we classify this molecule as being achiral (which it is in the **anti** conformation and the (Br, Br)-**eclipsed** conformation).

   (c) Prepare a model of butane and arrange it to illustrate the **anti,** the **gauche** and all of the **eclipsed** conformations. How many **gauche** conformations are possible? Are they identical? Is this molecule classified as being chiral or achiral? Explain. Which conformation has the lowest energy? Which has the highest energy?

7. (a) Depending upon the kind of models you are using, you may or may not be able to construct models of cyclopropane and cyclobutane. Do so, if possible, but do not break the linkages. Why do we say that these molecules are **strained?** What is the normal sp^3 bond angle? What are the (internal) C-C-C bond angles in cyclopropane?

(b) Make a model of cyclopentane. Does this molecule seem to have any ring strain in the sense that the three and four-membered rings do? The molecule is not completely strain-free because there is some eclipsing. Identify the eclipsed atoms or groups when the molecule is planar. In fact the molecule is not planar, but is puckered in either of two ways: (i) One C-atom is above or below the plane defined by the other four coplanar C-atoms defining an **envelope** conformation. (ii) Two adjacent C-atoms are arranged so that one is above and the other is below the plane defined by the other three C-atoms giving a **pseudo-chair** conformation. Try to make these conformations and note that eclipsing is diminished compared to the planar molecule. These conformations interconvert rapidly.

(c) Make both *cis*- and *trans*-1,2-dimethylcyclopentane. Classify each as being chiral or achiral. How can you be sure of your assignments? Identify any symmetry elements.

8. (a) Prepare a model of cyclohexane and make a boat conformation (see the illustration earlier). This is a high-energy conformation. Why? (Look for eclipsing and note the opposed 1,4-C-H bonds.) Now twist the model a little bit to give a **twist-boat** conformation. Note that eclipsing is somewhat relieved.

   (b) Make a chair conformation. Is there any eclipsing? Look at any four-carbon sequence and note that it has the same shape as a gauche conformation of butane. Next put white spheres on all six of the axial positions and leave the equatorial positions empty. Now do a chair-chair interconversion or "flip." To do so, you must twist the ring through a chair, half-chair, twist-boat, half-chair, chair sequence. The half-chair conformations represent energy maxima in the interconversions. What happens in a chair-chair flip? Repeat this interconversion a few times.

9. Prepare methylcyclohexane and generate both possible chair conformations. Which has the lower energy? Why? Replace the methyl group by a *tert*-butyl group and look at both possible chair conformations. Note any interactions which look important. In particular, look for **1,3-diaxial** interactions.

   The relative amounts of equatorial and axial conformers present at 25°C for alkyl-substituted cyclohexanes are:

   [Equatorial]/[Axial]: $-CH_3$, 15; $-CH_2CH_3$, 18; $-CH(CH_3)_2$, 35; $-C(CH_3)_3$, 4600.

   The *tert*-butyl group virtually "locks" the conformation to give equatorial *tert*-butyl. Why is this group so effective in this sense? Can you find a reason why isopropyl, $-CH(CH_3)_2$, is much less effective here?

10. (a) Prepare *trans*-1,2-dimethylcyclohexane and examine both possible chair conformations. Which is more stable (lower in energy)? Why? Is this molecule chiral? (Compare with the mirror image.)

   (b) Prepare *cis*-1,2-dimethylcyclohexane and examine both possible chair conformations. Do they differ in energy? Prepare a second model of *cis*-1,2-dimethylcyclohexane (best done by teaming up with a neighbor with the same kind of models you have). Arrange the chair conformations so that the models are surperposable. Then flip one model to give the other chair form. The two conformers should now be related as object and nonsuperposable mirror image, i.e., enantiomers. Note that the enantiomers are interchanged by "ring-flipping." Since this process is **extremely fast** at room temperature, the system represents a **dynamic racemate,** consisting of **rapidly interconverting conformational enantiomers.** As a consequence, we classify this molecule as being **achiral.**

(c) Prepare *cis*- and *trans*-1,3-dimethylcyclohexane. In each case look at both possible chair conformations to see if either one is preferred. Arrange the *cis* isomer to give the conformer with diaxial methyl groups. The *cis*-1,3-diaxial methyl interactions (repulsions) are especially high, destabilizing this conformation. Finally, classify each stereoisomer as being chiral or achiral.

(d) Prepare *cis*- and *trans*-1,4-dimethylcyclohexane and in each case determine whether or not any chair conformation is preferred. Is either isomer chiral? Explain. (Describe in terms of symmetry elements.)

11. Prepare a model of 2,3-pentadiene. Make the mirror image. Are the models superposable? Is 2,3-pentadiene chiral? Are there any chiral centers?

This molecule, called an allene, illustrates how chirality can arise even though there is no chiral center. We really are looking at a small segment of a helix, and a helix is chiral. (Refer to right and left-hand thread machine screws.) Molecules which are chiral without chiral centers are said to be **inherently dissymmetric.** Restricted rotation about single bonds due to steric hindrance can lead to chirality of this type. Consider the example in 10(b). What would happen if the methyl groups were replaced by much larger and bulkier groups which prevent chair-chair interconversion?

Prepare a model of 2,3,4-hexatriene. How many stereoisomers are possible. Are any chiral? Explain.

Advanced Exercises

12. Construct models of the two "decalins" (decahydronaphthalenes) shown below. In both systems the preferred conformers have chair conformations for the six-membered rings. Which stereoisomer is more flexible?

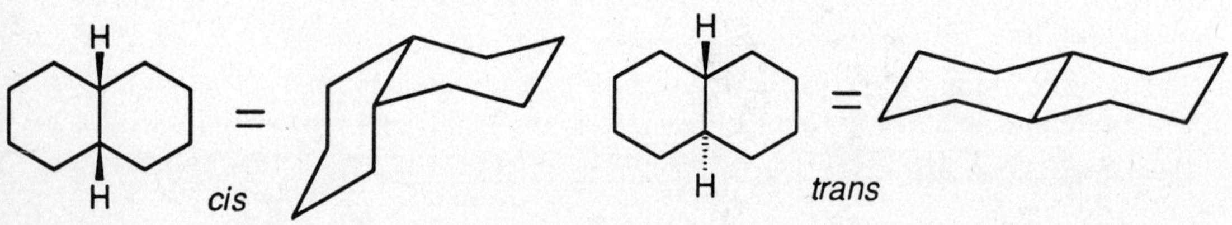

Can either stereoisomer undergo a chair-chair interconversion? Predict which isomer has the lower energy. Explain.

13. In spiro compounds, two rings share a single common atom. Construct a model of spiro[4.4]nonane, **8.** Is it chiral? Construct a model of the diketo derivative **9.** Is it chiral? Make the mirror image of **9.** Does **9** have any chiral centers? Explain the presence, or lack of, symmetry.

14. Construct models of adamantane, **10,** and bicyclo[2.2.2]octane, **11.** Does either model have any flexibility? Describe the preferred conformation of the 6-membered rings of each system. (Remember that a twist-boat conformation is lower in energy than a boat.) The carbon skeleton of adamantane has a special distinction; it is a basic subunit of the structure of diamond.

8

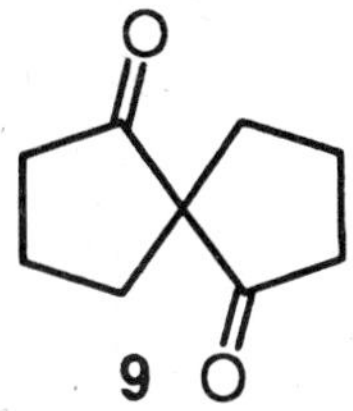

9

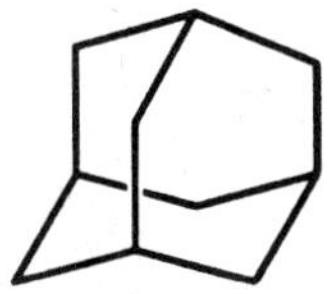

10

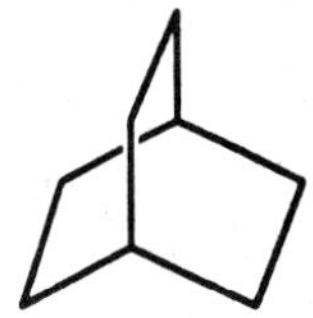

11

15. Construct a model of cyclooctane, a medium-ring system which illustrates the conformational complexities that arise as the ring size increases. While there are many possible conformations, a few are distinctive. Try to make conformations **12, 13, 14, 15,** and **16.** Slightly twisted versions of **14, 15,** and **16** are lower in energy than the untwisted conformations. Note how readily several conformations interconvert. The lowest energy conformation has been shown to be **12,** which minimizes repulsive interactions. However, it must be noted that the internal C-C-C ring angles have expanded from 109.5° (ideal tetrahedral angle) to *ca* 116° to minimize transannular (across the ring) interactions.

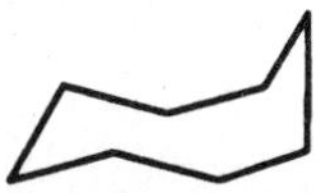

12

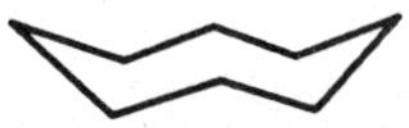

13

14

15

16

16. Construct a model of cyclohexene. Note that what you have made is actually *cis*-cyclohexene. Try to make a model of *trans*-cyclohexene. What can you conclude about *trans*-cyclohexene?

    Construct a model of *cis*-cyclooctene. Although there are some repulsive transannular interactions, this *cis* isomer is relatively strain-free. Now make a model of *trans*-cyclooctene. Although *trans*-cyclooctene is strained, it does exist. In fact it has been found to be chiral and the enantiomers have been separated. Examine your model to verify that is chiral; if possible, build the mirror image. Try to identify the structural features which cause chirality. What internal rotations are required to interconvert the enantiomers?

17. Try to construct models of cycloheptyne, cyclooctyne, and cyclononyne. Based upon your observations, explain the following facts: Cyclononyne is relatively stable; cyclooctyne can be prepared, but it is "abnormally" reactive; and cycloheptyne cannot be isolated (at least at room temperature).

10. Infrared Spectroscopy

A. Introduction

The purpose of this experiment is to obtain hands-on experience in the operation of an infrared (IR) spectrometer, to learn how to prepare samples for IR analysis, to practice the interpretation of IR spectra, and to relate spectroscopic results to chemical structure. Before starting, it is essential to study this section and related sections in your lecture text. Directions will be available from your instructor for the operation of the IR spectrometer.

Chemical bonds are similar to coil springs and, like springs, can be defined by force constants, which are related to bond strengths e.g., C≡C > C=C > C-C. A molecule can be envisioned as a three-dimensional array of balls of varying masses connected by springs. At any temperature, a molecule possesses energy which it stores in the form of vibrations. Increasing energy increases the amplitudes of the vibrations. The vibrations are dependent upon the strength of the chemical bonds, the distances between the atoms, the masses of the atoms, and the nature and positions of the atoms in the molecules. Since the atoms are all interconnected, the vibrations of one bond affect the vibrations of another; the vibrations are said to be "coupled." Vibrations are analyzed in terms of vibrational modes, such a stretching, bending, and wagging. The stretching mode, illustrated in Figure 1 is most important in organic chemistry because it can be related to functional groups.

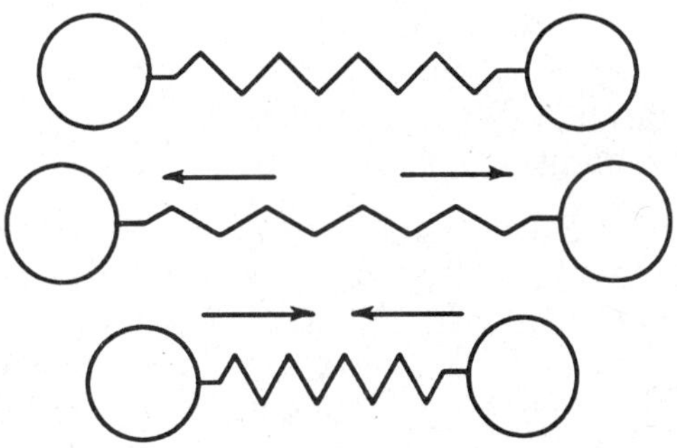

Figure 1. Illustration of the stretching mode of vibration between two atoms.

The vibrational frequencies of molecules are found in the infrared region of the electromagnetic spectrum. When infrared light passes through a molecule, if the frequency of light is the same as the frequency of a molecular vibrations (a condition called resonance) the molecule can absorb a photon and thereby increase the amplitude of the vibration; the vibration becomes "excited." Whether or not absorption of energy actually occurs, depends upon the nature of the vibration; some vibrations are more readily excited than others. However, vibrational excitation occurs **only** if there is a change in the permanent dipole moment of the molecule.

An infrared spectrum is produced by irradiating a sample of a compound with a range of infrared frequencies and determining the frequencies and intensities of the absorbed radiation. A plot of the percent of transmitted light (Y-axis) *vs.* the frequency of the radiation (X-axis) is called an infrared absorption spectrum. By convention, the baseline (100% transmission) is at the top and 0% transmission is at the bottom of the transmission scale. Thus IR "peaks" are inverted and are actually valleys. Typical IR spectra are shown at the end of this chapter.

IR peaks vary in width. Some are fairly sharp, others are broad. The peaks are usually referred to as "absorption bands" or simply "bands." The band widths reflect the fact that changes in closely-spaced rotational states are superimposed upon the vibrational states. In addition, other factors such as hydrogen-bonding, conformational differences, etc., can broaden the bands.

Each IR band arises from a particular type of vibration that the molecule is experiencing when that particular frequency of radiation is interacting with it. Some of these peaks represent vibrations of the molecule as a whole, but some of them represent vibrations of only particular parts of the molecule. Of special value to the organic chemist is the fact that many of these vibrations are vibrations of functional groups, such as the carbonyl and the hydroxyl groups. The vibrations are so characteristic of those particular functional groups, that a quick glance at an infrared spectrum tells immediately which functional groups are present and which are absent. Not only that, but the particular frequency of vibration will often allow distinctions to be made between different types of groups having similar structure. For example, the particular frequency of a carbonyl group will allow one to distinguish between aldehydes or ketones, esters, anhydrides, acid chlorides, amides, etc., all of which contain carbonyl groups.

By common convention, the frequency of the radiation (of the order of 10^{14} sec^{-1}) is expressed in reciprocal centimeters, cm^{-1}, called **wavenumbers,** a unit which is proportional to the energy of the infrared radiation. The higher the wavenumber for a peak, the greater the energy of radiation needed to excite the vibration. The infrared spectrum normally covers from 4000 cm^{-1} (on the left) to about 650 cm^{-1} (on the right). Because frequency and wavelength are reciprocally related, the wavelength scale in microns ($1\ \mu = 1\ \mu m = 10^{-6}$ m) is often included on IR spectra. However, the wavenumber scale is more generally used, and most spectra are linear in wavenumber. Wavelength and wavenumber are related by the expression: wavelength (μ) x wavenumber (cm^{-1}) = 10,000.

With some exceptions, no two compounds have identical infrared spectra. Thus, the structure of a particular compound can be verified by comparing the infrared spectrum of the compound with the infrared spectrum of an authentic sample of known structure. If the spectra are identical in every respect then the two compounds have a very high probability of being identical. If the spectra do not agree, then the structures are definitely not identical. Comparing spectra is a much better means for verifying structure than melting points, mixed melting points, boiling points, etc.

Infrared spectra are often used to assess the purity of compounds by observing the presence and intensity of peaks that are normally not present in the spectrum of the pure compound. By careful calibration and measurements, the degree of purity and the nature of the impurities can often be determined.

In analyzing infrared spectra, the functional group tables found in your lecture text and many handbooks may be consulted. With experience one learns the values of the functional group frequencies and can use this information to quickly interpret infrared spectra.

B. The Spectrometer

Two types of instruments are commonly used for obtaining IR spectra: dispersion spectrometers and Fourier Transform spectrometers. Both types of spectrometers send IR radiation (generated by a red-hot source) through a sample and then provide a plot of the intensity of absorption, *versus* the frequency of the radiation. However, the instruments obtain this information in different ways.

A dispersion spectrometer, using a diffraction grating or a prism, sends a continuously varied beam of infrared radiation through a sample. Using a double-beam technique, the absorption of radiation by the sample is compared to the absorption of radiation by an appropriate reference (usually the solvent). In this type of instrument an IR scan generally takes 3–10 minutes.

In a Fourier Transform infrared (FT-IR) spectrometer, a brief pulse of IR radiation covering the entire spectrum of interest is generated by an optical device called an interferometer. A sample placed in this pulsed beam absorbs radiation at its characteristic frequencies giving rise to an oscillatory pattern of intensity *versus* time called an interferogram. The interferogram is subjected to a mathematical process called a Fourier Transformation which converts the interferogram into a plot of intensity of absorption *versus* the frequency of the radiation. In practice, several pulses are averaged to improve the signal to noise ratio. An appropriate reference spectrum (stored in the computer) is subtracted automatically to correct for the background. Once operating parameters are set, the spectrum can be obtained in a few seconds. Printing of the spectrum takes a little longer, but stored data can be processed and printed at an auxiliary station. The high speed of the FT-IR method makes it feasible to use IR spectroscopy in large introductory laboratories classes.

C. Preparation of Samples for Infrared Analysis

Sample handling is very important in determining the IR spectra of organic compounds. In order to contain the sample and hold it in the beam of the infrared spectrometer, it is necessary to use cells made of materials that are transparent to the infrared radiation. One of the best cell materials is sodium chloride (NaCl) because it is transparent to infrared radiation throughout the 4000-650 cm^{-1} range of the spectrum. Thus, the cells used to contain the samples are made of large crystals of sodium chloride. NaCl is very soluble in water and so one must be very careful in handling the NaCl windows to keep all water away from them. **Water must never be allowed to contact the sample cells.** Water will dissolve the surface of the cell and render the cell unusable. **For that reason, all samples must be thoroughly dried to prevent damage to the cell windows.** In handling the cells hold them by the edges, never by the faces. The sample cells for use in this experiment will be NaCl disks (called plates), 25 mm in diameter, and 4-5 mm thick. These plates are particularly useful for liquid samples that do not evaporate too rapidly.

Handle the cells and plates carefully. Sodium chloride is easily broken. Do not drop the cells or plates because they will surely break!

Infrared spectra can be run on pure liquids, solids (or liquids) in solution, and solids dispersed in a medium that has few infrared bands to cause interference. A few of the techniques for sample preparation are given below.

Liquid samples. Obtaining a spectrum on a pure liquid sample is the most convenient procedure. If the sample is a liquid, the simplest way to run the spectrum is simply to place a drop or two on a NaCl plate, place another NaCl plate on top, and squeeze them together with a rotary motion. This process is called obtaining a spectrum on a "neat" sample (pure sample). The sandwich is then placed in the cell holder of the spectrometer. If you find that the peaks are too intense, remove a little of the liquid to achieve a thinner sample. To do so, separate the plates, gently touch one plate with a clean tissue and reassemble the sandwich. Repeat the process if necessary. After the spectrum is recorded, **clean the plates by washing both thoroughly with acetone or ethanol;** then dry them with a clean tissue and return them to their designated location.

Carbon Tetrachloride and Chloroform are excellent solvents and are widely used in IR and NMR spectroscopy because their absorption bands do not normally interfere with the functional group absorptions of organic compounds. However, these solvents should only be used in small quantities and then only where there is very good ventilation. Because both solvents are toxic (especially chloroform) and both are considered to be potential carcinogens, prolonged exposure to these solvents must be avoided. For that reason, sample preparation should be performed in a hood.

Samples soluble in CCl_4. Solid samples are often run in carbon tetrachloride solution using a solution cell. The solution cell is constructed of two 25-mm diameter NaCl disks, cemented together with a 0.1 mm spacer. A slot in the cell is used to introduce the sample. Commercial solution cells are too expensive (*ca* \$150 each) for routine use by students. However, a satisfactory solution cell can be easily constructed for about \$15, little more than the cost of two salt plates (see Winston, A., *J. Chem. Educ.* **1991,** *68 [May]*, A124-A125).

The sample is carefully dried and then dissolved in CCl_4, using *ca* 10 mg of sample in *ca* 10 drops of CCl_4. The solution is examined to see if there is any water present. Water will appear as small liquid droplets floating on top of the CCl_4. **If water is present, a fresh sample will have to be prepared from dry materials.** Using a special syringe with a #24 needle, the sample is introduced into the cell. The cell is placed in the spectrometer and the spectrum is recorded. When CCl_4 is used, the spectrum will only be useful from 4000 to 850 cm^{-1} because CCl_4 absorbs strongly below 850 cm^{-1} and the peaks of the sample will be obliterated in that region.

After use, clean the cells by flushing thoroughly with an appropriate solvent. The solvent should be completely evaporated from the cell by means of a vacuum line before the next use.

Mulls; Samples insoluble in CCl_4 and $CHCl_3$. If the compound is insoluble in CCl_4 and $CHCl_3$, it may be dispersed in a medium that is transparent to infrared radiation in the regions of greatest interest. This dispersion is called a "mull." To prepare a mull, the sample is crushed and ground on a watch glass with a spatula. A few milligrams of the powder is then placed on the center of a salt plate and a drop or two of the mulling liquid is added. A second salt plate is pressed on top and the two plates are rotated to disperse the solid in the liquid until a cloudy film is obtained. The spectrum is then recorded in the usual way. If the compound already consists of very fine crystals, the initial grinding step may not be necessary.

Some compounds become electrically charged upon grinding and tend to fly about. Grinding the solid on the watch glass along with the mulling liquid will avoid this problem, but the transfer to the salt plate may become somewhat more difficult. Alternatively, the compound can be ground while still wet with a volatile liquid and then allowed to dry before transferring to the salt plate.

Another method, is to dissolve the compound in a solvent such as methylene chloride, ether, etc., on the watch glass and then allow the solvent to evaporate. Most compounds will deposit very fine crystals which can then be transferred to the salt plate.

Nujol, a purified mixture of high molecular weight saturated hydrocarbons, is widely used for mulls. Aside from the strong absorption in the C-H region (2960-2850 cm^{-1}) and weak absorption at 1375 and 1460 cm^{-1}, Nujol is an excellent mulling medium as there is no interference with the important OH and C=O regions of the IR spectrum. Because of its high viscosity, it is not difficult to obtain an adequately thick dispersion between salt plates. Since Nujol is non-volatile, it will not evaporate while the spectrum is being recorded.

Carbon tetrachloride can also be used as a mulling agent. A small pile of the powdered crystals is placed on a salt plate and a drop of CCl_4 is added beside the solid. The second plate is immediately added on top and the two plates are rotated until the solid is evenly dispersed. The spectrum is immediately recorded before the CCl_4 has had time to evaporate. If the CCl_4 starts to evaporate, another drop of CCl_4 may be needed to get a good spectrum.

In order to produce stronger spectra, a slight depression, *ca* 1 cm diameter, can be ground into one face of a salt plate to hold more of the CCl_4 mull. The depression is formed by rubbing with a squared-off pencil eraser, a little grounding compound, and ethanol (or 50:50 ethanol-water, with care).

Samples dispersed in KBr Pellets: This technique is particularly useful for solids that are insoluble in CCl_4, but it has also become a very general method for preparing solid samples of all types. A small amount of the solid sample is thoroughly mixed with a small amount of dry KBr. The amounts are variable, but with experience the relative amounts can be easily estimated. The mixture is ground together to a fine powder, usually in a mechanical grinding device. A small amount of the mixture is then placed in a die and great pressure is applied to the powder, using either a hydraulic press, or by a bolts squeezed together with a long wrench. Under pressure the KBr mixture forms a continuous solid pellet with smooth faces. The assembly is taken apart and the die containing the KBr disk is used directly in the spectrometer. The advantage of this technique is that KBr is transparent to infrared radiation and has no interfering bands. Thus the spectrum of the sample is not confused by the presence of bands that are not derived from the sample. A disadvantage is that if the KBr, or the sample, is slightly wet, a broad band for water will appear in the OH region of the spectrum.

Polishing Salt Plates: After continued use, the faces of the salt plates become scratched and worn. They can easily be polished in the following way. Secure a sheet of heavy somewhat rough paper, such as construction paper. Wet a portion of the paper with 2 or 3 drops of a 50:50 mixture of ethanol-water from a small dropping bottle. Rub the face of the salt plate over the wet paper in a circular motion. After a few rubs, dry the plate by rubbing it over a dry portion of the paper. Inspect the face of the plate and repeat the process until the surface is smooth and clear.

If the plate is very badly damaged, the surface of the face may be restored by grinding the face on piece of plate glass using 95% ethanol and a grinding compound available from laboratory supply houses. The plates must then be polished as described above. New plates that have not had their faces polished should be conditioned by this same procedure of grinding, followed by polishing.

D. Recording Infrared Spectra

One or more unknown liquid samples will be provided in Chapter 12 for analysis by infrared spectroscopy. Obtain two NaCl plates and check them to make sure that the surfaces are smooth and not pitted. The plates should be reasonably transparent. If these conditions are not met, then the plates should be polished. Lay the plates on a clean flat surface and place one drop of the liquid unknown on the center of the face of one of the plates. Place the other plate on top and press firmly with a rotary motion to create a film between the two plates and to remove air bubbles. Place the sandwich in the cell holder in the chamber of the spectrometer and record the spectrum as indicated in the instructions for the operation of the spectrometer. It is possible that the intensity of the spectrum may be too great as indicated by some of the peaks bottoming out. The most intense peak should correspond to *ca* 10% transmission. If the spectrum is too intense and some bands bottom out, remove the sample cell from the spectrometer and remove some of the material to get a thinner layer. Then rerun the spectrum. Label the spectrum with your name, desk number, and the code designation of the unknown. **Clean the plates thoroughly** and return them to an appropriate location.

E. Interpretation of Infrared Spectra

There are two regions of the infrared spectrum that are of interest to the chemist: the functional group region and the fingerprint region. The functional group region extends from 4000 cm^{-1} to *ca* 1450 cm^{-1}. In this region are found absorption bands due to the stretching frequencies of various functional groups such as C=O, OH, NH, C=C, C-H, etc. This is the most important region for the practicing organic chemist because from the positions of the peaks one can tell quickly which functional groups are present and which are absent.

The fingerprint region extends from *ca* 1450 cm^{-1} to 600 cm^{-1}. Although certain functional group absorptions can sometimes be spotted in this region, it is most useful for comparison purposes. With few exceptions every organic compound will have a unique IR spectrum, one that is different from that of any other compound. Thus, if you have a spectrum of an unknown compound and you can match this spectrum peak for peak with the spectrum of a known compound, it is virtually certain that the unknown material is identical with the known material that gave the identical spectrum. If the spectra of the known substance and the unknown substance are different, then the two compounds are definitely different. This is where the fingerprint region is most useful. The peaks are usually sharp and it is easy to compare spectra quickly and spot any differences between them.

The most important functional group absorptions are listed in Table 1 and are described in more detail below. Examine the spectra of your unknown compounds and identify the peaks that are most likely caused by functional group absorptions. With the help of the following information, identify the functional group responsible for each functional group band in your spectra.

Table 1. Infrared Absorptions of Common Functional Groups

Functional Group	Wavenumber, cm^{-1}
O-H	3600 - 3350
N-H	3500 - 3300
C-H Aliphatic	3000 - 2800
C=C-H Alkene, Arene	3100 - 3000
C=O	1850 - 1650
C=C	1670-1600
C≡N	2260-2210
C≡C	2260-2100
CO_2*	2349

*Avoid exhaling in the sample chamber, or the intense band for CO_2 will be recorded.

It is important in the analysis of infrared spectra to get a feel for the general location of functional group bands. Then, when confronted with an IR spectrum, one can examine it quickly and immediately pick out bands that indicate the presence of certain types of functional groups. Conversely, the absence of bands in a particular region indicates immediately that certain functional groups that absorb in that region are absent. In order to assist you in visually scanning important functional group regions in the IR spectrum, the chart shown in Figure 2 has been prepared. The spectrum is linear in wavenumber, cm^{-1}, which is the most common unit of measurement. Note that there is a change in scale at 2000 cm^{-1}. This change has the advantage of expanding the 2000-800 cm^{-1} region and providing a greater **visual** separation of bands in the important carbonyl and double bond stretching region from 1800-1500 cm^{-1}. The wavelength scale is provided for reference at the top of the IR chart of Figure 2.

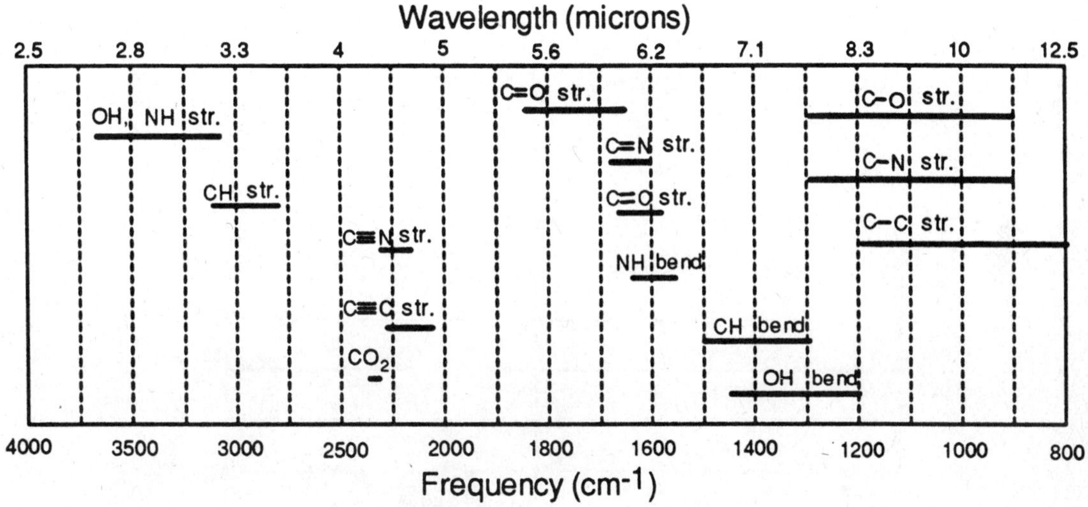

Figure 2. Common infrared group frequencies.

While Table 1 and Figure 2 give general ranges expected for the IR bands found for different groups, many features relating to each type of band have been noted. Some of the more important points concerning certain bands are outlined below.

Hydroxyl group, OH

Non-Hydrogen Bonded: Bands are sharp and appear at the higher end of the range. Characteristic of alcohols in dilute solution.

Hydrogen Bonded: Bands are broad and appear at the lower end of the range. Characteristic of alcohols run neat, in Nujol mulls, or in a KBr pellet.

Note: Under some conditions an alcohol will exhibit both a sharp band for a non-hydrogen bonded form and a broad band for the hydrogen bonded form.

Amine groups, NH, NH_2

Primary amines have two amine hydrogens and exhibit two N-H stretching bands, one for symmetric and for asymmetric modes of stretching vibration.

Secondary amines have only one amine hydrogen and exhibit only one N-H stretching band.

Tertiary amines have no bands in this region because there are no amine hydrogens atoms.

Carbon hydrogen, CH

Aliphatic carbon-hydrogen absorption occurs below 3000 cm^{-1}.

For alkene C-H and aromatic C-H, where the carbon is part of a double bond or aromatic system, the C=**C-H** stretching band will lie above 3000 cm^{-1}. This is one way to tell whether or not the C-H is part of a double bond, or aromatic system. Aliphatic C-H bands will fall below 3000 cm^{-1}

Terminal acetylenes will have a strong band at *ca* 3300 cm^{-1}.

Carbonyl group, C=O

The carbonyl region, 1850-1600 cm^{-1}, is one of the most important regions of the spectrum and a great deal of information can be learned from the exact placement and the appearance of the carbonyl bands. The various forms of the carbonyl group and the approximate placement of the absorption bands are listed in Table 2.

Table 2. Infrared Absorptions of Non-conjugated Carbonyl Groups

C=O Functional Group	Wavenumber, cm^{-1}
Ester	1750-1735
Ketone	1725-1705
Aldehyde	1735-1720
Carboxylic Acid	1725-1700
Anhydride (2 bands)	1850-1800 1790-1740
Acid chloride	1795-1810
Amide, Primary	1690-1650
Amide, Sec. and Tert.	1680-1630

Effect of Conjugation on the C=O Band. When a carbonyl group is in conjugation with a carbon-carbon double bond (C=C–C=O) the stretching frequency of the infrared absorption of the carbonyl group will be shifted about 30 cm^{-1} lower than the values for non-conjugated systems. Conjugation with an aromatic ring produces a similar lowering of the absorption frequency.

Carbon-Carbon Double Bond, C=C. The C=C stretching vibration of olefins and aromatic compounds is usually quite sharp. Conjugation with another carbon-carbon, or a carbon-oxygen double lowers the frequency by 20-30 cm^{-1}. If the double bond is symmetrical about a plane bisecting the double bond, the vibration is inactive and the band will not appear at all in the spectrum.

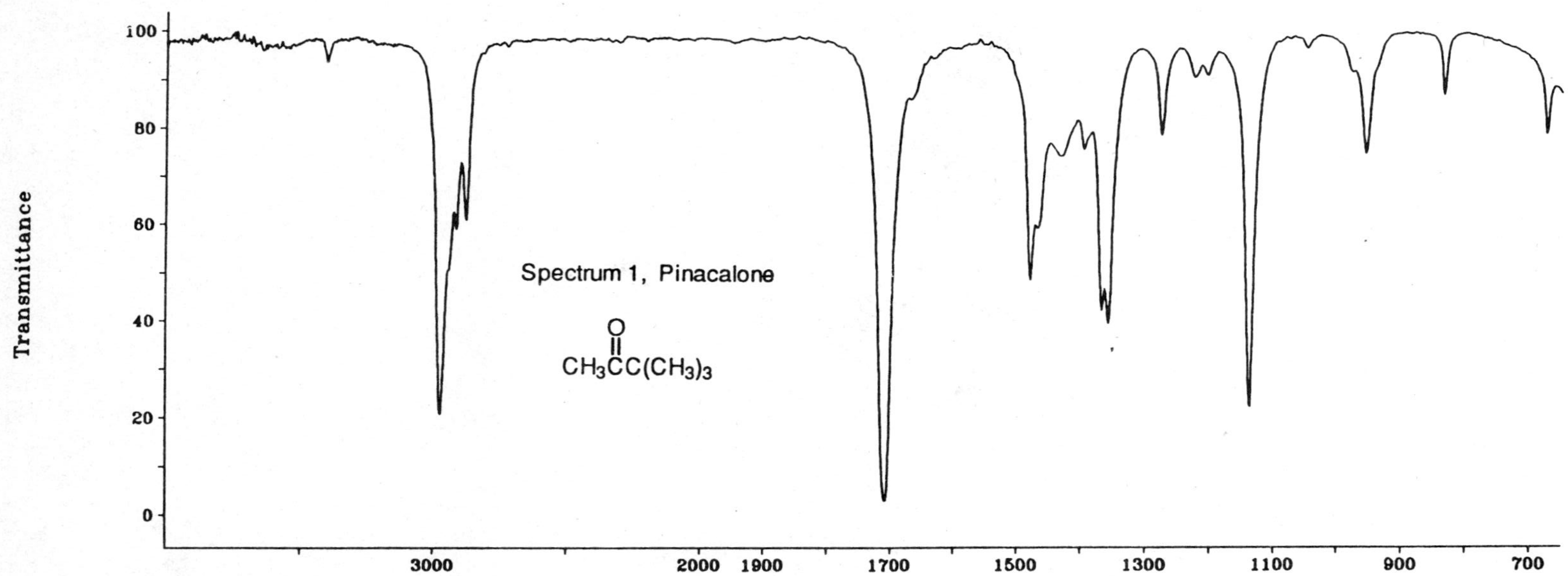
Transmittance
100
80
60
40
20
0
Spectrum 1, Pinacalone
$CH_3\overset{O}{\overset{\|}{C}}C(CH_3)_3$
3000
2000 1900
1700
1500
1300
1100
900
700

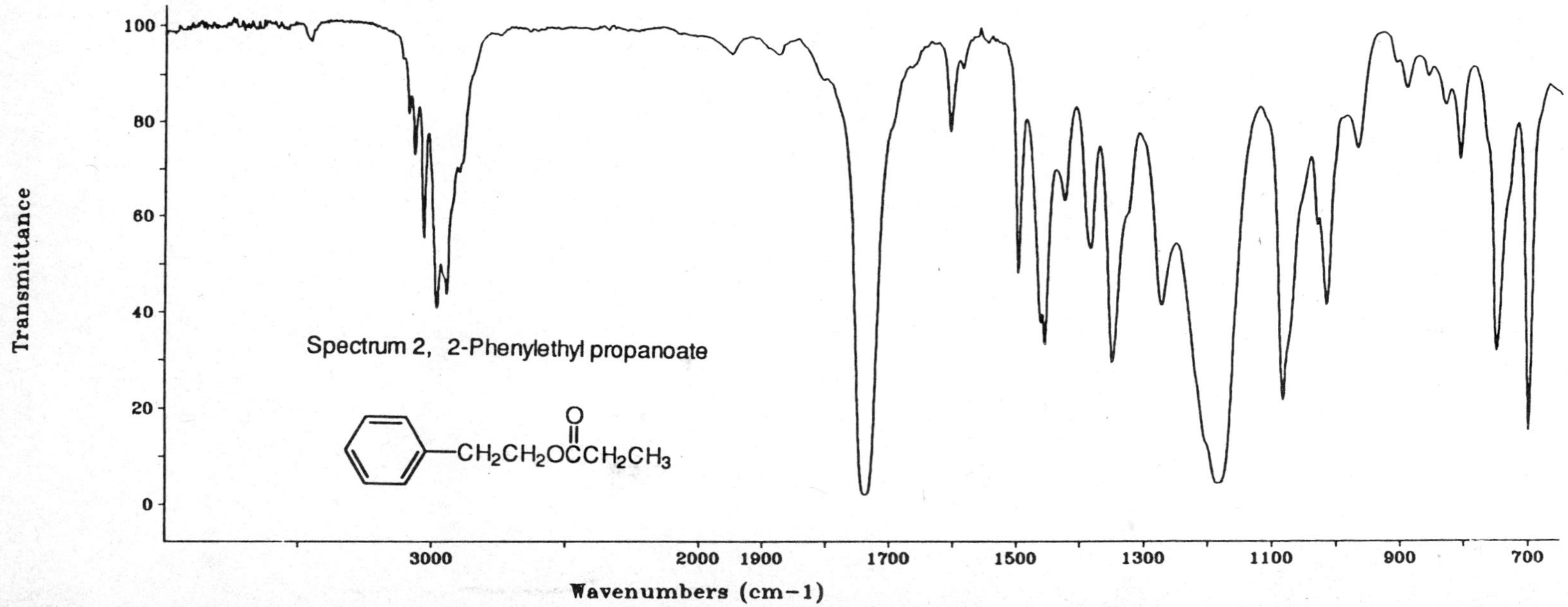
Transmittance
100
80
60
40
20
0
Spectrum 2, 2-Phenylethyl propanoate
$-CH_2CH_2O\overset{O}{\overset{\|}{C}}CH_2CH_3$
3000
2000 1900
1700
1500
1300
1100
900
700
Wavenumbers (cm−1)

11. Nuclear Magnetic Resonance Spectroscopy

A. Introduction

From the time nuclear magnetic resonance (NMR) spectroscopy came into general use in organic chemistry in the mid-1950s, the technique has progressed dramatically to become an indispensable tool for establishing structures of complex organic molecules. In NMR spectroscopy, nuclei in a strong magnetic field achieve a condition of resonance with a source of radiofrequency radiation; as a result, the nuclei absorb energy. Resonance refers to this condition of being frequency-tuned, not to valence bond structures. Recent advances in instrumentation have so simplified the acquisition of NMR spectra that is now possible to employ NMR methods in introductory organic chemistry laboratories.

For many years most NMR spectra were obtained with 60 megahertz (MHz) instruments with applied magnetic fields, H_0, of 14,100 gauss. Today some research instrument generate magnetic fields of 141,000 gauss and the spectra are obtained at 600 MHz. The use of more powerful instruments can simplify the interpretation of NMR spectra. The spectra shown in this chapter were obtained on a 270 MHz instrument.

Because modern organic chemistry texts have extensive discussions of NMR spectroscopy, including substantial introductions to NMR theory, the discussions that follow will focus mainly on principles as they are applied in the interpretation of NMR spectra. It is thus essential that you study the material on NMR spectroscopy in your lecture text in conjunction with the material in this and subsequent chapters.

B. Nuclear Spin, the Basis of NMR Spectroscopy

Certain nuclei, including ^{1}H, ^{13}C, ^{15}N, ^{17}O, ^{19}F, and ^{31}P possess a property called nuclear spin. When nuclei with spin are placed in a very strong magnetic field, they can absorb energy from a radiofrequency source, generating the signals which constitute NMR spectra. Other common nuclei such as ^{12}C and ^{16}O do not have nuclear spins and therefore do not generate NMR signals. In this chapter and the experiments which follow, we will focus on proton NMR, (^{1}H-NMR) exclusively. While ^{13}C-NMR is also important in determining structure, with many instruments the time involved in acquiring such spectra can be substantial and thus ^{13}C-NMR cannot yet be employed routinely in most large introductory organic laboratories.

The heart of the NMR spectrometer is a magnet which generates an extremely strong, homogeneous (uniform) magnetic field. The magnetic field presents no hazard to your health, but heed the precautions about bringing certain items too close to the magnet, such as mechanical watches which can be ruined and credit cards which can be erased. Figure 1 is a schematic diagram of an NMR spectrometer magnet and sample chamber (the probe). The sample of the compound is placed in the probe and the magnetic field strength is swept (scanned) by the spectrometer over a small range. As the field is swept, the sample is simultaneously subjected to a constant radio frequency (RF) radiation generated by an RF-source tuned to the field strength of the magnet. As the magnetic field is swept, the amount of energy absorbed from the RF-source is monitored. The output is a spectrum of signal intensities (absorption) versus increasing magnetic field strength. Because the RF frequency is directly proportional to the magnetic field strength, it is customary to present spectra in terms of intensity *versus* frequency shifts (in units of Hertz) relative to a standard.

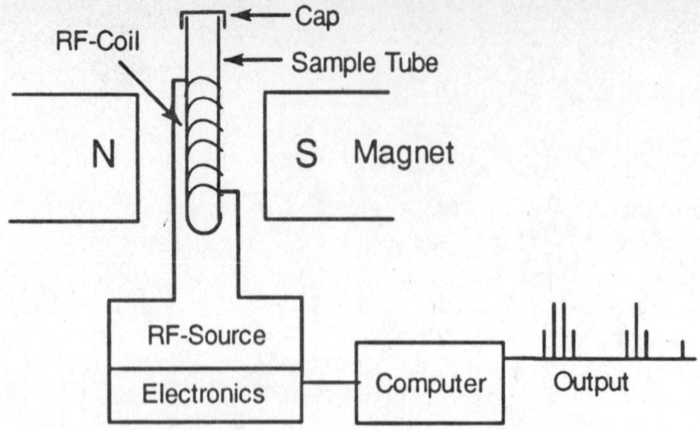

Figure 1. Schematic drawing of an NMR spectrometer.

Because of their nuclear spin, protons have magnetic moments which cause them to behave like tiny magnets. The magnetic moments must be aligned either in the direction of the applied field (α spin state, lower energy) or opposite to the applied field (β spin state, higher energy). When the sample absorbs energy from the RF-source, there is a net excitation (or flip) from the α to β spin states. The sweeping of the field brings different protons in the sample into resonance causing the sample to absorb energy at different applied field strengths as the protons change their magnetic alignments. The energy difference between the α and β spin states is directly proportional to the strength of the magnetic field experienced by each proton.

C. Sample Handling

Although NMR spectra can be obtained on pure liquids, generally *ca* 1-10% solutions in CCl_4 or $CDCl_3$ are employed. Occasionally deuteroacetone (CD_3COCD_3), D_2O, or other deuterated solvents are used (deuterium does not absorb in the ^1H region). The solvents provided for NMR generally have incorporated in them small amounts of tetramethylsilane (TMS) to serve as an internal reference. All lines of the NMR spectrum are measured in relation to the singlet of the TMS reference. The solutions should be clear of particles of undissolved material or foreign matter. Paramagnetic impurities broaden the NMR lines and make the spectrum interpretation more difficult.

The sample is placed in a 180 x 5 mm thin-walled precision-bore glass tubing. The tube is sealed with a plastic cap and then inserted into the probe centered between the pole faces of the magnet. Because sample tubes are spun rapidly in the probe, only tubes especially made for this purpose should be used. Spinning is necessary to average slight inhomogeneities of the field.

Great care must be exercised in inserting the sample tube into the spectrometer. Follow the instructions you are given. If a sample tube should break in the spectrometer, it can involve a great deal of time and expense to clean up the breakage and get the spectrometer back on line.

After the spectrum has been recorded and the integration performed, remove the NMR tube **CAREFULLY** and clean it out in preparation for the next sample.

D. Interpretation of NMR Spectra

The interpretation of ^1H NMR spectra involves three basic concepts: the **chemical shift,** which is the position of a signal (line) in ppm relative to the TMS reference; the **area of the lines** associated with each proton, or group of protons; and the **spin-spin splitting patterns** associated with the lines. Each of these measures provides important structural information. When combined, the information is often sufficient to assign a unique structure to the molecule

1. The Chemical Shift

The first analysis of the NMR spectrum is to examine the chemical shifts of the lines and try to relate them to the specific types of protons that could have given rise to these lines. The chemical shift of a proton is the position of the line in reference to the singlet line of TMS, which is assigned the value of $\delta = 0$ (δ is defined below). The chemical shift is influenced by the type of carbon atom to which the proton is attached, the position and nature of adjoining atoms, and the overall structure of the molecule.

Spectrum 1 shows the 270 MHz NMR spectrum of pinacalone, 3,3-dimethyl-2-butanone. (Spectra are shown at the end of this chapter.) The single line at δ 0 is the reference signal from TMS. The signal at δ 0.90 ppm arises from the nine protons of the $C(CH_3)_3$ group, while the signal at δ 2.10 ppm is due to the three protons of the CH_3CO group. These signals (in ppm) are the chemical shifts relative to the δ 0 singlet of the TMS standard. The vertical scale gives the relative intensities of signals on an arbitrary scale. The horizontal scale represents the applied magnetic field strength, H_0, increasing from left to right. To the left is referred to as downfield, to the right as upfield. The scale at the top represents the frequency shift in Hertz of a signal ($\Delta\nu$) downfield from the TMS signal. The numbers on this scale depend upon the frequency of the RF source used in the spectrometer (270 MHz in this case). The lower scale gives the chemical shifts in δ values which represent the fractional shift of a signal from that of TMS ($\Delta\nu$) relative to the applied radio frequency (ν) expressed in parts per million.

$$\delta = (\Delta\nu/\nu) \times 10^6 \text{ ppm}$$

For example: $\delta = (213\text{Hz}/270 \times 10^6 \text{ Hz}) \times 10^6 = 0.90$ ppm

Fortunately chemical shifts for "different kinds of protons" are remarkably reproducible. Consequently it has been possible to tabulate typical chemical shifts expected for a wide variety of substitution patterns (Table 1).

Table 1. Proton Chemical Shifts
Tetramethylsilane, $(CH_3)_4Si$, $\delta = 0$

Compound Type	Type of H	Typical Chemical Shift δ, ppm ±0.3
Primary Alkyl	RCH_3	0.9
Secondary Alkyl	R_2CH_2	1.3
Tertiary Alkyl	R_3CH	1.5
Allylic	$R_2C{=}C(R)CH_3$	1.7
Propargylic	$RC{\equiv}CCH_3$	2.0
Ester	CH_3CO_2R	2.1
Alkyl Ketone	$RCOCH_3$	2.2
Benzylic	$ArCH_3$	2.4
Alkyne	$RC{\equiv}CH$	2.5
Aryl Ketone	$ArCOCH_3$	2.6
Iodoalkane	RCH_2I	3.2
Bromoalkane	RCH_2Br	3.4
Chloroalkane	RCH_2Cl	3.6
Ether	$ROCH_3$	3.4
Ester	RCO_2CH_3	3.6
Alcohol	RCH_2OH	3.6
Terminal Alkene	$R_2C{=}CH_2$	4.9
Internal Alkene	$R_2C{=}CHR$	5.4
Aromatic	ArH	7.2
Aldehyde	RCHO	9.8
Carboxylic Acid	RCO_2H	11 ±1
Alcohol	ROH	1-6*
Amine	RNH_2	1-5*
Phenol	ArOH	4-8*

* Variable

The δ values are used to identify chemical shifts because δ values are ratios which are independent of the value of the frequency of the RF source and of the strength of the magnetic field of the spectrometer. Some of the δ values for protons of common structures are plotted in Figure 2 to aid in analyzing the problems presented in this manual.

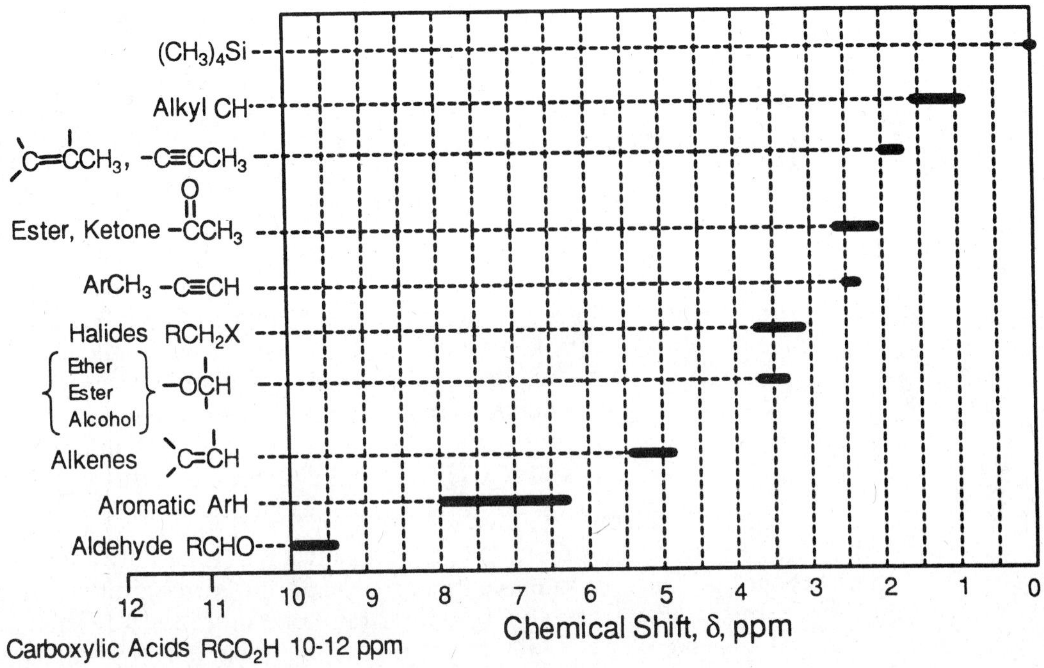

Figure 2. Chart relating structure to chemical shift.

2. Origin of Chemical Shifts

As examination of Table 1 and Figure 2 will show, protons in different molecular environments usually give different NMR signals; i.e., they have different chemical shifts. What causes chemical shifts? A bare proton would give a signal far downfield from the signals found for protons in molecules. All protons are diamagnetically shielded by the electrons in the molecules, especially the valence electrons. An electron moving in a magnetic field produces a small magnetic moment opposed to the applied field. Thus the magnetic field strength experienced by a proton in a molecule is slightly less that the applied field, H_0. Consequently H_0 must be increased a little in order to overcome this diamagnetic shielding and reach the field strength necessary to cause resonance. Fortunately the extent of diamagnetic shielding is dependent upon the immediate molecular environment of a proton. Because electronegative atoms and groups attract electrons, they reduce the general diamagnetic shielding. These atoms and groups are said to deshield the proton, giving rise to a chemical shift. The greater the deshielding the greater the chemical shift downfield from the TMS reference signal.

3. Predicting Chemical Shifts

Chemical shifts are usually approximately additive, meaning that the chemical shift for XCH_2Y can be predicted from the chemical shifts for RCH_2X and RCH_2Y. For example, the chemical shifts for CH_3Cl, CH_2Cl_2, and $CHCl_3$ show that there is approximately the same incremental shift as chlorine atoms are added.

	CH_4	CH_3CH_3	CH_3Cl	CH_2Cl_2	$CHCl_3$
δ	0.23	0.88	3.15	5.30	7.27

The chemical shifts for both methane and ethane are included. While methane might seem to be the appropriate reference compound, for our purposes the use of ethane as the reference gives better predictions of expected chemical shifts for XCH_2Y cases. Thus in the example above, the shift is *ca* 2.1 ppm per chlorine atom.

To keep the number of entries in Table 1 manageable, most of the chemical shifts are shown for methyl groups, CH_3, or methylene groups, $\mathbf{CH_2}$. The first three entries in Table 1 show that each additional alkyl group causes a chemical shift of ca 0.3 ppm. Thus, from the chemical shift for a methyl ketone, $RCOCH_3$, at δ 2.1, we would expect the chemical shift of $RCOCHR_2$ to be *ca* δ 2.7. Similarly, with the chemical shift of an alkyl methyl ether $ROCH_3$ at δ 3.4, we would expect the methylene protons of $ROCH_2CH_3$ to be at *ca* δ 3.7. The important point is to recognize that you can predict approximate chemical shifts for RCH_2X and R_2CHX from the chemical shift for CH_3X.

Figure 2 will be helpful in narrowing down the kinds of functionality which lead to observed chemical shifts. Keep in mind that the chemical shifts illustrated are for monofunctionality. In that regard, it is useful to make the empirical approximation that all functional group shifts reflect a shift from *ca* δ 0.9 (the first value in Table 1). Thus we would predict a chemical shift for $CH_2(CO_2R)_2$ based on the chemical shift of δ 2.1 for an ester CH_3CO_2R as follows: 2.1 + (2.1-0.9) =3.3, which is close to the observed value of δ 3.37 found for diethyl malonate.

Note that the signals for the protons of aldehydes, RCHO, and carboxylic acids, RCO_2H, are far downfield compared with most other chemical shifts. The absence of signals in this downfield region eliminates these groups from consideration when you are interpreting spectra.

4. Peak areas

It is an important fact that NMR peak areas are directly proportional to the number of protons generating the peaks. All NMR spectrometers have the capability of integrating the signals to generate relative peak areas. The integrals may be presented graphically on the spectrum as a rising curve consisting of a series of steps where the height of each step is proportional to the area of the peak producing it. In high-field instruments, the relative areas can also be printed for the peaks. The relative areas of the peaks will be the relative numbers of protons responsible for the peaks.

If the molecular formula is known for the compound generating the spectrum, it is an easy matter to transform relative peak areas into the actual numbers of protons. For pinacalone, Spectrum 1, the relative peak areas for the δ = 0.9 signal to the δ = 2.1 signal are in the ratio 75 to 25 or 3:1. Since the molecular formula shows a total of 12 protons, the corresponding proton ratios are 9:3. In providing spectra "to solve" in problems, the relative numbers of protons generating the signals may be indicated on the spectra.

5. Spin-Spin Splitting

In Spectrum 1, the signals for pinacalone are singlets, meaning that a single peak or line is observed for each set of equivalent protons. On the other hand, in the NMR spectrum of the ester 2-phenylethyl propanoate, Spectrum 3, four of the signals appear as clearly separated multiple peaks while the phenyl protons give rise to closely-spaced, overlapping peaks.

This phenomenon, called spin-spin splitting, is the splitting of the signals arising from nonequivalent sets of protons on adjacent carbons. **Magnetically equivalent sets of protons, i.e., protons which have the same chemical shift, do not give rise to mutual splitting.** Magnetically equivalent protons may have their signals split by nonequivalent protons, but not by each other. For example, the protons in the CH_3 group never give rise to mutual splitting; however, the signal from the CH_3 protons will be split by the protons in an adjacent CH_2 group.

In the spectrum of diethyl ether, Spectrum 2, there are two sets of equally spaced lines (peaks). The CH_2 protons give rise to a quartet of lines and the CH_3 protons generate a triplet of lines. Note that the spacing of the lines is the same in both sets of signals. This spacing is defined by the coupling constant, ***J***, and is measured in frequency units (Hz). It is an important observation that ***J*** is independent of the RF frequency of the spectrometer. Coupling constants vary widely depending upon the exact environment (including the relative stereochemistry) of coupled protons. Values of ***J*** range from less than 1 Hz to as much as 20 Hz. Many values are in the range 4–10 Hz.

What we are observing in the example of diethyl ether is called "first order splitting" and it is what is seen when the protons which are coupled have very different chemical shifts. The splitting patterns reflect the fact that the nuclear spins of nonequivalent adjacent protons interact to change the magnetic field strength at each set of protons. This interaction occurs through the valence electrons in the bonds and is most important for a two-bond sequence, H-C-H, and for a three-bond sequence H-C-C-H. In the H-C-H case, the protons are often equivalent, so no splitting is observed. Most of the splitting we see comes from examples of H-C-C-H coupling. Longer range coupling (through four or more bonds) is often seen superimposed on larger splitting patterns, but the splitting is usually small enough to be overlooked because it is still possible to recognize the major splitting patterns. However, long-range splitting can be useful; e.g., in the four-bond system H-C-C=C-H, *trans* coupling is *ca* 1–2 Hz while *cis* coupling is *ca* 0 Hz.

It is beyond the scope of this manual to discuss the theory of spin-spin splitting, but we can readily make use of observations made about first-order splitting. With the signal for a set of protons split into several lines, how do we assign the chemical shift? In first-order splitting the chemical shift can be assigned with confidence to the center of the splitting pattern. For example, the chemical shift is assigned to the center line of a triplet and to half way between the two center lines of a quartet. In more complex splitting, it may be less obvious where to assign the chemical shift, but the apparent center of the pattern will be adequate for our purposes. The related question concerning integration of the peaks in a splitting pattern is readily answered: the peak area must be summed over the entire set of lines in the splitting pattern.

6. Typical Splitting Patterns: The (n+1) Rule

The signal from a set of equivalent protons will be split into (n+1) lines by a second, nonequivalent set of n protons.

The number of lines in a signal reflects the number of coupled protons that are on adjacent carbons, not the number of protons generating the signal.

First-order splitting patterns of several molecules are shown in Figure 3. Study each case carefully to be sure you understand the origin of each splitting pattern.

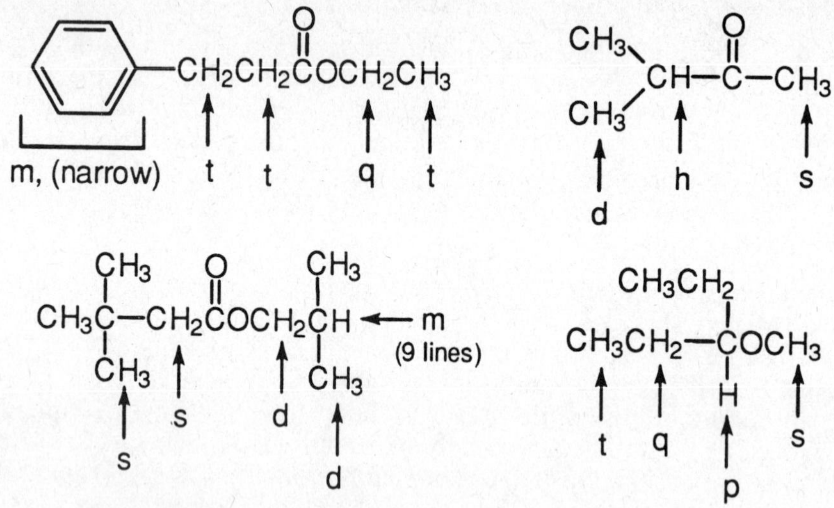

Figure 3. First-order splitting patterns. (s, singlet; d, doublet; t, triplet; q, quartet; p, pentuplet; sx, sextet; h, heptet; m, multiplet).

Although the relative intensities of the lines for first-order coupling are readily predicted (see your text), at least learn the intensity ratios for the following: doublet (1:1), triplet (1:2:1), quartet (1:3:3:1), pentuplet (1:4:6:4:1), sextet (1:5:10:10:5:1) (Figure 4).

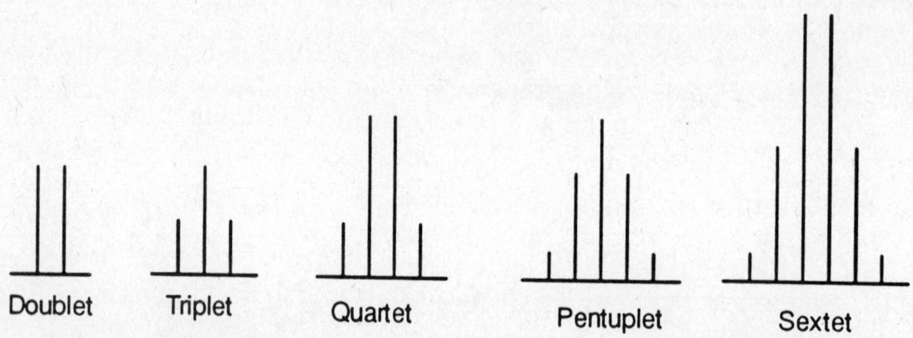

Figure 4. Relative intensities in first-order coupling.

7. The Spacing of Lines

Coupled protons split each other identically because there is a common coupling constant, *J*. Thus, the spacing between lines is the same in the signals for each set of protons. This fact can be very helpful in analyzing spectra. For example, a two-proton quartet near δ 4 indicates the presence of a methylene group. The splitting pattern suggests that there should be an upfield triplet with the same spacing near δ 1 arising from a methyl group.

8. Symmetry and Lack of Symmetry in Splitting Patterns

First order coupling predicts symmetrical splitting patterns, e.g., 1:3:3:1 intensities for the CH_2 quartet and 1:2:1 intensities for the CH_3 triplet, as shown for diethyl ether, Spectrum 2. If the chemical shift difference (Δ δ) between a coupled set of protons is small, this symmetry is lost and the pattern takes on what is sometimes called "leaning." In the combined pattern, the outer peaks diminish and the inner peaks grow as illustrated in Figure 5. As the chemical shift difference decreases, this loss of symmetry becomes increasingly pronounced.

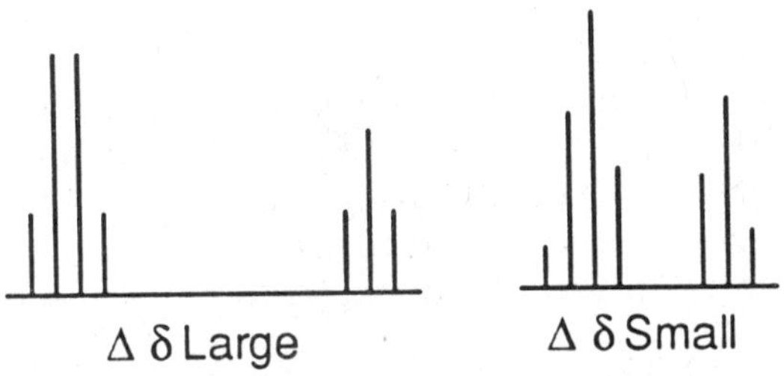

Figure 5. Effect of Δ δ on splitting pattern.

This effect is illustrated by comparing the NMR spectrum of 2-phenylethyl propanoate, $CH_3CH_2CO_2CH_2CH_2C_6H_5$, (**Spectrum 3**) with that of the isomer ethyl 3-phenylpropanoate, $C_6H_5CH_2CH_2CO_2CH_2CH_3$, (**Spectrum 4**). In each compound there is an "isolated" $-CH_2CH_2-$ system giving rise to the signals designated by **c** and **d**. In **Spectrum 3** the **c** and **d** signals are separated by *ca* 1.4 ppm and they show little leaning. But in **Spectrum 4** the separation is only *ca* 0.3 ppm with the result that substantial leaning occurs.

9. Chiral Centers and Diastereotopic Protons

In diethyl ether ($CH_3CH_2OCH_2CH_3$), Spectrum 2, and ethyl isopropyl ether, compound **1,** the methylene protons (CH_2) in each molecule are clearly equivalent. In an ether of the type **2**, the presence of the chiral center (*) means that the two methylene protons are now no longer equivalent; they are **diastereotopic.** As a consequence, they have nonidentical chemical shifts. The simple quartet for the equivalent CH_2 protons of compound **1** is replaced by a much more complex pattern from the nonequivalent protons of the CH_2 group of **2**. In a molecule of type **3**, the protons on each CH_2 group are also diastereotopic and can give rise to complex NMR signals.

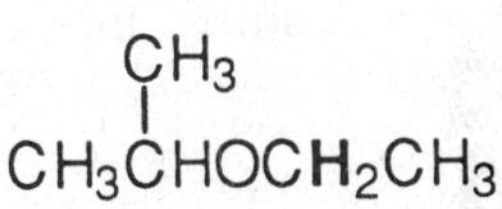

1

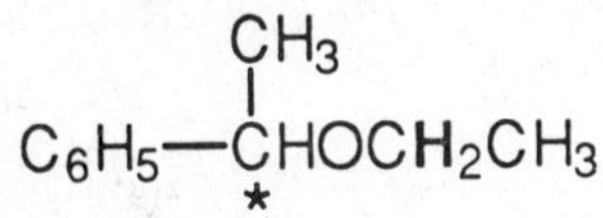

2

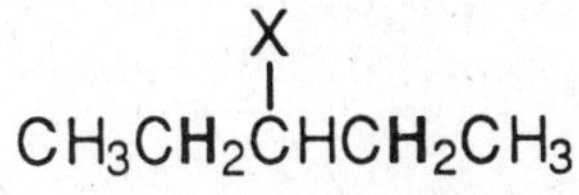

3

10. Signals from Alcohols and Amines

The OH group of an alcohol usually gives a singlet (which may be broad) even though there are adjacent protons; e.g., RCH_2OH would be expected to have both the OH and CH_2 signals mutually split, but no splitting is seen. This absence of splitting reflects very fast proton exchange between alcohol molecules which averages out the spin states of the OH proton. Because this process is catalyzed by a trace of acid or base, a very pure sample may show the expected splitting.

The protons of primary and secondary amines also give rise to broad singlets and do not split the signals of adjacent protons.

11. Aromatic Protons

Compounds with aromatic protons show chemical shifts in the range δ 7-8. Because phenyl groups, C_6H_5, are common, it helps to examine what kinds of signals they generate. If the substituent on the phenyl group is an alkyl side chain, then the phenyl protons usually give a 5-proton signal which is an apparent singlet at 60 MHz. At higher field strengths (e.g. 270 MHz) some splitting can be seen (refer to the phenyl protons of Spectra 3 and 4). Obviously the chemical shifts for the ring protons must be nearly identical. However, if the substituent on the ring is electronegative, then a much more complex spectrum is seen, often falling into a 3:2 pattern because now the ring protons have different chemical shifts.

E. Fourier Transform NMR Spectroscopy (FT-NMR)

The type of spectrometer described at the beginning of this chapter is called a continuous wave (CW) spectrometer. In recent years pulsed FT-NMR spectrometers having very powerful superconducting magnets (operating in liquid helium at 4 K) have been increasingly used. In an FT-NMR spectrometer the sample is placed in a very strong, constant magnetic field (the field is not swept). An intense pulse of RF energy (lasting a few microseconds) is sent to the sample as a continuum of RF frequencies covering a range wide enough to excite all of the protons in the sample. That is, the resonance condition is met simultaneously for all of the different kinds of protons in the molecule. Then, as a consequence of the way their magnetic moments behave in a magnetic field, the excited nuclei induce a current in a receiver coil. This current is monitored for a 1–5 seconds as it decreases while nuclei steadily "relax" (return to their initial state). The spectrometer output, a function of intensity *versus* time called a free induction decay (FID), is a steadily decreasing (damped) cosine wave. After a brief delay of 2–4 sec., the pulse sequence is repeated. The FIDs from a desired number of pulses are summed by a computer to improve the signal to noise ratio. Then the composite FID is subjected to a mathematical procedure called a Fourier Transformation by a powerful computer which converts the FID into a "conventional" spectrum of intensity *versus* frequency, i.e., the same kind of spectrum produced by a CW spectrometer.

Once operating parameters are established, spectra can be obtained quickly with an FT-NMR spectrometer. Also, the use of an increased field strength (e.g. 270 vs 60 MHz) leads to increased sensitivity, better resolution of signals with similar chemical shifts, and simpler spectra (more nearly first-order).

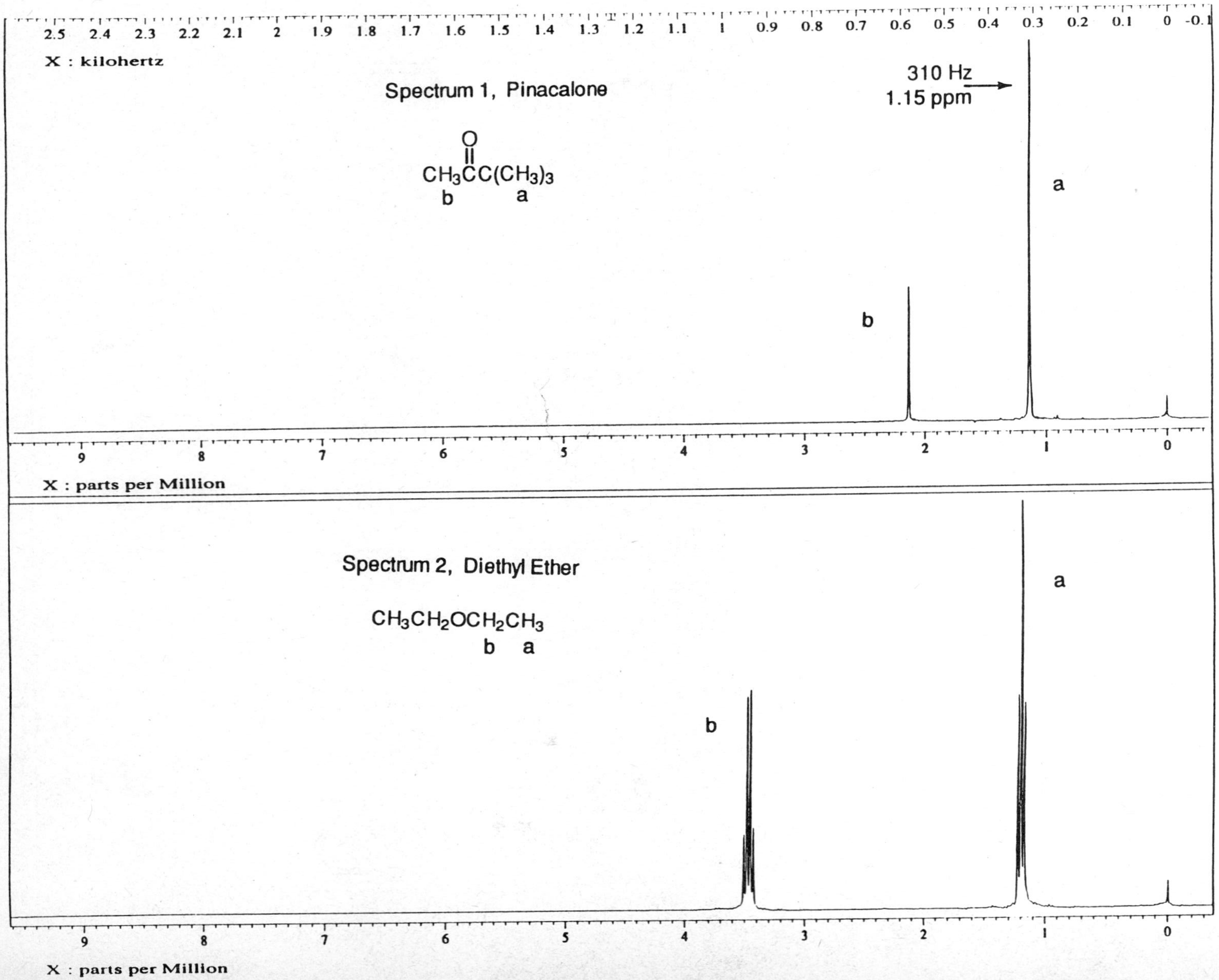

2.5 2.4 2.3 2.2 2.1 2 1.9 1.8 1.7 1.6 1.5 1.4 1.3 1.2 1.1 1 0.9 0.8 0.7 0.6 0.5 0.4 0.3 0.2 0.1 0 -0.1
X : kilohertz
Spectrum 1, Pinacalone
CH3CC(CH3)3
b a
O
310 Hz
1.15 ppm
a
b
9 8 7 6 5 4 3 2 1 0
X : parts per Million
Spectrum 2, Diethyl Ether
CH3CH2OCH2CH3
b a
a
b
9 8 7 6 5 4 3 2 1 0
X : parts per Million

Spectrum 3, 2-Phenylethyl propanoate

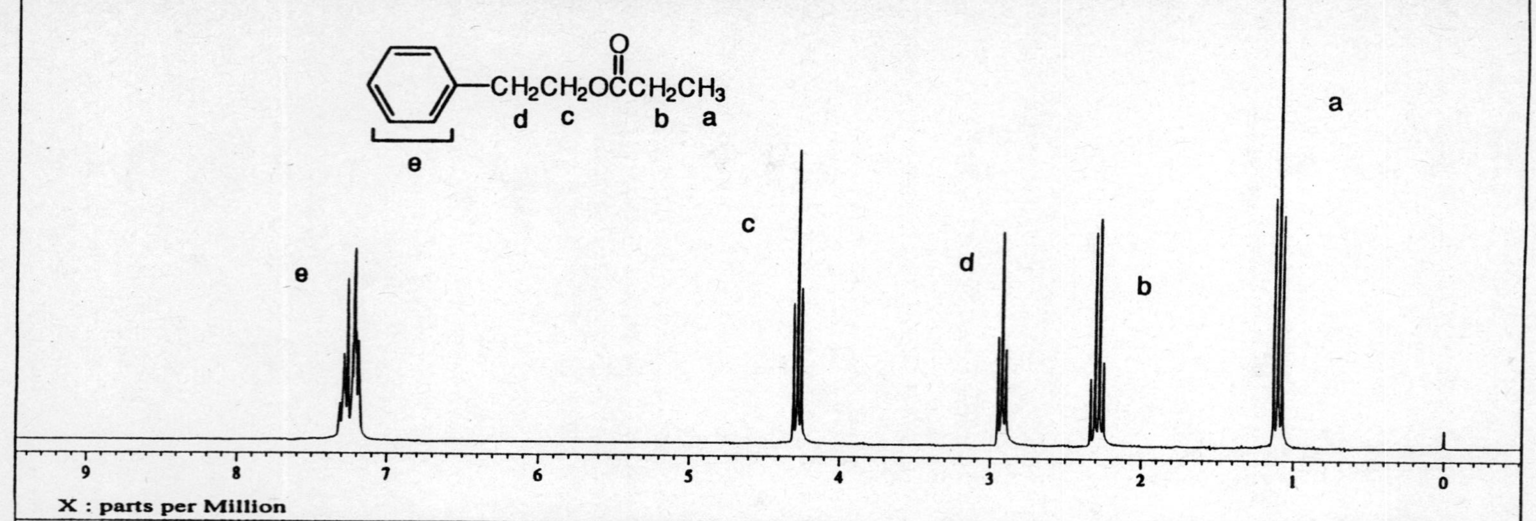

Spectrum 4, Ethyl 3-phenylpropanoate

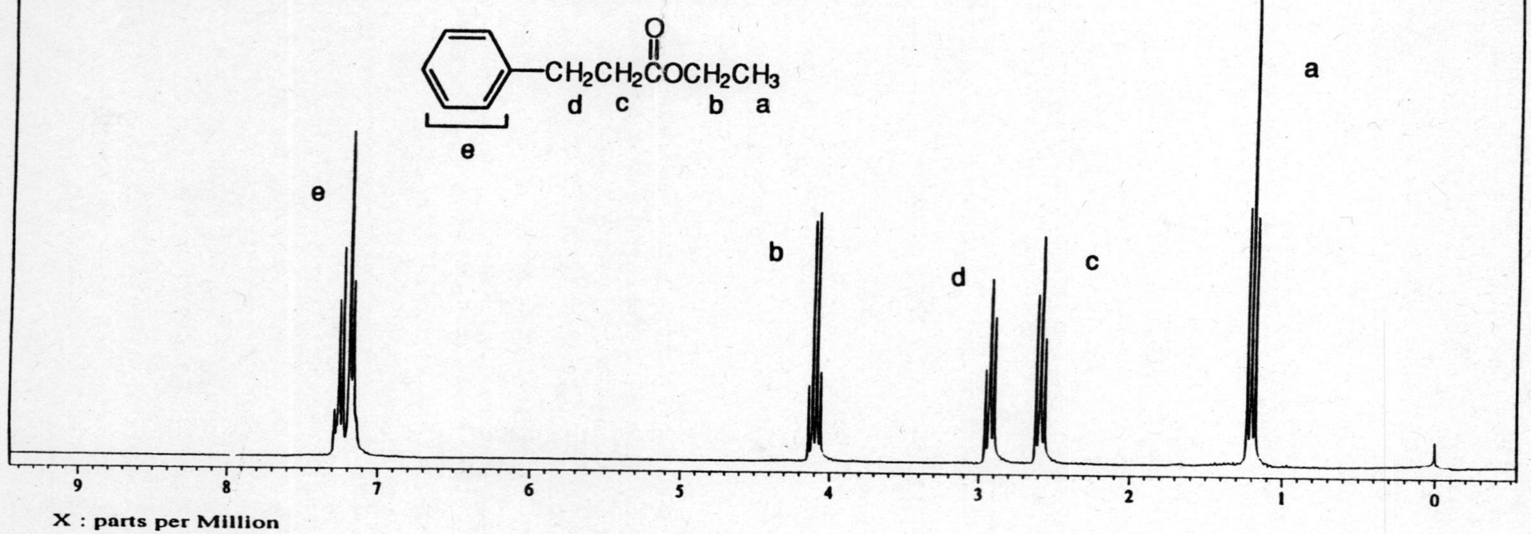

12. Assignment of Structure by IR and ^{1}H-NMR Spectroscopy

A. Introduction

> Before doing this experiment it is strongly advised that you review the sections on infrared and nuclear resonance spectroscopy in your textbook and this laboratory manual, especially as they pertain to the interpretation of spectra.

The purpose of this experiment is to determine the molecular structures of unknown substances by use of IR and NMR spectroscopy. Solving real problems is the best way to gain an understanding of applied spectroscopy. In the process you will learn to appreciate the enormous impact spectroscopy has had on modern organic chemistry. While the problems encountered in the experiment are designed to serve as an introduction to spectroscopy, the same kind of reasoning and logic is used to solve advanced problems in molecular structure. In each of the first two sections, you will record the IR spectrum of at least one unknown and then match the spectrum with one in a set of previously recorded spectra. Matching spectra to identify compounds is a widely used procedure. If the spectrum of an unknown compound matches peak for peak with the IR spectrum of a known substance, there is a very high probability that the unknown compound has the same structure as the known compound. If the spectra do not match, then the unknown compound is definitely not the same as the known compound. Once the IR spectra have been matched, the NMR spectrum of the unknown will be available for further spectral analysis. In the third section, you will be given a sample of an unknown compound and its molecular formula. The IR and NMR spectra are to be recorded and from these data, together with the molecular formula, the molecular structure is to be assigned.

In order to simplify the process of preparing samples for the recording of IR spectra, all of the unknown compounds for this experiment are liquids. It is relatively easy to obtain IR spectra on pure liquids by use of salt plates. Obtaining IR spectra of solids is somewhat more involved and the technique will be learned later in other experiments. The unknowns will be provided in small numbered bottles fitted with an attached micropipet. Do not remove the bottles from their designated location, and be sure to replace the droppers and caps.

Since all of the unknowns are liquids, IR spectra can be obtained on thin films of the liquids sandwiched between two salt plates. Be sure that the salt plates are clean and dry before use. A single drop of a liquid placed in the center of one plate should suffice. When the second plate is pressed onto the first plate, the liquid should spread to provide a film, without voids, between the plates.

Follow the protocol provided by your instructor for the particular IR instrument you are to use. If the spectrum is too intense, with strong absorptions that bottom out, separate the plates, gently touch one of the plates to a clean tissue, reassemble the plates, and record the spectrum again. After you have obtained a satisfactory spectrum, remove the salt plates from the spectrometer, separate the plates, wash both thoroughly with ethanol or acetone, and dry them with a clean tissue.

For NMR spectra, samples are run in carbon tetrachloride (CCl_4) or deuterochloroform ($DCCl_3$) solution in a an NMR tube. A sample size of 5–10 mg in 0.5 mL of solvent should be sufficient to produce a good NMR spectrum. NMR solvents commonly contain the proper amount of tetramethylsilane (TMS) reference. All chemical shifts are measured in ppm relative to the TMS line at δ 0.

You will have to take your turn at using the instruments. When you are not involved in preparing spectral samples or recording spectra, study the many other pairs of IR and NMR spectra provided in the laboratory with the aim of assigning molecular structures. Students are encouraged to discuss the spectra freely, but each student should make individual structural assignments based on a clear understanding of the information revealed by the spectral data.

B. What to Look For in Examining Spectra

First examine the infrared spectrum to see what characteristic absorptions are present and which are absent. For example, the presence of an absorption band in the 3600-3300 cm^{-1} region indicates that a compound is an alcohol (O-H) or an amine (N-H); the absence of absorption in this region indicates that N-H and O-H groups are absent. Look at the C-H stretching region around 2900 cm^{-1}. Is there absorption just above 3000 cm^{-1} to indicate aromatic, alkene, or alkyne groups? Are there bands in the triple bond region due to C≡C or C≡N triple bond stretching of alkynes or nitriles?

Examine the carbonyl region very carefully. Strong absorption in the 1800-1650 cm^{-1} region indicates the presence a carbonyl group (C=O stretching). The absence of any absorption in this region eliminates carbonyl compounds from consideration. Keep in mind that the exact position of a carbonyl band depends not only on the type of functional group (e.g., aldehydes, ketones, esters, acids, etc.) but also on other factors such as conjugation (conjugation lowers the frequency of an absorption band). Look at the rest of the spectrum. Does it show any bands around 1650 cm^{-1} which could be due to C=C stretching? While a detailed analysis of the fingerprint region is beyond our expectations, certain strong absorption bands can be helpful. For example, are there any strong bands in the 1000–700 cm^{-1} region which could be due to vinylic C-H bending?

By identifying the most salient features of an infrared spectrum and then looking at the structures of the compounds in Figures 1 and 2, you should be able to narrow the possibilities to two or three compounds even before you consider the related NMR spectrum.

When the interpretation of the IR spectrum is complete, examine the NMR spectrum. Three features will be important in the NMR analysis; the **chemical shifts** (the positions of the lines with respect to the TMS reference, the **integrated intensities** of the lines, and the **splitting patterns.** The chemical shifts indicate the types of protons responsible for the lines. The integrated intensities indicate the relative numbers of each type of proton. The splitting patterns give information concerning the location of the protons and the number of neighboring protons.

For example, look far downfield for signals due to carboxylic acid (COOH) or aldehyde (CHO) protons. Are there signals in the aromatic proton region? If so, how many protons cause the signals and what splitting pattern, if any, is seen? A singlet, which may be broad, suggests the presence of a C_6H_5 group. Look upfield; do any singlets stand out? A sharp 3-proton singlet at *ca* δ 3.5 could be a CH_3O group; a 3-proton singlet at *ca* δ 2.0 could be due to an acetyl group, CH_3CO; a 9-proton singlet at *ca* δ 0.9 indicates a *tert*-butyl group, $C(CH_3)_3$, etc. Look for splitting patterns from coupled protons; e.g., an OCH_2CH_3 will show a 2-proton quartet at *ca* δ 3.7 and a 3-proton triplet (with the same J-value) at *ca* δ 1.0.

Consider the data from the NMR analysis along with that from the IR analysis in order to generate partial structures, or components of the molecule. Then assemble these components to give a final structure consistent with all of the spectral features. We must emphasize, writing down the partial structures (components) is an essential part of the process which will enable you to build up the molecular structure logically.

C. Matching Spectra with Structural Formulas

1. Determining the Structure of an Unknown

Some, or all, of a set of IR and matching NMR spectra for 30 compounds will be available in the laboratory. Their structural formulas are shown in Figures 1 and 2. These compounds have been selected to illustrate how a variety of spectral features can be used in the identification of compounds. The actual compounds corresponding to these spectra will be available as a set of unknowns. Each student will be assigned at least one of the numbered unknowns. The IR spectrum of the unknown is to be recorded and then matched peak for peak with one of the IR spectra in the set of IR-NMR unknown spectra. Then the IR and NMR spectra for the unknown are to be analyzed and the results should be examined in relation to the unknowns listed in Figures 1 and 2. From this information, determine the structural formula of the unknown that is consistent with the data.

2. Deducing Structural Formulas from Spectral Data and Molecular Formulas

In this experiment, a second set of matched IR and NMR spectra will be made available together with the molecular formulas of the compounds which gave rise to the spectra. These compounds will be available in the laboratory as a second set of unknowns. Record the IR spectrum of at least one assigned unknown, match this spectrum with a spectrum in the set of available spectra, and then proceed to assign a structural formula to the unknown based on your analysis of the IR and NMR spectra and the molecular formula of the compound. Note that in this case you do not have a list of possible compounds to choose from.

Use the time when you are not recording spectra to analyze the spectra in the complete set and assign structures to as many of the compounds as possible.

D. Determining Structure from IR and NMR Data

To gain additional experience in interpreting spectra, study all of the spectra in the set of 30 unknowns and assign each pair of IR and NMR spectra to one of the compounds listed in Figures 1 and 2. Before coming to the laboratory, you should have predicted the important IR and NMR features expected for each of these compounds (see Prelab Question #1). This preliminary analysis will greatly facilitate the interpretation of the spectra in the laboratory.

E. A General Unknown (Optional)

In this section it is assumed that a proton NMR spectrometer is available for student use. If not, your instructor may give you the NMR spectrum and molecular formula of your unknown.

In this experiment you will be given an unknown liquid compound and the molecular formula of the compound. Record both the IR and NMR spectrum of the unknown. Then based on your analyses of your spectra, and the molecular formula, assign a structural formula to the compound. Record the structure and provide in your report a summary of the spectral features that led to this assignment.

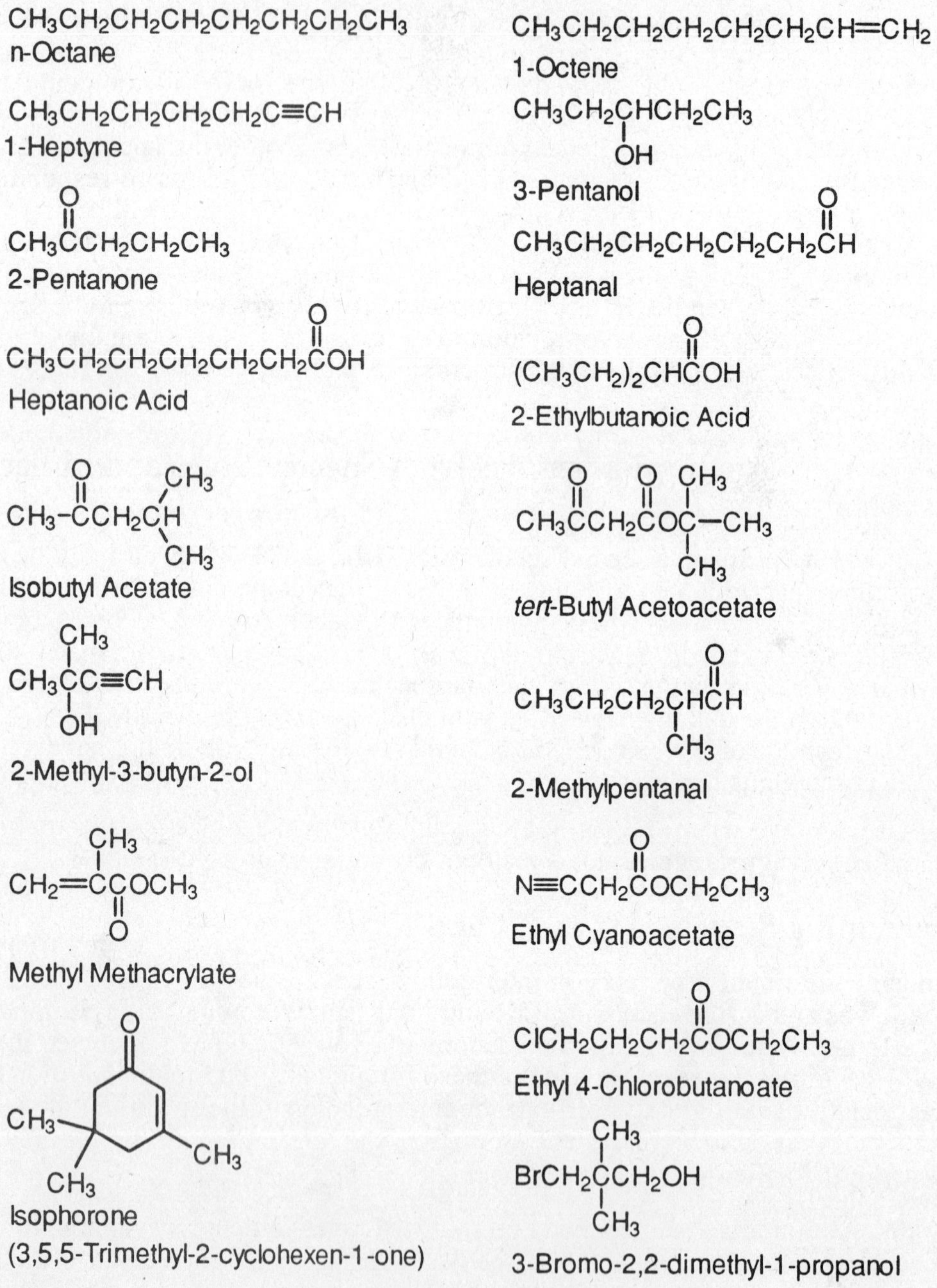

Figure 1. Aliphatic compounds available for unknowns.

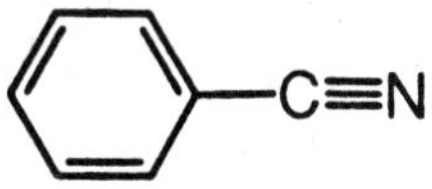

Benzonitrile

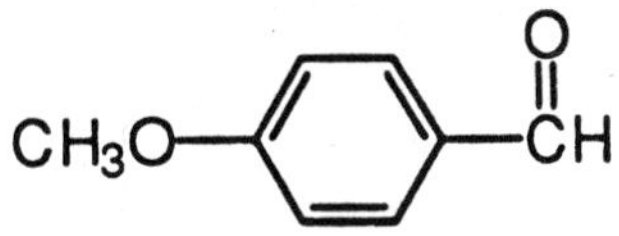

4-Methoxybenzaldehyde
(*para*-Anisaldehyde)

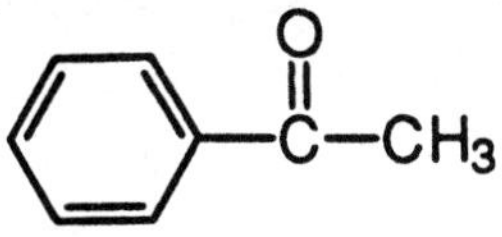

Acetophenone
(Phenyl Methyl Ketone)

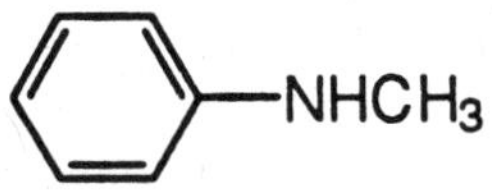

N-Methylaniline

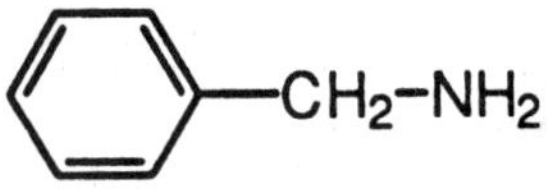

Benzylamine

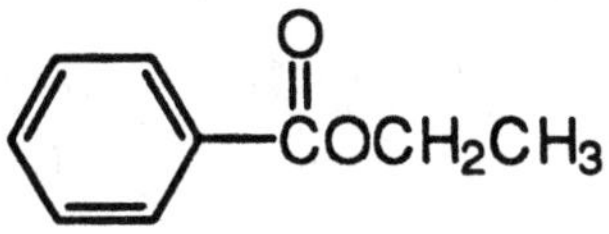

Ethyl Benzoate

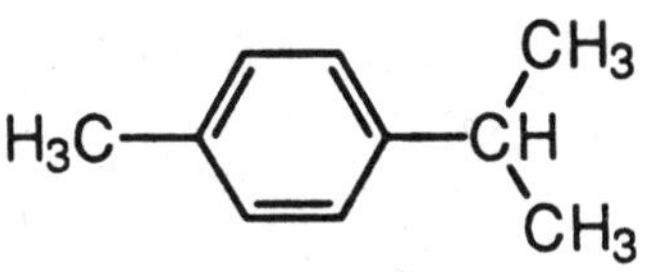

para-Cymene
(4-Isopropyltoluene)

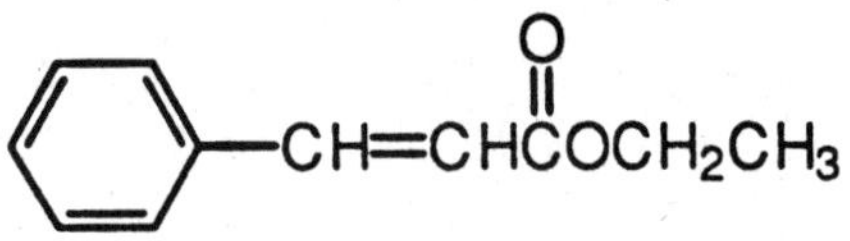

Ethyl *trans*-Cinnamate

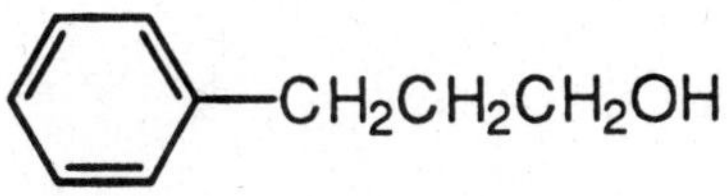

3-Phenyl-1-propanol

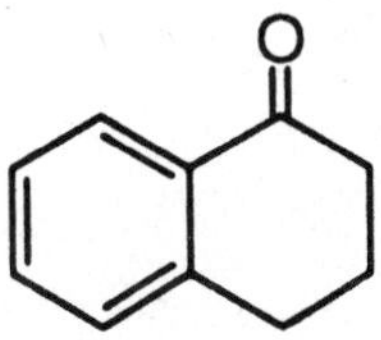

α-Tetralone

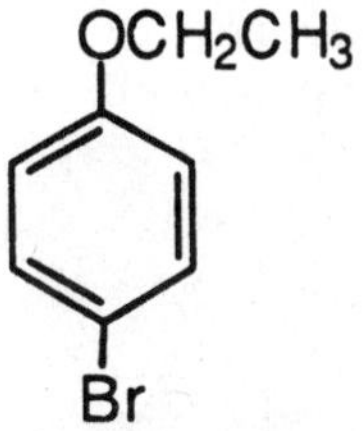

4-Bromophenyl Ethyl Ether
(4-Bromophenetole)

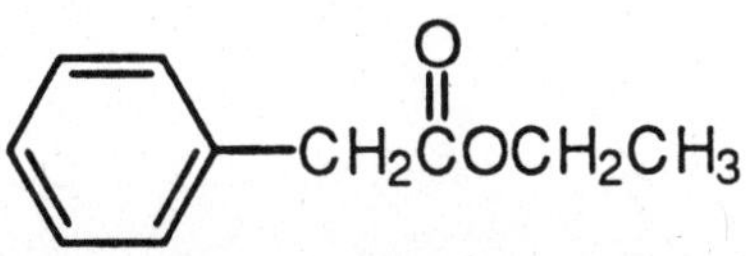

Ethyl phenylacetate

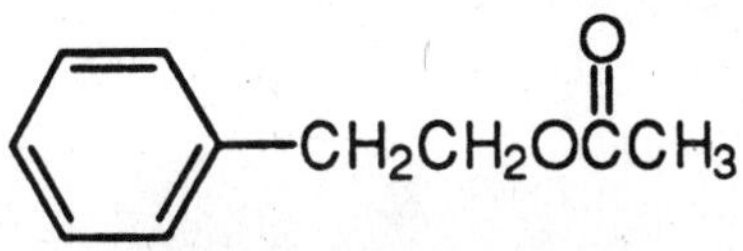

2-Phenylethyl Acetate
(Phenethyl Acetate)

Figure 2. Aromatic compounds available for unknowns.

F. Prelab Questions

1. For each compound in Table 1, predict the IR bands and the NMR signals which are expected to be useful in assigning the molecular structure. List only the most important IR bands expected, e.g. OH stretching, C=O stretching, etc. In anticipating the NMR spectrum, predict the approximate chemical shift and the splitting pattern

expected for each different type of proton, or set of equivalent protons. Draw each structure in your notebook and indicate the expected spectral features as illustrated in the example below.

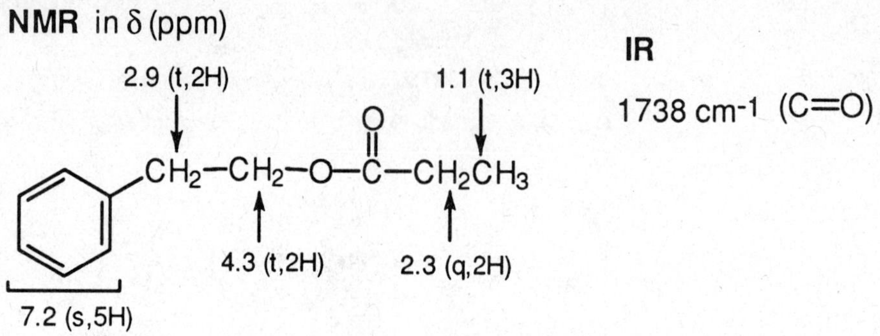

Note that predicting these spectral features will greatly facilitate correlating the molecular structures and spectra when you come to the laboratory.

G. Postlab Questions

1. In each case, compare the spectral predictions you made in the prelab question with what you found on examining the actual spectrum of that compound in the laboratory. Note any significant differences.

13. Preparative Scale GLC. Separation of Carvone and Limonene from Caroway Seed Oil

A. Introduction

Caroway seed oil is essentially a mixture of limonene and (S)-(+)-carvone. Spearmint oil is a mixture of limonene and (R)-(-)-carvone. The different odors of the two substances are due mainly to the difference in the chirality of the two optically active carvones. From this fact we can infer that human olfactory receptor sites are also chiral and respond differently to each of the two enantiomers. In this experiment caroway seed oil will be separated into its components, limonene and (S)-(+)-carvone, by preparative scale GLC. Identification of the two components will be made by infrared analysis. The structures of limonene and the two enantiomers of carvone are shown in Figure 1.

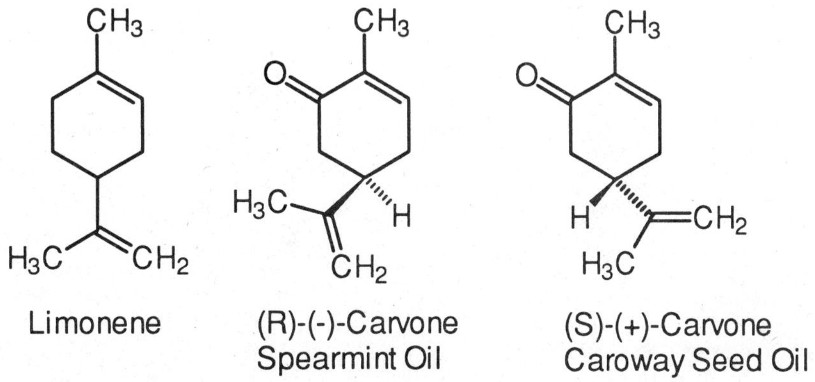

Figure 1. Structures of limonene and carvone.

B. Procedure:

The GLC column for this experiment is 4 feet of 1/4 inch copper tubing packed with 20% Carbowax 20M on Chromosorb PAW-DMCS, 80/100 mesh. The temperature is maintained at 200-225°C. The exit port should be heated to prevent premature condensation of the effluents. If the gas chromatograph does not have a heated exit port, heating tape coiled about it may be used instead.

Before starting on the preparative scale GLC, an analytical run should be performed first to see if the nature of the column and the conditions of the gas chromatograph do indeed effect a separation of the components of caroway seed oil. To do this, simply inject 2-4 μL of caroway seed oil and observe the separation of the components. After each peak appears, the pen should return to the baseline and the two peaks should be well separated. If the peaks overlap, or are not separated sufficiently, the components emerging during the preparative scale run will overlap and a clean separation will not be possible. If this occurs, the column and/or the conditions of the chromatograph will have to be changed in order to give a good separation.

Take two GLC collection tubes, number them with a marking pen, and record their weights. One type of collection tube (Figure 2) consists of a glass tube with two bulbs and a ground glass joint at one end. The joint is used to connect the collection tube to the exit port of the chromatograph through a special metal adapter attached to the exit port. Because the components of caroway seed oil are high-boiling, the collection tubes do not have to be cooled during the collection procedure. If desired, cooling can be achieved by wrapping a short strip of cloth around the tube with the free end of the cloth dipped into a small beaker of ice-water.

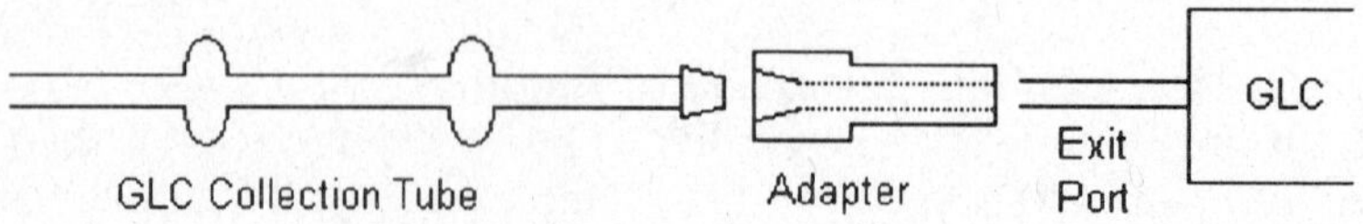

Figure 2. Preparative GLC collection tube and adapter assembly.

Using a 25 μL syringe, inject 25 μL of caroway seed oil into the gas chromatograph. When the large peak of the major component of the caroway seed oil starts to emerge, connect one of the collection tubes to the exit port of the chromatograph. When the sample has completely emerged, remove the collection tube. Wait until the second major component starts to emerge, and connect the second collection tube to the chromatograph. Collect the second component. When both samples have been collected, weigh the collection tubes and determine the percent recovery. (Density: Limonene 0.840 g/mL, Carvone 0.959 g/mL; bp: Limonene 177°C, Carvone 230°C.) Check the odor of each compound. Which component has the odor characteristic of caroway seed oil?

Rinse out the tubes with a few drops of carbon tetrachloride, collecting the washings in small screw-cap vials. Run the infrared spectra on both components using solution cells. By means of the spectra, identify each of the two components and assign IR bands to the functional groups.

For comparison, spearmint oil may also be separated and analysed by the same procedure. Collection of the first component may be difficult, as the concentration is not as great as in caroway seed oil. Compare the odors of the components of spearmint oil with the odors of the components of caroway seed oil. Remember that the difference in odor is due primarily to the difference in chirality of the two carvones.

Dispose of waste chemicals in labeled waste containers provided in the laboratory.

C. Prelab Questions

1. How would the IR spectrum of limonene differ from that of carvone?
2. How would the IR spectra of R-carvone and S-carvone compare?

D. Postlab Questions

1. Look up the definition of a terpene and show how limonene and carvone are both examples of terpenes. Circle each isoprene unit of the terpenes.
2. How could the conditions of the gas chromatograph be changed to slow up the passage of limonene and carvone through the GLC and increase the resolution of the compounds?

14. S_N1 and S_N2 Reactions

A. Introduction

Substitution reactions of alkyl halides are classified as S_N1 (substitution, nucleophilic, unimolecular) or S_N2 (substitution, nucleophilic, bimolecular). The mechanisms of these two processes are compared in Figures 1 and 2.

The **S_N1** reaction proceeds in two discrete steps. The first is an initial slow step in which the organic reactant dissociates to give two ions, a positive carbocation and a negative ion, X^- (the leaving group). The second step is a **fast** addition of the nucleophile, Y^-, to the carbocation to give the product. Since the second step cannot occur until the first step takes place, the rate of the reaction depends on only the slow first step. The carbocation, is trigonal and planar; thus the nucleophile Y^- can add to the carbocation from either side (Figure 1).

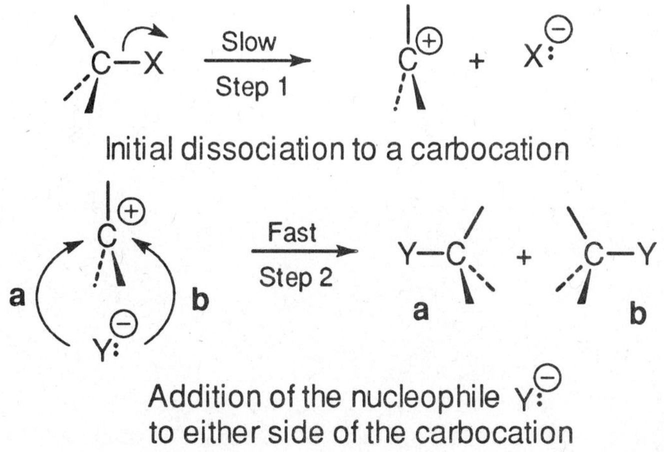

Figure 1. The reaction path (mechanism) of the S_N1 reaction.

The **S_N2** reaction takes place in a single-step, direct attack of the nucleophile Y^- at the carbon nucleus and the simultaneous loss of a leaving group X^- (Figure 2).

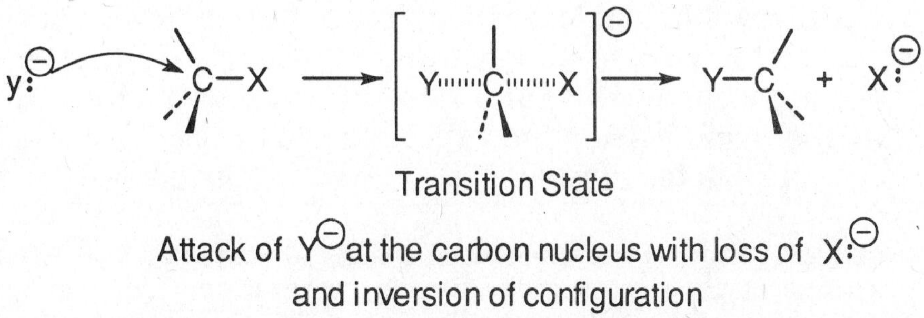

Figure 2. The reaction path (mechanism) of the S_N2 reaction.

The reaction proceeds through a transition state in which all five groups are attached to carbon. Note that in this case only one configuration of the product is allowed. This configuration is opposite to the configuration of the starting compound; i.e., the reaction occurs with inversion of configuration. (This point is demonstrated if the substitution occurs at a chiral center.)

In this experiment, we carry out a synthetic reaction which follows the S_N2 mechanism, and in a second part we compare the S_N1 and S_N2 reactivities of a series of organic halides by performing qualitative tests.

B. Synthesis of an Alkylthiocyanate

1. Discussion

The reaction involves the conversion of *p*-nitrobenzyl bromide to the thiocyanate by the S_N2 displacement of the bromide substituent by thiocyanate ion.

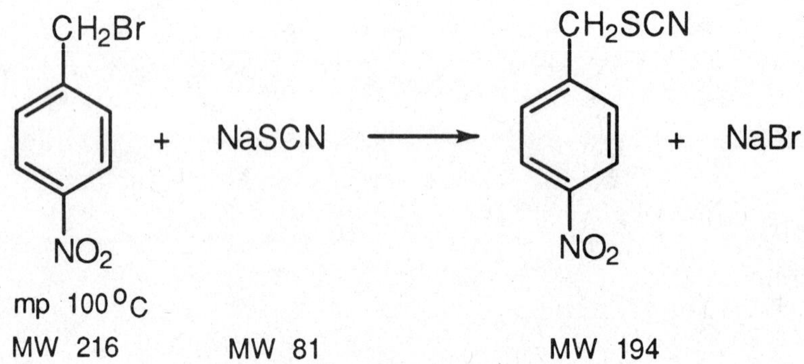

Even though the thiocyanate ion, ^-SCN, is essentially neutral, thiocyanate is a very strong nucleophile, which is characteristic of a negative sulfur atom. Sodium thiocyanate, NaSCN, is remarkably soluble in certain organic solvents such as ethanol, so it is possible to carry out many S_N2 reactions with ^-SCN in alcohol. This particular reaction is convenient to carry out because the *p*-nitrobenzyl bromide is quite reactive toward displacement of the bromide in an S_N2 process and the product, *p*-nitrobenzylthiocyanate, is a crystalline solid and can be readily isolated and purified.

2. Procedure

_p_-Nitrobenzyl bromide is a skin irritant; keep it off your hands and be careful to avoid spilling it while weighing. Many reactive volatile halides are lachrymators and some are vesicants. Keep them off your skin and avoid getting the vapors in your eyes. Work in the hood. Your instructor may suggest that you wear gloves.

In a clean, dry Craig tube, weigh out *ca* 100 mg (0.46 mmol) of *p*-nitrobenzyl bromide. (It is not necessary to have exactly 100 mg, but try to be within 4 or 5 mg and record the weight to ±1 mg.) Support the Craig tube in a small Erlenmeyer flask while weighing.

In a small weighing cup, weigh out *ca* 80 mg (1.0 mmol) of sodium thiocyanate. Carefully transfer all of this material to the Craig tube containing the *p*-nitrobenzyl bromide.

Add 1.2 mL of 95% ethanol to the mixture. Then heat the mixture on a steam bath, or an 80°C water bath, to **near boiling,** with gentle shaking (or tapping with a finger) until the material all dissolves. Continue heating for 5 minutes but **do not** boil off any of the ethanol.

Add 0.6 mL of water to the solution in the Craig tube. The water will keep NaBr from precipitating, and will cause the product to separate as a crystalline solid. Shake or tap the tube to ensure complete mixing. The product may start to separate as a white "cloud." If so, warm the tube to put everything back in solution. Then loosely stopper the tube (or close with aluminum foil) and allow the reaction mixture to cool slowly. Since the product tends to become supersaturated in solution, it may be necessary to add seed crystals. Seed crystals can be obtained by dipping a microspatula into the warm solution, removing it, and letting the solvent evaporate. Small seed crystals should form on the spatula. Scratching the walls of the tube may also initiate crystal formation.

The product should crystallize as fine white needles. If the material separates too quickly as a mass of very fine crystals, rewarm the tube carefully until almost all of the crystals have dissolved. **Leave a few crystals undissolved!** Then allow the tube to cool slowly.

When crystallization seems to be complete, cool the tube for a few minutes in an ice bath. Then add the Teflon insert and collect the crystals by centrifugation. Rinse the crystals with about 1 mL of cold water (to ensure removal of inorganic salts) and centrifuge again to remove the water. Remove a few seed crystals and add *ca* 1 mL of ethanol. Heat to dissolve the solid, then add 0.5 mL of water, and mix well. Allow the solution to cool slowly (seed it if necessary). Finally cool the tube in an ice bath and then collect the crystals by centrifugation.

Transfer the crystals to a tared (weighed) watch glass to air dry. When dry, record the weight and melting point. Calculate the percent yield. Transfer your crystals to a clean vial bearing a label with your name and the structure of the product. Cap the vial and turn it in to your instructor.

Calculation of the percent yield. Review the section in the introduction chapter regarding the calculation of the percent yield. Follow the procedure exactly as described. To emphasize the importance of yield calculations, the steps are summarized below.

1. Write a balanced equation for the reaction.
2. Under each reagent write its molecular weight.
3. Under each molecular weight write the weight of each reagent that you actually used, in mg.
4. Convert each weight to mmol.
5. Determine the limiting reagent and underline it.
6. Determine the theoretical yield in mmol based on the limiting reagent. Write this under the product.
7. Convert mmol theoretical yield to mg.
8. Calculate the percent yield from the theoretical yield and the actual yield of product from the reaction.

C. Relative Reactivities of Organic Halides by S_N2 and S_N1 Reactions

A series of ten organic halides will be compared to determine their reactivities under S_N1 and S_N2 conditions (Table 1).

Table 1. Organic Halides

Structure	Common Name	IUPAC Name	bp, °C
$CH_3CH_2CH_2CH_2Cl$	*n*-Butyl Chloride	1-Chlorobutane	78
$CH_3CH_2CH_2CH_2Br$	*n*-Butyl Bromide	1-Bromobutane	104
$CH_3CHBrCH_2CH_3$	*sec*-Butyl Bromide	2-Bromobutane	91
$(CH_3)_2CHCH_2Br$	Isobutyl Bromide	1-Bromo-2-methylpropane	92
$(CH_3)_3CCl$	*tert*-Butyl Chloride	2-Chloro-2-methylpropane	51
$CH_3CH{=}CHBr$	Propenyl Bromide	1-Bromopropene	60
$C_6H_5CH_2Cl$	Benzyl Chloride		181
$CH_2{=}CHCH_2Br$	Allyl Bromide	3-Bromopropene	70
C_6H_5Br	Bromobenzene	Bromobenzene	156
$C_6H_5C(CH_3)_2CH_2Cl$	Neophyl Chloride	1-Chloro-2-methyl-2-phenylpropane	96/10 mm

Your instructor may elect to give you an unknown. You should determine the S_N1 and S_N2 reactivity of this unknown by comparing its reactivity with the reactivities of the compounds listed in Table 1.

1. S_N2 Reactivity

Discussion: Sodium iodide in acetone solution reacts with organic halides by an S_N2 process.

$$Na^{\oplus}\ I^{\ominus} + R{-}X \xrightarrow{\text{Acetone}} R{-}I + \underset{\text{Ppt.}}{NaX}$$

$$X = Br \text{ or } Cl$$

Acetone is a good solvent for many organic substrates and is a remarkably good solvent for sodium iodide. On the other hand, acetone is not a good solvent for sodium chloride or sodium bromide. When the iodide displaces the chloride (or bromide) from the substrate, one of the products is sodium chloride (or sodium bromide). Since sodium chloride and sodium bromide are insoluble in acetone, the salt will immediately precipitate. Thus, a precipitate of the salt indicates that a reaction has occurred. The time that it takes for the precipitate to form is a measure of the rate of the reaction. A comparison of these rates for different organic halides is a measure of their relative reactivities.

Note: The reactivity tests of organic halides depend upon observing the formation of a precipitate, but do not be misled by a "false positive" test. While the organic halides should be pure, they are commercially available samples and may contain small amounts of impurities, which give precipitates more rapidly than the pure halide. However, you can spot such a false positive test because only a small amount of precipitate will form; i.e. **there will not be a steady increase in the amount of precipitate, as occurs in the case of a true positive test.**

Procedure: Label ten small test tubes to identify each halide. (You could number each compound in your notebook and simply put the code number on each label.) Place *ca* 0.5 mL of 18% sodium iodide in acetone solution in each test tube (ca 8 mm deep in a 3-inch test tube).

Add 4 microdrops of an organic halide to a test tube containing the NaI/acetone solution. Note the time of addition. Stopper the tube and shake it by tapping to effect thorough mixing. Then note the time of the first **definite** appearance of a precipitate. If no reaction occurs within 5 minutes, loosely stopper the tube and heat it in a 45°C warm water bath (use a beaker on your hot plate) for at least 5 min. Note the time heating is started and the time when a **definite** precipitate (if any) appears. (Avoid evaporating acetone because NaI might precipitate giving a false positive test.)

2. S_N1 Reactivity

Discussion: Silver nitrate ($AgNO_3$) in ethanol is a reagent useful for studying the S_N1 reactivity of organic halides. Silver ion has a great affinity for halides. In solution, the silver coordinates with the halogen atom of the organic halide and helps to pull off the halide ion. The result is a dissociation of the organic halide to form a carbocation; in essence this is an S_N1 reaction. The carbocation then reacts with the solvent ethanol to form an ether. The silver halide that is formed is very insoluble and precipitates as AgX. If a substantial amount of precipitate forms, then we know that a reaction has taken place: i.e., we have a positive test (rather than a false positive test due to impurities). The reaction is shown in Figure 3.

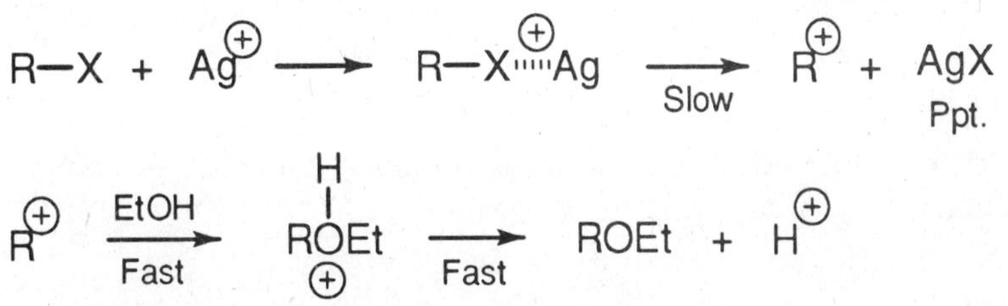

Figure 3. An S_N1 reaction induced by Ag^+.

The nature of the leaving group, X^-, and especially the structure of the carbocation determine how fast the reaction occurs. By comparing how fast AgX is formed, we can gain an understanding of the S_N1 reactivity of the organic halides. The time that it takes a precipitate to form is a measure of the rate of the reaction. A comparison of the rates for different organic halides, is a measure of the **relative reactivities of the halides.**

Procedure: Empty the test tubes, rinse with water, rinse twice with ethanol, and then place *ca* 0.5 mL of 1% silver nitrate in ethanol in each tube. Then add 4 microdrops of an organic halide to a test tube containing the silver nitrate/ethanol solution. Note the time of addition, mix well and then note when the first **definite** precipitate appears. If no precipitate appears in 5 minutes, heat the tube in a 45°C warm water bath for 5 minutes. Note the time heating is started and the time when a **definite** precipitate appears.

Analysis of the data should include the following factors for both the S_N1 and S_N2 conditions:

(a) The nature of the leaving group (Cl vs. Br) for the 1-halobutanes.
(b) The effect of structure: primary, secondary, tertiary.
(c) Steric hindrance: primary *versus* hindered primary.
(d) Allylic and benzylic *versus* primary and tertiary.
(e) Alkenyl halide (1-bromopropene) and aromatic halide (bromobenzene *versus* aliphatic halides.

Dispose of waste chemicals in labeled waste containers provided in the laboratory.

D. Prelab Questions

1. Draw the **transition state** for the reaction of RCH_2Br with SCN^- in ethanol.

2. Draw the two major resonance structures for SCN^-. Which structure seems to be more important? Why?

3. Suppose we **doubled the volume** of the solution (by adding more solvent) while keeping the **quantities** of reactants constant in an S_N2 reaction. How would the reaction rate change? Would this dilution change the time it would take to get a fixed **percentage** (e.g., 95%) of starting material to react? Explain briefly.

4. Consider the same dilution of an S_N1 reaction. Would the time required for 95% of the starting material to react change as a result of this dilution? Explain briefly.

E. Postlab Questions

1. Compare **your reactivity results** to those predicted by "theory." Refer to your lecture text and/or lecture notes for the "theoretical" predictions.

2. How do organic iodides compare in reactivity with bromides? Could we determine the relative reactivity of alkyl iodides by the methods employed in this experiment? Explain briefly.

15. E2 Elimination Reactions, Dehydrobromination of the Butyl Bromides, C_4H_9Br

A. Introduction

Elimination reactions are designated by the terms E1 (unimolecular) and E2 (bimolecular). In a nucleophilic displacement (S_N1 or S_N2) the nucleophile (base) replaces a leaving group, such as bromide. If, instead of replacing the bromide, the nucleophile removes HBr to generate an alkene, the process is a β-elimination. The β- refers to the fact that the proton is removed from the β-carbon, which is the carbon next to the carbon bearing the bromide (the α-carbon).

When an alkyl bromide is treated with a strong base such as an alkoxide, RO^-, two competing reaction pathways can be followed: nucleophilic displacement, or β-elimination. A comparison of an S_N2 displacement with an E2 elimination is shown in Figure 1.

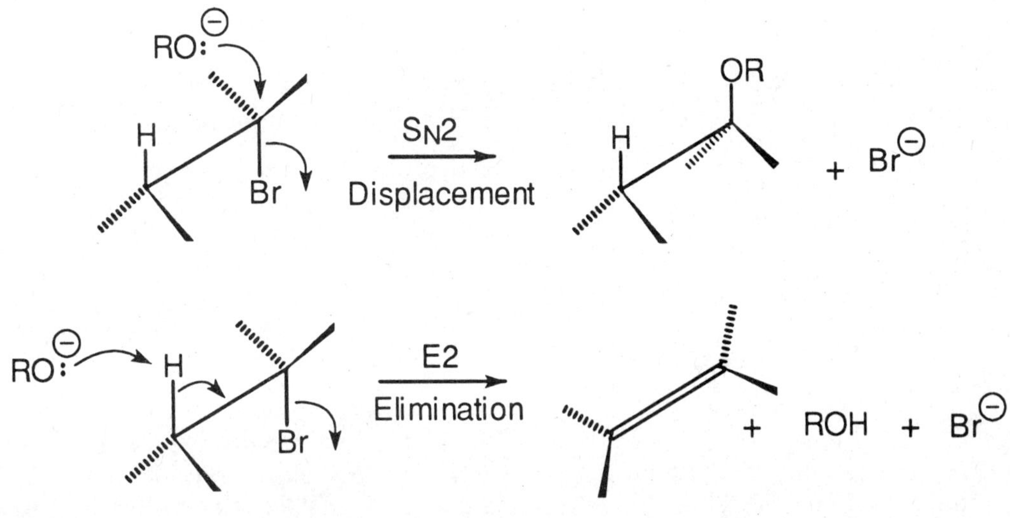

Figure 1. Comparison of the mechanisms of the S_N2 nucleophilic displacement and the E2 elimination.

The relative amounts of displacement and elimination will depend upon several factors: the nature of the base (nucleophile), the concentrations of the reactants, the temperature, the solvent, and the structure of the organic bromide.

To ensure that elimination occurs and not displacement, a **strong base** is employed, such as potassium *tert*-butoxide, $(CH_3)_3O^-K^+$). The large *tert*-butyl group in the *tert*-butoxide ion, $(CH_3)_3CO^-$, causes this ion to be **sterically hindered** inhibiting S_N2 displacement. The size of this reagent is so great that it has great difficulty in approaching the nucleus of a carbon atom to cause displacement. On the other hand the size of this atom does not impede its ability to "attack" a proton on the β-carbon of the alkyl bromide.

A high concentration of base promotes a second order process, either S_N2 or E2. Heating accelerates all reactions, but the rates of E2 eliminations are increased more than the competing S_N2 displacement reactions.

The solvent for the elimination reactions is dimethyl sulfoxide, (DMSO, $(CH_3)_2SO$, bp 189°C). DMSO has remarkable solvent properties because of its high polarity. This property, and the fact that DMSO is aprotic (it has no -OH groups), make it a particularly effective solvent for enhancing the basicity of the alkoxide ion and thus promoting E2 elimination. Since DMSO has a high boiling point, it can be heated to fairly high temperatures without loss by evaporation.

The four isomeric butyl bromides shown in Table 1 will be subjected to the conditions described above that favor E2 elimination.

Table 1. Butyl Bromides

Structure	Common Name	IUPAC Name	bp, °C
$CH_3CH_2CH_2CH_2Br$	*n*-Butyl Bromide	1-Bromobutane	104
$CH_3CHBrCH_2CH_3$	*sec*-Butyl Bromide	2-Bromobutane	91
$(CH_3)_2CHCH_2Br$	Isobutyl Bromide	1-Bromo-2-methylpropane	92
$(CH_3)_3CBr$	*tert*-Butyl Bromide	2-Bromo-2-methyl-propane	73

Because of the conditions selected for this reaction, even *n*-butyl bromide, which normally undergoes S_N2 displacement, follows the E2 elimination path. *tert*-Butyl bromide normally undergoes S_N1 displacement, but again the conditions force the reaction to go by E2 elimination.

A significant feature of E2 eliminations is that **no rearrangement of the carbon skeleton occurs.** Consequently the **double bond will be formed at the site where the bromide leaving group was attached.** Thus, elimination of HBr from *n*-butyl bromide, *tert*-butyl bromide, and isobutyl bromide should give a single alkene in each case. Elimination from 2-bromobutane, on the other hand, can generate 1-butene, *trans*-2-butene and *cis*-2-butene.

By determining the composition of the products of these reactions, we can learn something of the nature of E2 eliminations. The alkenes expected from E2 elimination of the four bromides are shown in Table 2. The products are all gases which will be collected and then analyzed by gas chromatography to determine the compositions. Side products arising from S_N1 or S_N2 processes will be liquids and will not contaminate the gaseous products.

Table 2. Products arising from the E2 Elimination of the Butyl Bromides

Structure	Common Name	IUPAC Name	bp, °C
$CH_3CH_2CH{=}CH_2$	n-Butene	1-Butene	-6.3
$CH_3CH{=}CHCH_3$		*cis*-2-Butene	3.7
$CH_3CH{=}CHCH_3$		*trans*-2-Butene	0.9
$(CH_3)_2CH{=}CH_2$	Isobutene	2-Methylpropene	-6.9

Collecting gases on a small scale is easily carried out by displacement of water in an inverted test tube. Gas samples are then injected into the gas chromatograph using 1-mL syringes.

Before conducting the experiment, observe the following precautions.

Precautionary Notes

All alkyl halides should be handled with care. Do not spill them on your skin and avoid inhaling the vapors. Please be careful to avoid contaminating the reagents. Use only a clean, dry syringe for taking aliquots.

Potassium *tert*-butoxide is a white solid obtained from the reaction of potassium with *tert*-butyl alcohol. It will be supplied as a solution in DMSO, in which it is quite soluble. The solution is easily deactivated by water; therefore be sure to close the container after obtaining the reagent.

Handle the potassium *tert*-butoxide/DMSO reagent with care. DMSO must be kept off the skin because it can carry potentially harmful substances through the skin, including pathogens, into the underlying tissue. If you should spill any on your skin, wash it off immediately with a steady stream of running water. Clean up any spills immediately.

Handle hypodermic needles with care. The points on all needles should be cut off and each needle ground to give a blunt end to reduce the chance of puncture wounds. If you find a sharp (unmodified) needle in your desk, turn it in to your instructor who will exchange it for a blunt needle.

B. Experimental Procedure

1. The Reactions

Each student should dehydrobrominate each of the four bromides. The order in which this is done is not important, but each time you start a new reaction, make sure that the equipment is clean and not contaminated with the reactant bromide or product alkene from a prior run.

The apparatus to be used is illustrated in Figure 2.

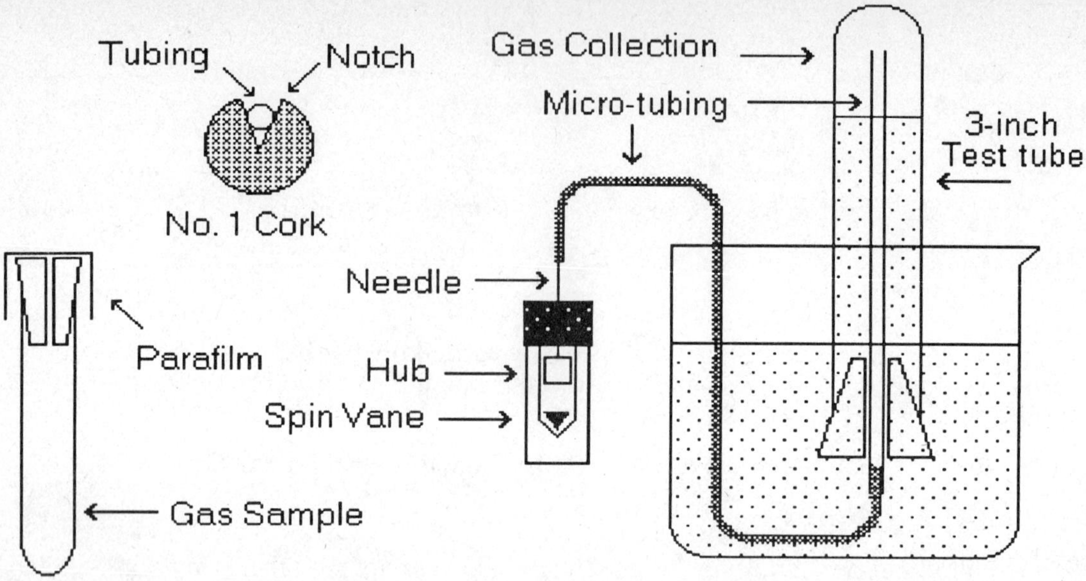

Figure 2. Diagram of the equipment used for the dehydrobromination reaction.

Select a new solid septum. Insert a blunt #20 gauge hypodermic needle (yellow hub) through the center of the blue side of the septum. Then carefully attach a 40-cm length of polyethylene "microtubing" to the needle and slide it up to the septum. If the needle hangs up, make another try at attaching the tubing.

With a triangular file, cut a small notch in the edge of a #1 cork. This size cork fits snugly into a 3-inch test tube. The notch should be large enough to allow the poly-ethylene tubing to slide through the notch freely when the cork is in the test tube, leaving some free space for water to be pushed out during gas collection. (Avoid an excessively large notch, or the tubing might slip out of the test tube when in use.) Fill a 3-inch test tube with water and insert the notched cork.

Center the aluminum block on the hot-plate and fit a thermometer into the block; be sure that the thermometer is loosely supported by a stand. Heat the block to about 80°C (the final heater setting should be close to "low") and leave the block at this temperature until all of the reactions have been run.

Place the spin vane in a **dry, clean** 5-mL conical reaction vial. Then add *ca* 1.0 mL of 1.6 M potassium *tert*-butoxide/DMSO reagent to the vial (use the markings on the vial to set the volume) and promptly attach the septum, with needle and tubing attached, to the vial using the screw cap to give a **gas-tight fit.** (Do not use an O-ring.)

Place the vial in the heating block (80°C) and turn on the stirrer to give gentle stir-ring. Allow at least five minutes for the vial to come up to the temperature of the block. Then insert the free end of the microtubing through the notch in the cork and into the test tube filled with water. **Push the tubing all the way to the closed end of the test tube.** Then invert the test tube in a small beaker which is two-thirds filled with water.

AT THIS POINT HAVE YOUR INSTRUCTOR CHECK YOUR SET-UP.

Note: to avoid contaminating the alkyl halides provided in the reagent bottles, it is essential that your syringe be clean and dry before you remove a sample. Your instructor may choose to have you obtain your sample from a small test tube attached to the reagent bottle. Transfer a small volume of the halide to the test tube with the pipet provided, then draw a sample into your syringe.

A syringe must always by carried in the laboratory in a 6-inch test tube (needle down).

In your **clean, dry,** 1-mL syringe draw up 0.10 mL of the bromide. You do **not** need to expel the air bubble from the syringe. Bring the liquid level (the meniscus) to the 0.10 mL mark with the air bubble above the liquid. (Later, when you add the reagent by pushing the plunger all the way down, you will deliver 0.10 mL of the liquid.)

Insert the needle through the septum and push it down so that the tip is just over the solution in the vial. Then, **over a period of 5 to 10 seconds,** inject the alkyl bromide into the *tert*-butoxide/DMSO solution. The reaction should be rapid with an abrupt evolution of gas, which will enter the test tube through the tubing, forcing out the water.

Remove the tubing from the gas-filled test tube and cover the notched cork with Parafilm or aluminum foil. Remove the reaction vial from the heating block (hold the vial by the cap) and record the appearance of the reaction mixture.

It is advisable to analyze each sample by GLC as soon as it is obtained to stagger the use of the gas chromatographs. However, gas samples can be kept almost indefinitely if the Parafilm-sealed tube is inverted, corked end down, in a beaker of water. Be sure to use the same gas chromatograph for all runs.

To prepare for a run with another bromide, **flush the reaction vial and spin vane thoroughly with tap water.** (To avoid losing the spin vane, it is advisable to empty the washings into a beaker before disposing of wastes.) After flushing, rinse several times with small volumes of acetone. Then dry the vial and spin vane by butting the end of the vial against the square-cut end of a heavy hose attached to the house vacuum system. Flush air in and out of the vial repeatedly until it is dry.

Thoroughly flush the needle/septum/tubing unit to remove all traces of the previous gas sample. **Rinse the test tube and notched cork thoroughly with water** (or use a new test tube fitted with a notched cork if the GLC has not been run on a previous sample).

Remove the plunger from the 1-mL syringe and rinse the syringe and plunger thoroughly with acetone to remove all traces of the organic bromide. Dry the plunger with a clean towel and dry the barrel of the syringe by pulling air through it. **Be sure your syringe is clean and dry before you proceed.**

Add 1.0 mL of the potassium *tert*-butoxide solution to the reaction vial and assemble the apparatus as before. When the vial has reached reaction temperature on the hot-plate, add the next alkyl bromide and proceed as above. Continue until all four of the alkyl bromides have been run.

2. Analysis of Gas Samples by Gas Chromatography

Since the gaseous products of the reaction are nonpolar, a nonpolar column (silicone oil as stationary phase) will be used at room temperature. Gas-tight 1-mL syringes will be provided for injection of the products into the gas chromatograph. **To remove a previous gas sample, flush the syringe several times with air.**

Insert the needle through the Parafilm and the notch in the cork, withdraw 0.5 mL of gas sample, and inject it into the gas chromatograph. The instructor will indicate the proper **attenuator** setting. Since the samples contain a great deal of air, the air peak is generally large. It is all right for the air peak to go off scale, but the sample peaks should remain on the chart paper and fill up a good portion of the paper. If the peaks are too small, run the analysis again with more sample, or at a higher sensitivity setting of the attenuator. If the peaks go off scale, use less sample or less sensitivity.

Record **retention times** for your peaks **measured from the air peak.** Calculate the **percentage composition** (ignoring the air peak) and identify the peaks in all samples. To determine peak areas, refer to the discussion given in the experiment on gas chromatography. **Peak height times half-band width** or **peak height times retention time** (measured from the air peak) will give a good estimate of the relative peak areas. With narrow peaks the latter method is preferable.

Dispose of waste chemicals in labeled waste containers provided in the laboratory.

C. Prelab Questions

1. Give two valence-bond structures for DMSO. Show formal charges where appropriate.
2. Draw the transition state for the E2 reaction of *tert*-BuO$^-$ with *tert*-butyl bromide. Why is the S_N2 reaction unfavorable?
3. Why does a hindered base favor E2 elimination over S_N2 displacement?
4. Give the structural formulas for the expected product(s) of E2 elimination for each of the bromides to be studied in this experiment.
5. If we use exactly 0.100 mL of *tert*-butyl bromide and 1.0 mL of 1.6 **M** potassium *tert*-butoxide in DMSO:

   (a) What is the limiting reagent? Show the reasoning.

   (b) Calculate the theoretical yield in milliliters of the butene produced at 740 mm and 25°C when it is collected over water. R = 0.0821 liter atm/mol K and the vapor pressure of water is 24 mm at 25°C. (Recall, PV = nRT)

D. Postlab Questions

1. Do the compositions of the gases from the four alkyl bromides agree with the expected compositions based on the E2 mechanism of the reaction? Explain briefly.
2. What was the solid formed in these reactions?
3. Suppose we measured the rates of elimination for the four bromides, but used a lower temperature and more dilute conditions to slow reactions enough to permit measurements. Predict the relative reactivity of these bromides. Explain your answers briefly.

16. E1 Elimination Reactions, Dehydration of Butyl Alcohols

A. Introduction

In this experiment, we will examine the acid-catalyzed dehydration of the four butyl alcohols, C_4H_9OH, Table 1, to give butenes.

Table 1. Butyl Alcohols

Structure	Common Name	IUPAC Name	bp, °C
$CH_3CH_2CH_2CH_2OH$	*n*-Butyl Alcohol	1-Butanol	118
$CH_3CHOHCH_2CH_3$	*sec*-Butyl Alcohol	2-Butanol	100
$(CH_3)_2CHCH_2OH$	Isobutyl Alcohol	2-Methyl-1-propanol	108
$(CH_3)_3COH$	*tert*-Butyl Alcohol	2-Methyl-2-propanol	82

All four alcohols can be dehydrated by heating with concentrated sulfuric acid. However, the ease of dehydration varies strikingly with the structure of the alcohol.

The general course of dehydration of an alcohol with sulfuric acid is shown in Figure 1.

(1) $ROH + H_2SO_4 \longrightarrow ROH_2^{\oplus} + HSO_4^{\ominus}$

$ROH_2^{\oplus}$ is called an oxonium ion

(2) $ROH_2^{\oplus} \longrightarrow R^{\oplus} + H_2O$

$R^{\oplus}$ is called a carbocation

(3) $R^{\oplus} \xrightarrow{-H^{\oplus}}$ alkene(s)

(4) $R^{\oplus} \xrightarrow{\text{rearrange}}$ new $R^{\oplus} \xrightarrow{-H^{\oplus}}$ alkenes

Figure 1. Mechanism of the dehydration of an alcohol with sulfuric acid.

In the first step, the alcohol reacts with sulfuric acid to form an oxonium ion, analogous to H_3O^+. Here the alcohol behaves as a base toward the sulfuric acid. The equilibrium is established instantly and the position of the equilibrium lies far to the right.

In the second step, the protonated alcohol (the oxonium ion) loses a molecule of water by dissociation to form a carbocation, R^+. **Carbocations are key intermediates in the sequence.** The fate of the carbocation determines what product(s) that are formed.

In the third step, the base extracts a proton from the carbocation to form an alkene. What bases are present in this system? Alcohol, water, and hydrogen sulfate ion all can act as bases. As long as the alcohol is in excess, it probably is the main proton acceptor, forming more ROH_2^+. Since the carbocation may have more than one way to lose a proton, a mixture of alkenes may be formed.

The fourth step recognizes an important feature of carbocation behavior: **a carbocation can and will rearrange to a more stable carbocation.** Since the stability of carbocations is in the order of tertiary > secondary > primary, we can anticipate rearrangement, if possible, to proceed in the order: primary to secondary to tertiary. The new (rearranged) carbocation can then lose a proton to generate one or more alkenes.

Under appropriate conditions several of these reactions (Figure 1) are reversible, but here we stress the forward reactions leading to alkenes.

In addition to the steps above, several other reactions can occur. We might consider them to be "side reactions": although that outlook is purely arbitrary. These reactions are shown in Figure 2.

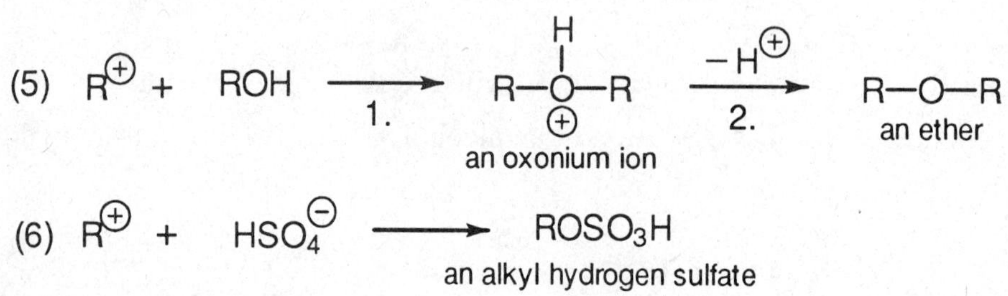

Figure 2. Possible side reactions in the dehydration of alcohols.

In reaction 5, the carbocation is "captured" by a molecule of the alcohol generating a new oxonium ion, a protonated ether. The protonated ether then rapidly transfers a proton to another base (water, alcohol, HSO_4^-). Depending upon the experimental conditions, the ether may be the major product. However, since the two steps of rection 5 are reversible, the ether that is formed initially may subsequently undergo the reverse reactions, ultimately generating alkenes from the carbocation R^+.

Reaction 6 shows that a carbocation can be captured by the HSO_4^- ion to form an alkyl sulfate. However this reaction is readily reversed by heating, so that under our conditions alkyl sulfates will be decomposed.

It should be evident that heating an alcohol with sulfuric acid is a fairly complex process. Other "side reactions" can occur, leading to tar-like side products. We will not seek to establish optimum conditions for dehydrating each alcohol, but rather will follow a common procedure, increasing the reaction temperature as required until alkenes are produced.

First, consider the dehydration of *tert*-butyl alcohol (Figure 3) which undergoes dehydration at the fastest rate. The oxonium ion formed upon protonation by sulfuric acid undergoes facile loss of a water molecule because a **tertiary carbocation,** the *tert*-butyl carbocation, is formed. In this case the reaction is quite "clean": loss of a proton gives isobutene in excellent yield.

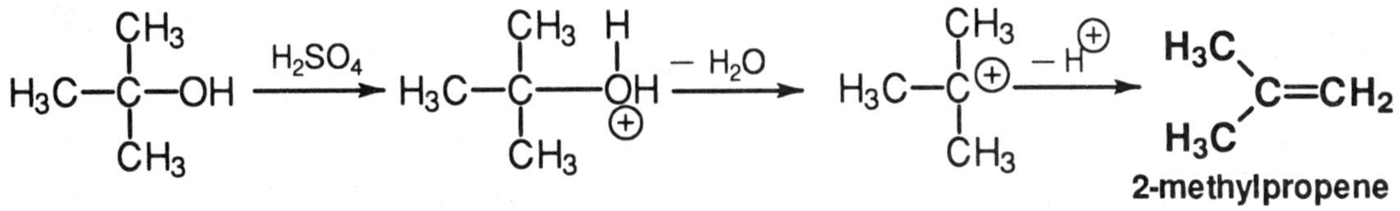

Figure 3. Acid-catalyzed dehydration of *tert*-butyl alcohol.

Next consider the dehydration of *sec*-butyl alcohol (Figure 4).

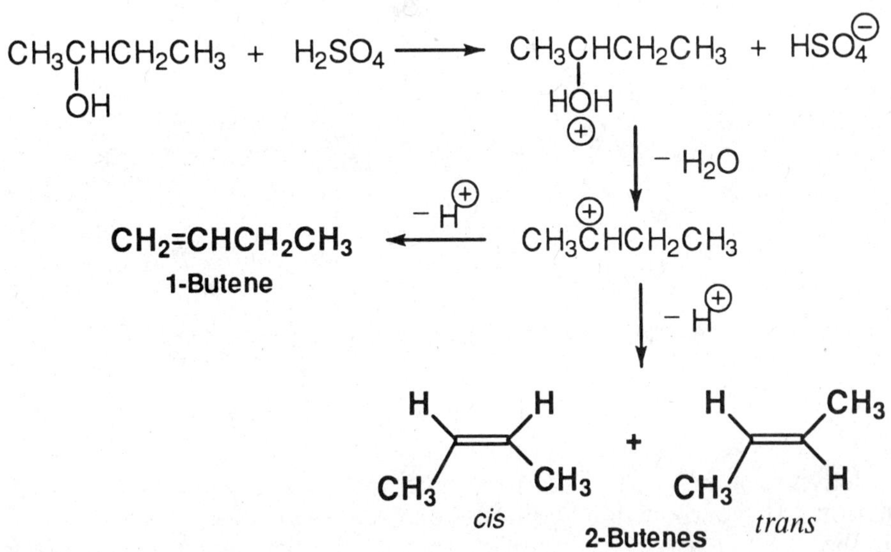

Figure 4. Acid-catalyzed dehydration of *sec*-butyl alcohol.

Although the oxonium ion is formed readily, the loss of water is slower than in the first case and thus the reaction mixture must be heated to a higher temperature in order to generate the 2-butyl carbocation, a **secondary carbocation.** This carbocation can then lose a proton from either the terminal methyl group to give 1-butene or from the methylene group at C-3 to give a mixture of *cis* and *trans*-2-butene.

Dehydration of isobutyl alcohol (Figure 5), is still more difficult and requires a substantially higher temperature because a **primary carbocation** must be generated. While this carbocation could lose a proton to give isobutene, it can also undergo **rearrangement** to a more stable carbocation. In fact, the carbocation **might undergo rearrangement as it is formed** to yield a more stable species. A hydride ion shift (⁻:H) would give the *tert*-butyl carbocation, while a methyl shift (⁻:CH_3) would give the 2-butyl carbocation. Both carbocations would then lose a proton, to give rise to a mixture of all four C_4H_8 alkenes:

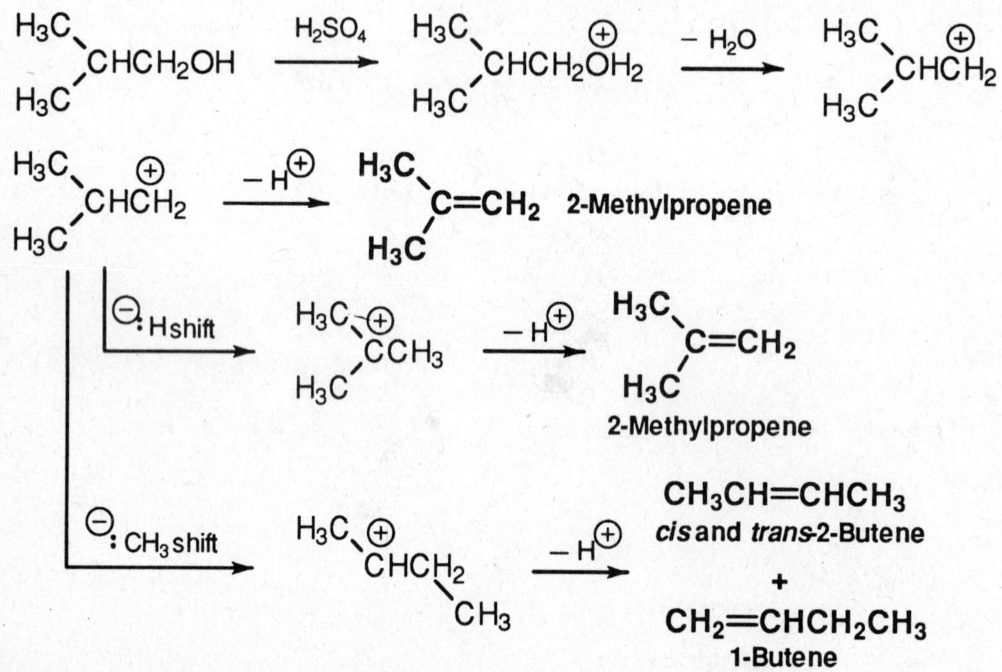

Figure 5. Acid-catalyzed dehydration of isobutyl alcohol.

Finally, dehydration of 1-butanol requires a still higher temperature to generate a **primary carbocation.** This carbocation could lose a proton to generate 1-butene. However, in this case (as in the case above), the carbocation actually may be undergoing rearrangement as it is formed, generating the 2-butyl carbocation which would subsequently lose a proton by the competing pathways outlined earlier:

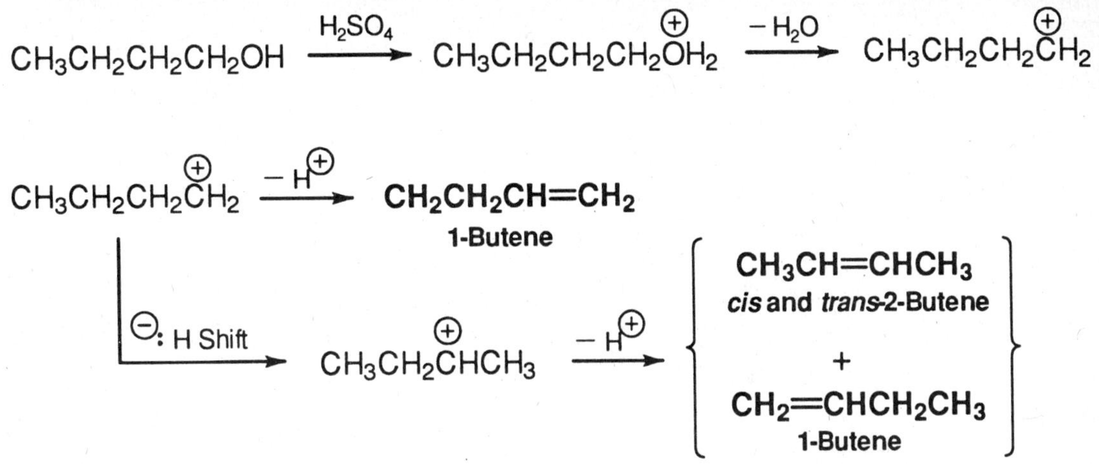

Figure 6. Acid-catalyzed dehydration of *n*-butyl alcohol.

In this experiment, as in the dehydrobromination of the butyl bromides, we will be dealing with mixtures of the isomeric butenes. (The boiling points of the butenes are given in the dehydrobromination experiment.)

All students will dehydrate both *tert*-butyl alcohol and *sec*-butyl alcohol. Then each student should dehydrate either isobutyl alcohol or *n*-butyl alcohol; your instructor will tell you which to do. If time permits, you are encouraged to do both.

B. Experimental Procedure

The apparatus is identical to that used in the dehydrobromination of the butyl bromides (see E2 Elimination Reactions (Figure 2). The same procedure is used for each butanol, the only differences being the temperatures employed to effect dehydration.

Using micropipets, place 24 microdrops (*ca* 0.25 mL) of the alcohol and 3 microdrops of concentrated sulfuric acid in a clean, dry 5-mL conical vial containing a magnetic spin vane.

Using the screw cap, attach to the vial the septum bearing the inverted needle with microtubing. Do **not** use an O-ring. Tighten the cap securely to ensure a gas-tight fit. Then place the vial in the aluminum heating block centered on the stirrer hot-plate. Mount a thermometer in the aluminum block and support it loosely near the top with a clamp.

Fit the notched cork into a 3-inch test tube **filled** with water and insert the polyethylene tubing well into the test tube. Invert the test tube and immediately submerge the corked end in a 100-mL beaker two-thirds filled with water.

AT THIS POINT HAVE YOUR INSTRUCTOR CHECK YOUR SET-UP.

When your set-up has been approved, start the stirrer to give a brisk stirring rate and turn on the heater. The heater should be adjusted so that the heating block will attain the temperature at which dehydration should be expected to occur. The approximate temperatures required for dehydrating the different alcohols are given in Table 2. To save time, initially use a heater setting higher than that needed to maintain the desired temperature. To avoid overshooting, monitor the temperature carefully and decrease the heater setting as the desired temperature is approached.

Table 2. Approximate Temperatures Required for Dehydration of Butanols

Alcohol	Temperature of Dehydration
tert-Butyl Alcohol	90°C
sec-Butyl Alcohol	110°C
Isobutyl Alcohol	140-150°C
n-Butyl Alcohol	160-170°C

On heating the reaction vial, the air expands and some bubbles will be seen to rise in the test tube. This is not the start of the reaction. When the reaction begins, the rate of gas evolution will increase rapidly and the test tube will fill with gas. **Record the temperature** when steady and rapid displacement occurs. Then keep the temperature adjusted to maintain a steady rate of gas evolution.

Allow the gas collection tube to fill completely. The tube will contain some air forced over from the reaction vial. The concentration of butene in the gas sample can be increased by simply allowing the collector to flush until the bubble rate slows down. Then remove the gas collection tube from the water bath and withdraw the tubing. **Promptly seal** the corked end of the test tube with a small piece of Parafilm. Turn off the stirrer, but leave the heater on in preparation for the next run. Carefully remove the reaction vial from the aluminum block and allow it to cool.

Analyze the gas sample by GLC. (Review the GLC procedures in the experiment on E2 elimination reactions.) Be sure to use the same gas chromatograph for all runs. Record **retention times** for your peaks **measured from the air peak.** Calculate the **percentage composition** (ignoring the air peak) and identify the peaks in all samples. To determine peak areas, refer to the discussion given in the experiment on gas chromatography. **Peak height times half-band width**, or preferably **peak height times retention time** (measured from the air peak) will give a good estimate of the relative peak areas.

When the reaction vial has cooled, fill it with water (to dilute the sulfuric acid), remove the spin vane, and wash the vial and the spin vane thoroughly with water. Then wash with acetone to dissolve any dark residue. **Dry the equipment thoroughly before starting the next reaction.**

The tubing and needle contain gaseous products from the reaction and must be flushed with air, or else the gases will contaminate the sample obtained from the next reaction, leading to an erroneous GLC analysis. The syringe for injecting the samples into the chromatograph must always be flushed several times with air before use to remove any remaining gases that may contaminate the sample.

When clean-up is completed, the next reaction may be started. Follow the temperature chart given in Table 2. In order to save time, the reactions should be run in order of increasing reaction temperatures (in which case you do not need to allow the hot-plate to cool).

Dispose of waste chemicals in labeled waste containers provided in the laboratory.

C. Prelab Questions

1. Why is a **strong** acid such as sulfuric acid used in these dehydrations instead of a weak acid?

2. Suppose a 120-mg sample of *tert*-butyl alcohol is dehydrated. Calculate the **volume** in milliliters of isobutene which would be formed if the yield is 100%. Assume that the gas is collected at 25°C and that it is saturated with water vapor (which has a vapor pressure of 24 mm Hg at 25°C) and that the barometric pressure is 740 mm. The gas constant R has a value of 0.0821 L atm mol^{-1} K^{-1}.

3. Predict the **major** organic products formed by acid-catalyzed dehydration of each of the following alcohols.

   (a) 2-Hexanol
   (b) 3,3-Dimethyl-1-butanol
   (c) 1-Methylcyclohexanol
   (d) 2-Methylcyclohexanol

D. Postlab Questions

1. Summarize your results and include those of a neighbor who dehydrated the alcohol which you did not dehydrate (use only your data if you dehydrated all four alcohols).

2. Draw conclusions about the rearrangement of carbocations in this experiment. What apparently happened? For example, compare 2-butanol and 1-butanol; compare isobutyl alcohol and 2-butanol; compare *tert*-butyl alcohol and isobutyl alcohol; and make any other comparisons which seem appropriate.

3. The relative stabilities of the butenes at 298°K are as follows. The numbers indicate the relative amounts present at **equilibrium**, compared to 1-butene as the standard.

1-Butene	1.0
cis-2-Butene	7.2
trans-2-Butene	25
Isobutene	85

   (a) Explain this trend qualitatively.
   (b) Find how these data may relate to the results you (and your fellow students) obtained in this experiment. There is no "correct" answer, but you should be able to draw some conclusion.

4. Compare the results of the acid-catalyzed dehydration of the butyl alcohols with the results of the E2 dehydrobromination of the corresponding butyl bromides (see prior experiment):

   tert-Butyl alcohol *versus* *tert*-butyl bromide
   2-Butanol versus 2-bromobutane
   Isobutyl alcohol versus isobutyl bromide
   1-Butanol versus 1-bromobutane

   What conclusions can you draw about **E1** *vs.* **E2** eliminations?

17. Hydroboration-Oxidation of Alkenes

A. Introduction

Diborane, B_2H_6, is a toxic, very reactive, colorless gas. Dissolved in tetrahydrofuran (THF), diborane dissociates to give a borane-tetrahydrofuran complex (BH_3-THF), which is stable in the absence of water and other hydroxylic species. THF solutions of borane are commercially available, providing a very convenient source of BH_3 for use in a variety of synthetic reactions.

One of the most useful reactions of borane is the addition to alkenes to form organoboranes (R_3B). Reaction of these organoboranes with hydrogen peroxide and sodium hydroxide leads to the formation of alcohols. Although there are many ways to synthesize alcohols from alkenes, the unique result from the use of borane is that the borane reacts with the alkene in such a way as to place the hydroxyl group on the **least substituted** carbon. Thus, borane reacts with alkenes by anti-Markovnikov addition. As a result of this regioselectivity, alcohols prepared by "hydroboration-oxidation" have structures that could not have been realized by other means. The course of hydroboration-oxidation is summarized in Figure 1.

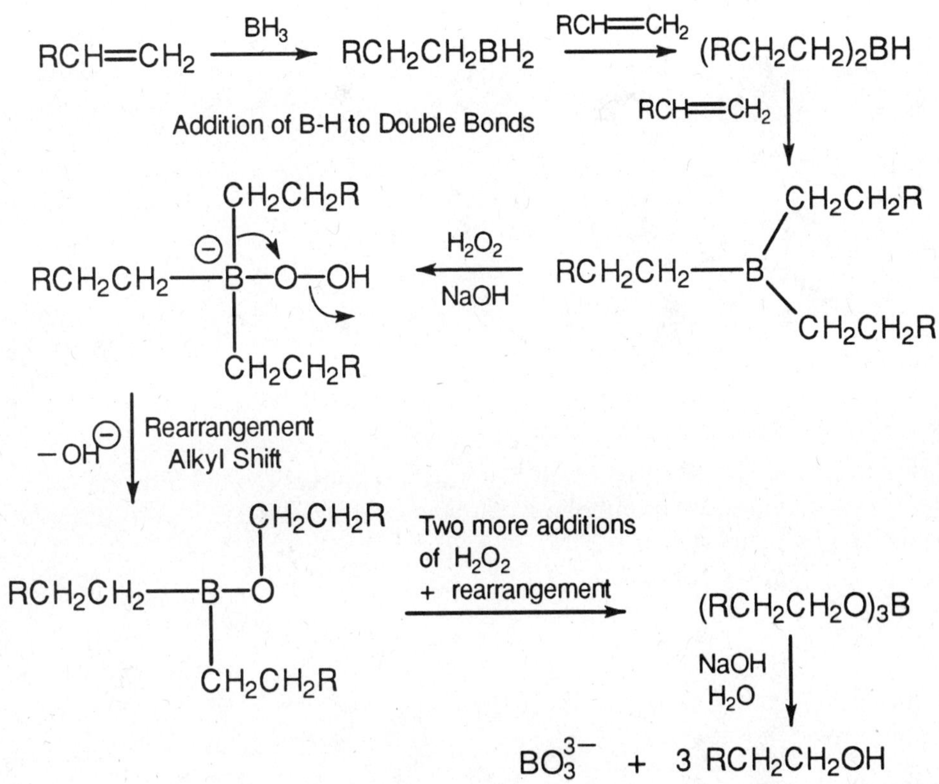

Figure 1. The course of the hydroboration-oxidation reaction.

The first step in the reaction is the addition of B-H of the BH_3 across the double bond of the alkene with the boron adding to the terminal carbon of the double bond and the hydrogen adding to the other carbon. This addition is preferred because of electronic effects where the boron adds to the double bond in the "normal" sense of an electrophile and because of steric effects where the bulky nature of BH_3, and the BR_3 that is formed, direct the addition to the less hindered terminal carbon.

The initially formed alkyl borane, $RCH_2CH_2BH_2$, reacts with a second molecule of alkene to form a dialkylborane, $(RCH_2CH_2)_2BH$. Reaction of the latter with a third molecule of the alkene gives a trialkylborane, $(RCH_2CH_2)_3B$.

After the borane addition is complete (about 5 min), hydrogen peroxide and base are added to the reaction mixture. The hydrogen peroxide adds to the boron as shown and then there is a rearrangement in which an alkyl group attached to the boron migrates to the oxygen of the hydrogen peroxide addition complex and displaces the hydroxide ion. Finally the base hydrolyzes the alkoxyborane to give the product, in this case a **primary alcohol** and boric acid (as the borate salt) (Figure 1).

The degree of stereospecificity of this reaction depends on the degree of bulkiness of the borane. In the example shown, BH_3 is not very bulky; thus one can expect both possible products: the primary alcohol from terminal addition of boron which will be the major product, accompanied by a minor amount of a secondary alcohol from addition of boron to the secondary carbon. There are other boranes that are much more bulky. One in particular is 9-borabicyclo[3.3.1]nonane (9-BBN-H).

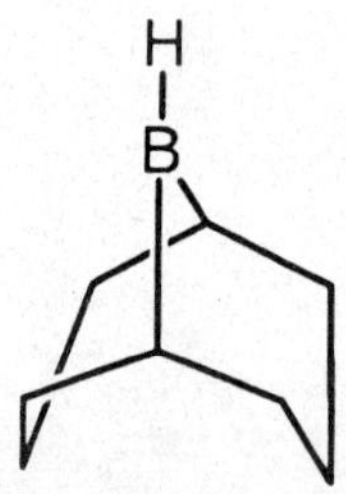

9-Borabicyclo[3.3.1]nonane (9-BBN-H)

9-BBN-H is extremely bulky and adds almost **exclusively** to the terminal carbon of olefins with the result that primary alcohols can be synthesized in very pure state without contamination by secondary alcohols.

In this experiment we will examine the hydroboration of 1-octene and/or styrene to form alcohols (Figure 2).

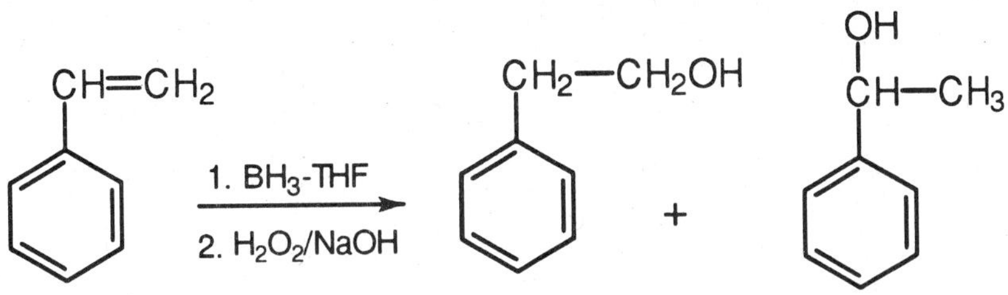

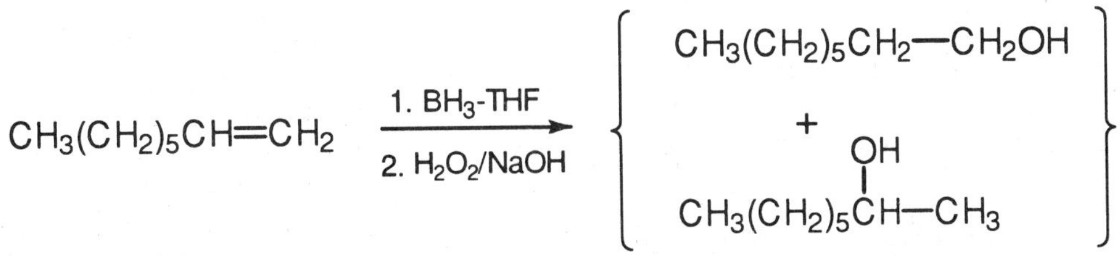

Figure 2. The hydroboration of styrene and 1-octene.

The products of the reactions will be examined by gas chromatography to determine the product ratio of primary to secondary alcohol. The reactions of 1-octene and styrene are carried out in an identical manner. Your instructor will tell you which alkene to use. Some students may do one and some the other so that the results of the GLC analyses can be compared. There is a significant difference in the product ratios in the two cases. The properties of the reagents and products are given in Table 1.

Table 1. Properties of Organic Reagents and Products

Compound	Mol. Wt.	bp, °C	Density, g/mL
1-Octene	112	122	0.715
1-Octanol	130	195	0.827
2-Octanol	130	180	0.819
Styrene	104	145	0.909
2-Phenylethanol	122	220	1.020
1-Phenylethanol	122	204	1.018

B. Experimental Procedure

Notes: The reaction vial, spin vane, and syringe must be dry. It is not necessary to oven-dry the equipment provided that it is "desk-dry." If styrene is to be used in the reaction, test for the presence of polystyrene. Add a few drops of the styrene to a small beaker of methanol. If a precipitate forms, polystyrene is present and the reagent should not be used. Secure fresh reagent and test again.

Place 200 μL (143 mg, 1.28 mmol) of 1-octene (or 200 μL, 182 mg, 1.75 mmol of styrene) into a dry 5-mL reaction vial with a spin vane. Place the vial in the aluminum block centered on the magnetic stirrer. Attach a screw cap with septum and commence stirring.

HANDLE THE BH_3-THF REAGENT WITH CARE. IT REACTS VIOLENTLY WITH WATER.

Using a **dry** 1-mL syringe fitted with a blunt needle, obtain about 0.8 mL of the BH_3-THF reagent. (Note: The BH_3-THF reagent is available as a 1-M solution, but the concentration diminishes in time due to hydrolysis from moisture that may enter the bottle. If the BH_3-THF is old, more reagent may be needed. The amount taken will depend upon the actual concentration of BH_3 in the reagent)

Inject the BH_3-THF reagent dropwise into the stirred 1-octene over a period of about 30 seconds. Then allow the solution to **stir for 5 minutes**. Using a capillary pipet, add 12 microdrops of acetone and allow the reaction mixture to stir for 2 minutes. The acetone reacts with excess BH_3 to form isopropyl alcohol. **IMPORTANT — Do not overlook this step. The acetone destroys excess BH_3 which otherwise would react violently with water in the next step.**

With continued stirring of the reaction mixture, cautiously add 3 or 4 microdrops of water (there should be no apparent reaction) followed by 0.25 mL of 3 **M** NaOH. Then add **cautiously, dropwise,** 0.25 mL of 30% H_2O_2 over a 30-second period and stir for *ca* 1 min. The oxidation is exothermic, and the mixture will become warm. Then replace the aluminum block by a 50-mL beaker half-filled with water and heat with continued stirring at 50-60°C for 5 minutes. The reaction should be complete and the mixture may now be worked up.

Add 1 mL of saturated aqueous NaCl solution to the vial. At this point you should have a two-phase system. The upper layer is organic and the lower layer is aqueous. There may be some gelatinous material on the bottom of the aqueous layer giving off bubbles. (This may be a hydrogen peroxide-boric acid complex.) If so, continue heating at 50-60°C with stirring to eliminate this material and give two clear phases.

> Note: Saturated NaCl is often used in washing the organic layers rather than plain water. The reason for this is that organic compounds are often slightly soluble in water and if plain water were used in the washing, then some of the product would be lost by dissolving in the water layer. If a salt (NaCl) is dissolved in the water, the salt renders the water more polar and decreases the solubility of the organic material in the water. Thus, less material will be lost by solubility by using a salt solution rather than plain water.

Cool the reaction mixture to room temperature and add *ca* 1 mL of diethyl ether. Stir rapidly with the spin vane for about a minute, but avoid splashing. Stop the stirring, allow the layers to settle, and carefully remove the **lower aqueous layer** with a capillary pipet. Add 0.5 mL of saturated NaCl solution, stir again, allow the layers to settle, and remove the **lower aqueous layer.** Repeat this washing with a second 0.5-mL portion of saturated NaCl. Discard the aqueous layers.

Prepare a drying column in a micropipet. Add a small plug of cotton, 2-3 mm of sand, and 2-3 cm of anhydrous sodium sulfate. Using a micropipet, transfer the ether solution of the products to the column and collect the effluent in a **weighed** 3-mL conical vial. (If you carefully flush the ether solution into and out of the micropipet a few times, the bulb will become saturated with ether vapor and the solution will not tend to leak out.) Rinse the vial and spin vane twice with small volumes of ether and pass these ether washes through the filter also.

Carefully evaporate the ether and other solvents using a **warm (not hot)** hot-plate (low heat) and a stream of nitrogen. Remove as much solvent as possible until only a small amount of liquid residue remains. Cool and weigh. Then continue the evaporation until a constant weight is obtained. Record the final weight and determine the crude yield.

Analyze the product by gas chromatography using a Carbowax column. For the hydroboration of 1-octene, the column temperature should be about 150°C; for hydroboration of styrene, 175°C. Authentic mixtures of the expected alcohols will be available for comparison by GLC analysis. Measure the area of the peaks due to the two alcohols and calculate the product ratio of primary to secondary alcohol in the product. (**Note:** Use peak height times retention time measured from the air peak to determine relative peak areas.) Run the infrared spectrum of the product **neat** (between salt plates).

Dispose of waste chemicals in labeled waste containers provided in the laboratory.

C. Prelab Questions

1. Explain why the primary alcohol predominates in this reaction.
2. What product(s) would be expected if the mixture of alcohols you obtained were to be heated with a catalytic amount of concentrated sulfuric acid?
3. Write the reaction which occurs when BH_3 encounters water. Hint: a gas is evolved and boric acid is formed.
4. Draw a valence bond structure for the BH_3-THF complex showing formal charges (the complex has a B-O covalent bond).

D. Postlab Questions

1. What other sequence of reactions could be used to convert 1-octene to 1-octanol?
2. How does acetone react with BH_3?
3. Why are rearrangements **not** observed when BH_3 adds to alkenes?
4. Give the structure of the product when 3,3-dimethyl-1-butene is treated with BH_3-THF followed by H_2O_2/NaOH.
5. Compare the relative retention times and the relative boiling points of the isomeric alcohols you prepared. Do you see any correlation? Explain briefly.

18. Reduction of Ketones to Alcohols with Sodium Borohydride

A. Introduction and Theoretical Discussion

Sodium borohydride ($NaBH_4$) and lithium aluminum hydride ($LiAlH_4$) are important reagents in organic synthesis because they readily reduce aldehydes and ketones to alcohols. However, they differ greatly in reactivity and selectivity as well as in the nature of the reaction conditions required for reduction.

$NaBH_4$ is much less reactive and much more selective than $LiAlH_4$. While $NaBH_4$ reduces only aldehydes and ketones to alcohols, $LiAlH_4$ not only reduces aldehydes and ketones, but also reduces the carbonyl groups of esters, amides, and carboxylic acids to alcohols. $LiAlH_4$ is dangerous to handle because it reacts violently with water to liberate hydrogen and can ignite spontaneously in air when crushed in a mortar. Thus $LiAlH_4$ must be used in anhydrous aprotic solvents such as diethyl ether. In contrast, although $NaBH_4$ reacts slowly with "neutral" water, it is stable in strongly basic solution (4.5 **M** $NaBH_4$ in 14 **M** aqueous NaOH is commercially available). Because most organic compounds are not water soluble, methanol and ethanol have been widely employed as solvents to carry out reductions with $NaBH_4$. However, $NaBH_4$ is more stable in ethanol and therefore ethanol is the preferred solvent.

The reaction of the complex metal hydrides with carbonyl groups results in the addition of the **equivalent** of a **hydride ion** ($:H^-$) to a carbonyl group. However, **no free** hydride ion is involved; instead, these reagents serve as **hydride transfer agents** (Figure 1A). All four hydride hydrogen atoms of $NaBH_4$ are "active", and participate in the reduction process as is shown in the balanced equation for reaction of $NaBH_4$ with a carbonyl group (Figure 1B). Thus, one mole of $NaBH_4$ will reduce 4 moles of an aldehyde or a ketone.

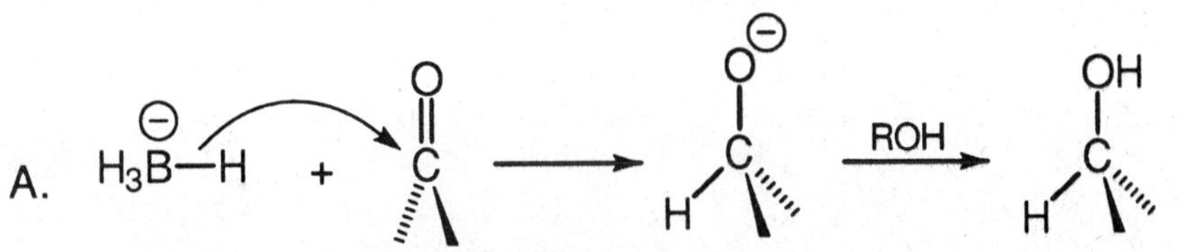

B. $NaBH_4 + 4\,R_2C{=}O + 4\,C_2H_5OH \longrightarrow 4\,R_2CHOH + NaB(OC_2H_5)_4$

Figure 1. Reaction of sodium borohydride with ketones.

The $NaBH_4$ reduction will be described for two ketones, 4-*tert*-butylcyclohexanone and benzoin.

4-*tert*-Butylcyclohexanone (Mol. Wt. 154, mp 47°C) is reduced by $NaBH_4$ to a mixture of *cis*- and *trans*-4-*tert*-butylcyclohexanol (Mol. Wt. 156). These alcohols are solids having nearly identical melting points and are not easily separated by recrystallization. However, gas chromatography readily shows the two isomers due to their different retention times. From GLC analyses, the ratio of the two isomers can be determined. The major product is the more stable *trans*- isomer, indicating that **axial attack** is preferred with BH_4^- (Figure 2). On the other hand, reagents that are more bulky than BH_4^- attack from the equatorial position to form the *cis*- isomer as the major product.

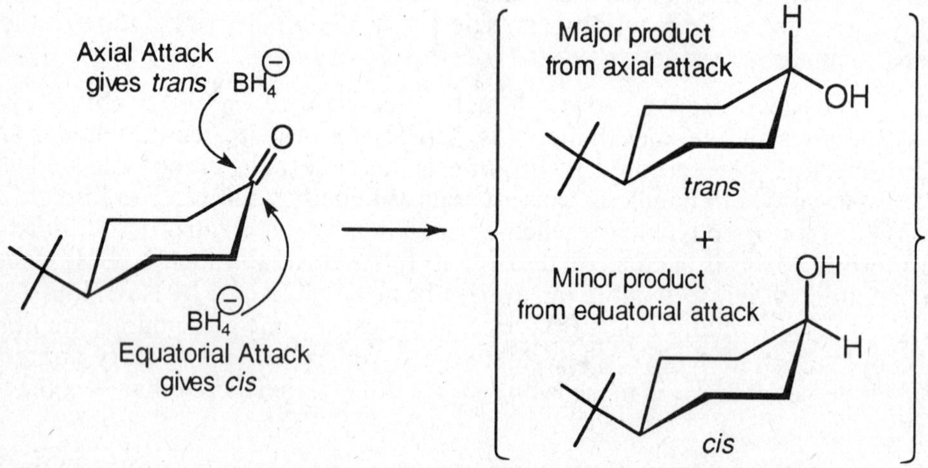

Figure 2. $NaBH_4$ reduction of 4-*tert*-butylcyclohexanone.

The reduction of benzoin (2-hydroxy-1,2-diphenylethanone; mol. wt. 212, mp 135°C) with $NaBH_4$ gives hydrobenzoin (1,2-diphenyl-1,2-ethanediol; mol. wt. 214) as a mixture of stereoisomers. One of the stereoisomers, *meso*-hydrobenzoin, is the major product and is readily isolated in good yield. The *dl*-hydrobenzoin that is formed remains in the "mother liquor."

The formation of the of the *meso*- isomer can be understood by application of Cram's Rule (Figure 3). This rule explains how the chiral center adjacent to the carbonyl group in benzoin controls the stereochemistry which results when the ketone is reduced to generate a new chiral center. The molecule is oriented as shown in the Newman projection (Figure 3), with the chiral center (bearing H, OH, and C_6H_5 groups) away from the viewer. To minimize transition state interactions, the oxygen of the carbonyl group lies between the **smallest group (H)** and the **medium-sized group (OH)**; it is directed away from the **largest group, (C_6H_5)**. The hydride attack then occurs preferentially from the **less hindered face** of the carbonyl group, which is on the side of the **smallest group (H)** on the chiral center. The result is the formation of the *meso* isomer.

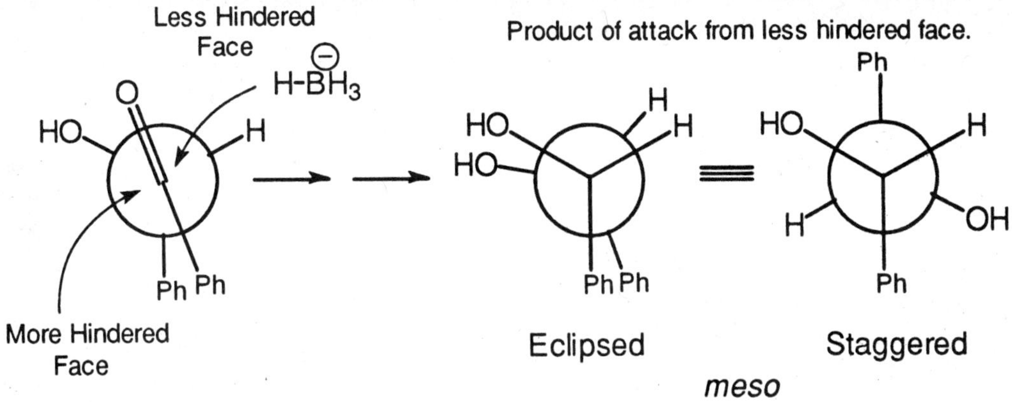

Figure 3. Stereoselective reduction of benzoin following Cram's Rule.

For the reductions, $NaBH_4$ will be used as a solution in 95% ethanol to which NaOH has been added. The NaOH gives a reagent that is stable for several days. The concentration should be made up to be about 0.6 **M** in $NaBH_4$ and 0.4 **M** in NaOH. In cases where it is not desirable to have a strongly basic solution (which can cause other reactions), freshly prepared $NaBH_4$ should be used in neutral ethanol.

In each case a substantial excess of $NaBH_4$ will be used to ensure that all of the ketone is reduced. The reduction of the ketone is rapid, but a few extra minutes will ensure that the reaction is complete. The excess $NaBH_4$ is subsequently destroyed by the cautious addition of dilute hydrochloric acid; note the evolution of hydrogen.

B. Procedure

1. Reduction of 4-*tert*-butylcyclohexanone

Dissolve *ca* 80 mg (0.52 mmol) of 4-*tert*-butylcyclohexanone in *ca* 0.3 mL of ethanol contained in a 5-mL reaction vial (use the graduations on the vial). While stirring with a spin vane, add 0.3 mL of the $NaBH_4$ solution in ethanol. A small amount of white material may separate. Continue to stir for 10 min.

At the end of the reaction period, using a micropipet, add dropwise 1–2 drops of 3 **M** hydrochloric acid. **CAUTION! The excess $NaBH_4$ reacts to form hydrogen with considerable foaming.** Note that this foaming indicates that an excess of $NaBH_4$ was present and that the reduction of the ketone should have gone to completion. When the foaming has stopped, add an additional 0.1 mL (about 10 microdrops) of hydrochloric acid. Then add 1.0 mL of water and 0.5 mL of methylene chloride. Stir briskly with the spin vane. Allow the layers to separate and then **carefully remove the lower methylene chloride layer with a micropipet.** Repeat the extraction with two more 0.5 mL portions of methylene chloride. Note: **Be careful to remove only the methylene chloride layer without any of the aqueous layer.** Pass each methylene chloride extract through a cotton-plugged pipet containing a 2 to 3-cm column of anhydrous Na_2SO_4. Collect the effluent in a **weighed** 5-mL conical vial.

Evaporate the solvent using heat and a stream of nitrogen. If a steam bath is used, be particularly careful that no water gets into the flask. Attaching an air condenser to the flask during the evaporation will reduce the chance of water entering the system. If a hot-plate is used, do not heat it over 80°C. Continue heating until a constant weight is reached. On cooling, the product should crystallize. Weigh the flask and contents to obtain a "crude" yield. If any liquid remains after evaporation, water has gotten into the sample. If this happens it must be removed, see your instructor. With a spatula, scrap down the walls of the flask so that all of the product is collected at the bottom of the 5-mL vial. Your instructor will indicate if the product should be sublimed (below).

Purification by Sublimation: Review the section on sublimation in the introduction chapter. Insert the sublimator into the 5-mL vial clamped in the aluminum block on the hot-plate. If an adjustable sublimator is used, adjust the cold finger so that the bottom of the cold finger is at the level of the 1-mL mark. Seal using a plastic cap and O-ring. Attach a vacuum hose to the side arm of the sublimator and the other end to the vacuum source. The vacuum may be produced by a pump, a house vacuum system, or by a water aspirator. If an aspirator is used, make sure that there is a water trap between the sublimator and the aspirator. With a micropipet, add some ice-water to the cold finger. Apply vacuum and heat the vial on the hot-plate. The material will sublime onto the cold finger at a block temperature of from 100-150°C, depending upon the vacuum produced. Change the water in the cold finger periodically during the sublimation. When the sublimation is complete, break the vacuum, remove the rubber hose, remove the sublimator **carefully,** and scrape off the product onto a weighing dish. Determine the weight, percent yield, and melting point. Place the solid in a small screw cap vial and make sure that the cap is on tightly, otherwise the material will all evaporate in a few days.

Analysis by GLC. Note: A four foot 1/4" column packed with 20% Carbowax on 80/100 mesh Chromosorb will separate the *cis-* and *trans-* isomers, provided the column is fairly new. With continued use, the column may not separate the isomers well. If separation cannot be achieved, a freshly prepared column will be necessary.

Transfer *ca* 10 mg of your product to a small vial and dissolve the solid in *ca* 100 μL of methylene chloride. Keep the vial tightly capped to avoid evaporation of the methylene chloride. If you did not sublime your product, then **show your solution to your instructor,** who will check to be sure that you do not have any water in it. If water is present, it must be removed before running the GLC analysis. Analyze the product by gas chromatography using a Carbowax column at about 150–170°C. Locate the peaks for the *cis-* and *trans-* isomers and determine the *cis/trans* ratio, expressed as percent *cis-* and percent *trans-*. (By using a substantial excess of $NaBH_4$, we ensured that reduction was complete and thus no ketone should remain in the product.) **Immediately after injecting your sample, flush the GLC syringe thoroughly with acetone to remove all traces of product. Syringes which are not cleaned are apt to become plugged.**

Analysis by Infrared Spectroscopy: Place a few drops of the methylene chloride solution of the product on the surface of a salt plate and allow the methylene chloride to evaporate. During the evaporation, angle the salt plate this way and that so that the salt plate becomes evenly coated with the product. When the solvent has evaporated, add one or two drops of carbon tetrachloride. Immediately cover the plate with a second salt plate, rotate one plate against the other, and then record the IR spectrum. If a solution cell is available, a better spectrum may be obtained using a carbon tetrachloride solution of the compound. The presence of a carbonyl band in the IR spectrum indicates that the reduction of the ketone was not carried to completion. Locate the hydroxyl group band and mark it on the spectrum.

2. Reduction of Benzoin

This experiment should be started when you have a break in the above procedure. Weigh 50 mg of benzoin into a Craig tube supported in a small Erlenmeyer flask or beaker. Add 0.5 mL of 95% ethanol and stir with a micro stirring rod to suspend the solid. Then add 0.3 mL of the $NaBH_4$ solution and stir thoroughly. The solution initially may turn yellow and the solid should dissolve. Allow the reaction mixture to stand for 10 minutes, at which time it should be clear and colorless.

Cautiously add 3 **M** hydrochloric acid **dropwise** until the foaming stops. Then add another 5 to 10 microdrops followed by 0.9 mL of water. Heat to near boiling on the steam bath (or boiling water bath) whereupon all the solid material should dissolve leaving a clear solution. Crystallization starts readily upon cooling. The best crystals result if the mixture is heated until almost all of the crystals dissolve. Stopper the tube and allow the mixture to cool slowly and finally cool it in an ice bath. Collect the crystals by centrifugation and wash with 0.5 mL of ice-cold 1:1 ethanol-water. Centrifuge again to remove the wash liquid, remove the crystals, and allow the product to dry in air. Store the hydrobenzoin in a screw cap vial. Determine the yield and melting point.

Dispose of waste chemicals in labeled waste containers provided in the laboratory.

C. Prelab Questions

1. Draw stereo-formulas for the more stable chair conformation for both *cis*- and *trans*-4-*tert*-butylcyclohexanol and explain why each conformation is preferred.
2. If you could find a way to equilibrate *cis*- and *trans*-4-*tert*-butylcyclohexanol (there are ways to do this), which isomer would predominate? Explain.
3. The reduction in Figure 3 is illustrated with (*R*)-Benzoin. Using stereoformulas show that (*S*)-benzoin also reduces to *meso*-hydrobenzoin.
4. Using stereoformulas, show that reduction of(*R*)-benzoin from the more hindered face (Figure 3) would give (*R*,*R*)-hydrobenzoin.

D. Postlab Questions

1. How could the *meso*- and *dl*- forms of hydrobenzoin be differentiated experimentally?
2. Recognizing that both *meso*- and *dl*-forms are optically inactive, what fundamental experimental approach can be employed to distinguish between such stereoisomers?
3. Suppose we reduced benzil, $C_6H_5COCOC_6H_5$, instead of benzoin with an excess of $NaBH_4$. What would the major product be? Explain.
4. Predict the major product expected from the $NaBH_4$ reduction of 3-*tert*-butylcyclohexanone. Explain.

19. Stereospecific Bromination of *cis*- and *trans*-Stilbene

A. Introduction

The bromination of alkenes proceeds in a stereospecific sense in that the two bromine atoms add from opposite sides of the double bond. The reaction is believed to proceed through a three-centered "bromonium ion" intermediate, which then undergoes a displacement of the bromine from the rear side by a bromide ion to give overall *anti*, or *trans*, additions. The process is illustrated for the bromination of *cis*- and *trans*-stilbene, (*cis*- and *trans*-1,2-diphenylethene) (Figures 1 and 2).

In the case of *trans*-stilbene (Figure 1), attack of the bromide ion by path **a** gives structure **a**. Attack by path **b** gives structure **b.** However, examination of these structures (or molecular models) will show that **a** and **b** are identical and thus only one compound is produced, the *meso* isomer (see Prelab Question 2a). Note that with either **a** or **b**, rotation about the central bond by 180° gives a conformation having a plane of symmetry passing through, and perpendicular to, the central bond.

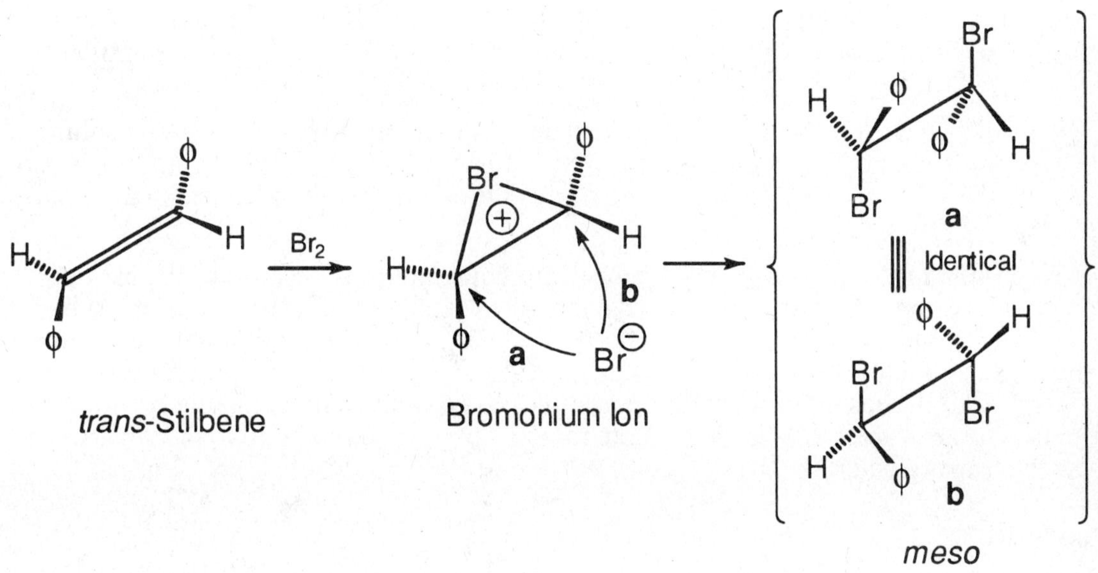

Figure 1. Bromination of *trans*-stilbene.

For *cis*-stilbene (Figure 2), reaction by path **c** gives structure **c** and by path **d** gives structure **d.** In this case, however, the two structures **c** and **d** are **not** identical, but are enantiomers (non-superposable mirror images). Since the bromonium ion is symmetrical, the reaction to form the dibromide goes with equal probability by path **c** and by path **d.** Thus, compounds **c** and **d** are be produced in equal amounts to give an optically inactive racemic mixture of the two enantiomers. We call the mixture a *dl*-pair (see Prelab Question 2b).

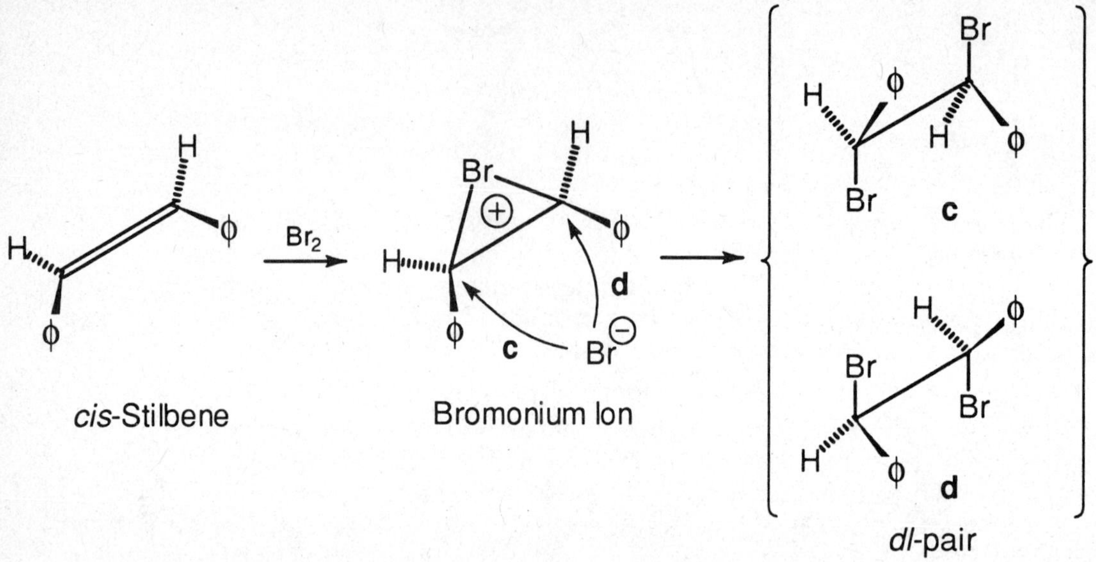

Figure 2. Bromination of *cis*-stilbene.

In this experiment we brominate *trans*- and *cis*-stilbene. The two dibromides, the *meso* and the *dl*-pair, differ dramatically in melting points and solubilities and thus are easily differentiated.

trans-Stilbene can be brominated conveniently in methylene chloride using a solution of bromine. Although the reaction is not instantaneous, it is quite rapid. The product, *meso*-stilbene dibromide, is not very soluble in most solvents and readily crystallizes in a virtually pure state.

The bromination of *cis*-stilbene requires different conditions because of the ease of isomerism of *cis*- to *trans*-stilbene under acid conditions. A trace of HBr, often found in solutions of bromine, is sufficient to cause isomerization. The bromination will be conducted in acetic acid to which **pyridine** has been added to neutralize any HBr in the bromine solution. Because of the ease of isomerization, samples of commercially available *cis*-stilbene may contain some of the *trans*- isomer. Note that the *cis*- isomer costs 20 to 30 times as much as the *trans*- isomer.

B. Procedure

1. *trans*-Stilbene Dibromide (*meso*-1,2-Dibromo-1,2-diphenylethane)

NOTE: All work must be done in the hood. Handle bromine with care. Bromine is toxic and it can cause nasty chemical burns. If it is spilled on the skin, wash it off immediately with running water. Then see your instructor.

Weigh about 200 mg (1.11 mmol) of *trans*-stilbene into a 5-mL conical reaction vial and add a spin vane. Then add about 2.0 mL of methylene chloride (use the markings on the vial). Place the vial in the aluminum block on the stirring hot-plate (no heat is used) and stir vigorously until the stilbene has dissolved.

In a clean, dry 3-inch test tube supported in a small Erlenmeyer flask, obtain about 1.4 mL of 5% bromine (by volume) in methylene chloride (this reagent is about 1.0 **M** in Br_2).

Add the bromine solution dropwise with a micropipet to the well-stirred solution of *trans*-stilbene until a light red-brown color persists. On standing, the color will fade and soon the product will begin to separate as a solid. As the color fades, periodically add a few drops of the bromine solution. Keep the mixture well-stirred to suspend the solid and enable the bromine to become distributed throughout the solution. This periodic addition may take a few minutes. You can begin working on the bromination of *cis*-stilbene between additions.

When a **light** red-brown color persists for several minutes without fading, the reaction is over. Destroy the excess bromine by adding **one microdrop of cyclohexene**. The cyclohexene reacts with the excess bromine to form 1,2-dibromocyclohexane which is easily removed in the work-up. Stir well to ensure complete removal of all free bromine. The mixture should be nearly colorless with a white precipitate. Cap the vial and place it in an ice bath for at least 10 minutes. Then collect the product on a **Hirsch** funnel. Rinse out the vial and wash the product with about 1-2 mL of ice-cold methylene chloride. Turn off the suction while the wash liquid is added, and use a pipet to distribute the wash liquid over the entire solid. Turn on the suction and let air pass through the product for 5 or 10 min. Collect the product and allow the solid to air-dry for a few minutes. Determine the weight and melting point.

2. *cis*-Stilbene Dibromide (*dl*-1,2-Dibromo-1,2-diphenylethane)

Weigh a clean, dry, 3-inch test tube. Add **4 microdrops** of *cis*-stilbene to the test tube and reweigh to determine the weight of *cis*-stilbene.

In another test tube supported in a small Erlenmeyer flask, place 1.5 mL of acetic acid followed by 0.30 mL of pyridine. Then add 0.30 mL of 5% bromine (by volume) in acetic acid. Mix the solution thoroughly.

Work in a hood. Handle the bromine solutions with care.

To the test tube containing the *cis*-stilbene, add about one-fourth of the bromine/acetic acid/pyridine solution and heat the test tube in the steam bath or in a beaker of boiling water. As the color fades, periodically add more of the bromine/acetic acid/pyridine solution (with continued heating) until a distinct yellow color persists.

Cool the test tube and add one or two microdrops of 10% sodium bisulfite (sodium hydrogen sulfite) solution to destroy the excess bromine and mix thoroughly. The yellow color should be discharged giving a colorless, or possibly a faintly colored solution. Then add 1.5 to 2 mL of water and the product will separate, probably as a white oil or a semi-solid. Cool the test tube in an ice bath for 15 min. When the oil has solidified, centrifuge the test tube to settle-out all of the product. (You may need to place a #4 cork in the centrifuge tube-holder to support the 3-inch test tube. Be sure to counterbalance the test tube and to remove the corks when you are through.)

Carefully decant the aqueous layer from the product, which should adhere to the test tube. Then wash the product with about 1 mL of water, centrifuge again, and remove as much water as possible. Then add about 0.8 mL of 95% ethanol and heat the mixture in a beaker of boiling water to dissolve the product. Additional ethanol may be added if necessary. However, if any *meso*-dibromide has been formed (from *trans*-stilbene as an impurity) it will not dissolve in the hot alcohol. If, on addition of a few more drops of ethanol, the remaining solid does not dissolve, do not add more ethanol. The insoluble solid is the dibromide of *trans*-stilbene and must be removed by filtration.

Using a metal rod or wooden stick, tightly pack a small cotton plug into the tapered end of a micropipet to make a microfilter. The filter should be flushed with a little solvent to remove any stray fibers. If the liquid flows too slowly, it can be forced through the filter by squeezing a pipet bulb butted against (or attached to) the end of the micropipet. Place the microfilter in a Craig tube supported in a 25-mL Erlenmeyer flask, which can be placed on a hot-plate (Figure 3).

Heat the alcohol solution of the *dl*-dibromide to boiling and heat up a micropipet by running the hot solution up and down in the micropipet a few times. Then quickly transfer the alcohol solution to the filter and, with the aid of a pipet bulb, collect the filtrate in the Craig tube. Pass a few drops of hot ethanol through the filter to flush out any product remaining in the filter. If you have much more than 1 mL of solution, carefully boil off some of the ethanol. Cool the solution, slowly at first, and then in an ice bath for 15-20 min and collect the product by centrifugation. Remove the product from the Craig tube, allow it to dry, and determine the weight and melting point.

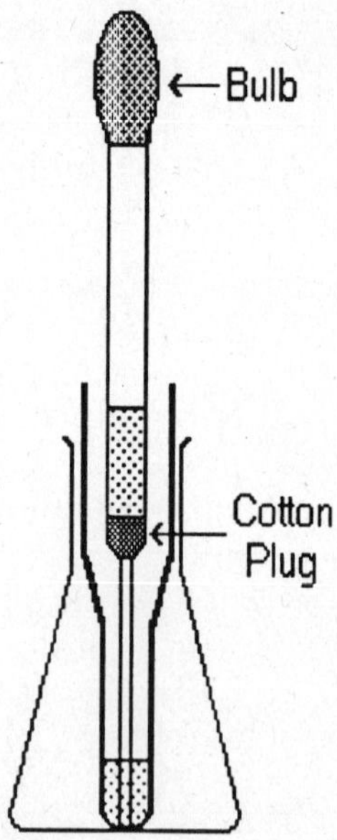

Figure 3. Filtration assembly.

Dispose of waste chemicals in labeled waste containers provided in the laboratory.

C. Prelab Questions

1. Write an equation for the reaction which occurs when cyclohexene is added at the end of the reaction of *trans*-stilbene with bromine.

2. (a) Assign the absolute configuration (R or S) to each chiral center of structures **a** and **b** in Figure 1 and explain how with these assignments you can identify the structure as being *meso*.

   (b) In the same way, assign the absolute configuration of the chiral centers for structures **c** and **d** in Figure 2. How do these assignments confirm that this is a *dl*-pair?

3. Pyridine is a heterocyclic aromatic compound similar in structure to benzene. It is widely used as an organic base because it reacts with acids, but is inert to most reagents. Write an equation with structural formulas to show how pyridine reacts with HBr.

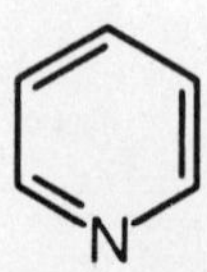

Pyridine

4. Addition of bromine to 130 mg of *trans*-stilbene gave 200 mg of *meso*-dibromide. Calculate the percentage yield.

D. Postlab Questions

1. Why is *trans*-stilbene more stable than the *cis*- isomer?
2. Propose a mechanism for the acid-catalyzed isomerization of *cis*- to *trans*-stilbene.
3. *meso*-Stilbene dibromide can be used to make diphenylacetylene. Suggest how this conversion can be achieved and write the reaction.
4. Is the *dl*-dibromide obtained in this experiment certain to be racemic? Explain.
5. Why is the *meso*-stilbene dibromide much less soluble in all solvents than the *dl*- isomer? (Hint: consider the melting points.)

20. Diphenylacetylene and (*E*)-1,2-Dibromostilbene

A. Introduction

The removal of two moles of HBr from a stilbene dibromide generates diphenylacetylene. For this dehydrobromination we will use *meso*-stilbene dibromide, the major product prepared in the previous experiment. It should be noted that this is the usual precursor for the synthesis of diphenylacetylene (the *dl* isomer is not used). In addition, we will do a qualitative debromination (removal of Br_2) of *meso*-dibromide and investigate the addition of bromine to diphenylacetylene.

1. Dehydrobromination of *meso*-Stilbene Dibromide

This experiment illustrates consecutive eliminations of HBr from a dibromide.

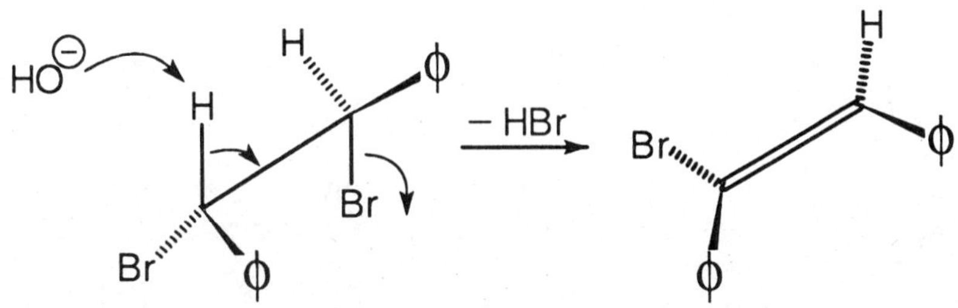

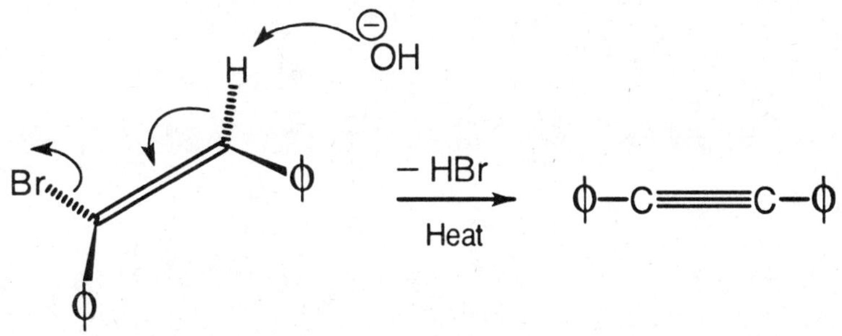

The first step is a "conventional" E2 reaction with *anti* elimination (removal of HBr). This step is relatively facile, giving rise to (*E*)-1-bromostilbene as shown. Elimination of HBr from this bromoalkene to give diphenylacetylene also follows the E2 mechanism, but it is much more difficult than the first step because: (a) the elimination involves removing HBr from a carbon-carbon double bond to generate a triple bond, a process which always has a high activation energy; and (b) the elimination must be *syn* rather than *anti*. Very harsh conditions (potassium hydroxide, 190°C) are required for the second step. If milder conditions are used, only the first step occurs to form 1-bromostilbene.

2. Debromination of *meso*-Stilbene Dibromide

The debromination of *meso*-stilbene dibromide is a stereospecific E2-type elimination initiated by iodide ion. It is a general reaction for 1,2-dibromides and follows the *anti* pathway if possible.

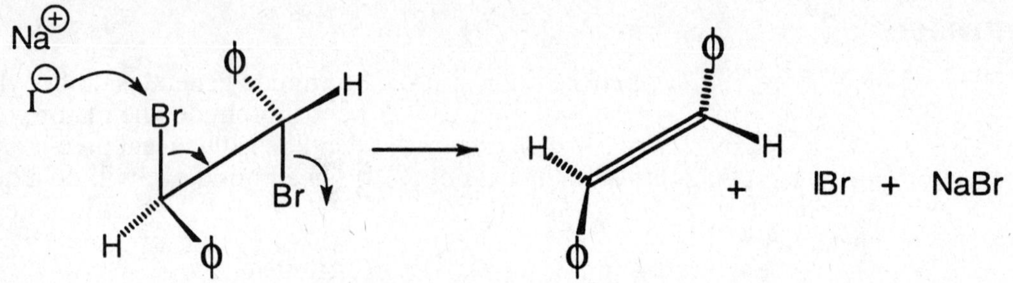

Then IBr + NaI ⟶ I_2 + NaBr

3. Addition of Bromine to Diphenylacetylene

Diphenylacetylene is a typical alkyne to illustrate the addition of Br_2 to the carbon-carbon triple bond. *Anti* addition occurs to give the *trans*- adduct, (*E*)-1,2-dibromostilbene.

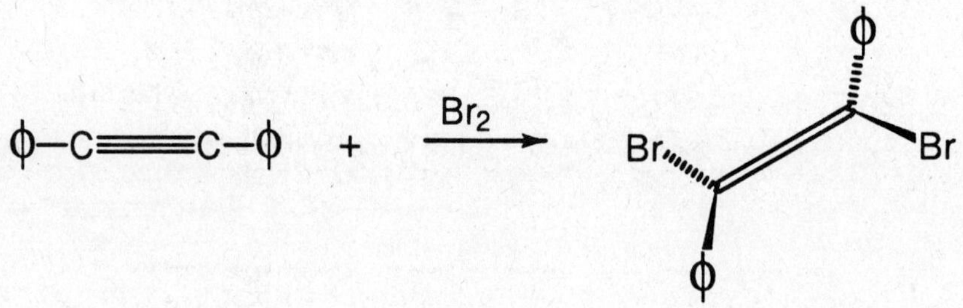

Addition of a second mole of Br_2 requires more forcing conditions because the two bromine atoms on the carbon-carbon double bond deactivate it toward reaction with electrophilic reagents such as Br_2.

B. Procedure

1. Diphenylacetylene

Weigh at least 200 mg of *meso*- stilbene dibromide, obtained from the previous experiment, into a 5-mL reaction vial. (Save a few mg for the qualitative reaction described below.) A large excess of KOH, supplied in the form of pellets, is used. Place two pellets (about 0.20 g) at the center of the full-sized weighing paper placed on the lab bench and fold the paper over the pellets several times. Then crush the pellets by tapping with a makeshift hammer; the aluminum heating blocks work well for this purpose. The powdered KOH should be used promptly because it is hygroscopic.

Caution: KOH causes chemical burns. If you get it on your skin, flush it off with water immediately.

Carefully add the powdered KOH to the reaction vial. Then add about 0.8 mL of triethylene glycol, which serves as an inert, high-boiling solvent. Connect an air condenser to the vial securing it with a lock nut and an O-ring. Then heat the vial to 190°C using the aluminum heating block. At this point, while holding on to the air condenser, remove the vial from the heating block and gently swirl it for a few seconds to ensure that the solids have dissolved. Return the vial to the heating block for at least 5 min.

After the heating period is over, cool the vial first in air and then in a water bath to approximately 50°C (it should feel warm but not hot) and add 2.0 mL of water. Stir the viscous dark oil thoroughly with a microspatula to induce solidification of the product. Then cool the vial in an ice bath for 10 min. Collect the solid product on a Hirsch funnel using suction filtration and wash it with water to remove inorganic salts.

Recrystallization of Diphenylacetylene — The crude diphenylacetylene tends to be contaminated by some dark impurities which can be removed by filtration. Prepare a microfilter made from a micropipet packed fairly tightly with 3 or 4 mm of cotton. Place the microfilter in a Craig tube and put the assembly in a hot water bath contained in a small Erlenmeyer flask (see Figure 3, previous experiment).

Place the crude diphenylacetylene in a 3-inch test tube and dissolve it in 0.5 mL of hot ethanol. Quickly transfer this hot solution with a micropipet to the hot microfilter and, using a bulb if necessary, filter the solution into the Craig tube. Rinse the filter with a **few** microdrops of hot ethanol. The solution may be colored, but it should be clear.

The volume of the alcohol solution should be less than 0.7 mL. Concentrate if necessary. Crystallization can usually be induced by briefly touching the bottom of the tube to ice. If the product oils out, warm the solution and re-cool. When crystallization has become well advanced, cool the tube in an ice bath for 10 or 15 min. Collect the crystals by centrifugation, wash with a few microdrops of ice-cold ethanol, and centrifuge again, giving adequate time to remove all of the ethanol possible. Collect the crystals on a weighed watch glass and air-dry to constant weight (10–15 min. usually). Determine the weight, melting point, and infrared spectrum.

2. Debromination of *meso*-Stilbene Dibromide

Place several milligrams of the *meso*-dibromide in a 3-inch test tube and dissolve the solid in a few drops of acetone with warming. Then add about 0.2–0.3 mL of 15% NaI in acetone. Warm the solution to near reflux on the steam bath. Describe what happens. Explain any color change and/or formation of a precipitate. Warm the solution until no further color change occurs. Then add an equal volume of water and a few drops of aqueous $NaHSO_3$ solution. Explain the results.

3. (*E*)-1,2-Dibromostilbene [(*E*)-1,2-Dibromo-1,2-diphenylethene]

Weigh about 3/4 of your diphenylacetylene (save some to turn in to your instructor) into a Craig tube and dissolve it in a few microdrops of methylene chloride. Add dropwise a solution of bromine in methylene chloride until a light bromine color persists for several minutes. Then discharge the excess bromine with a microdrop of cyclohexene. The volume of the solution should be about 0.5 mL (concentrate if necessary). The product is only slightly soluble and usually crystallizes readily. Cool in an ice-bath and centrifuge to remove solvent. Wash the crystals by adding a few drops of cold methylene chloride and re-centrifuging. Collect the crystals and, after air-drying, determine the weight, yield, and melting point.

Dispose of waste chemicals in labeled waste containers provided in the laboratory.

C. Prelab Questions

1. Give stepwise reactions showing how *dl*-stilbene dibromide would react with hot KOH in triethylene glycol.
2. Suggest a reason for using the *meso*-dibromide instead of the *dl*-dibromide to prepare diphenylacetylene.
3. Triethylene glycol (TEG), $HO(CH_2CH_2O)_3H$, has a boiling point of 285°C. Would you expect it to be soluble in water? Explain. Why is it used as a solvent in this experiment?
4. Why is a strong base such as KOH used in this experiment instead of a weak base such as potassium acetate?

D. Postlab Questions

1. How could dipole moment measurements be used to distinguish between the (*Z*) and (*E*) stereoisomers of 1,2-dibromostilbene?
2. Zinc causes elimination of bromine from 1,2-dibromides. Suggest a mechanism for this reaction.
3. A student proposed the synthesis of dibenzylacetylene by adding bromine to 1,4-diphenyl-2-butene followed by heating the resultant dibromide with KOH in TEG. Would this synthesis be successful? What would be the most likely product of the reaction with KOH?
4. What organic product would be formed upon treatment of *dl*-stilbene dibromide with NaI in acetone? How could you prove that this product is formed?

21. Addition Polymerization

A. Introduction

Polymers are high molecular weight materials formed by the successive addition of monomers to form long chains often consisting of hundreds of thousands of similar units joined together. Polymer means units (mer) taken many times (poly). The compound from which the polymer is constructed is called the "monomer" since there is only one unit. Polymers may be formed by a number of different processes, but one of the more common types is called "addition polymerization." An example of addition polymerization is the polymerization of compounds containing carbon-carbon double bonds. Such monomers react readily with free radicals generated by the decomposition of a chemical initiator. These free radicals are extremely reactive and readily add to unsaturated compounds to promote the successive addition of monomers to form polymer chains. The process is extremely rapid, with thousands of units adding to the growing polymer chain in only a few seconds.

Free radical polymerization proceeds in four stages: 1. Generation of radicals; 2. Initiation; 3. Propagation; and 4. Termination. **Free radicals may be generated** by decomposition of an initiator. One example is benzoyl peroxide, which on heating decomposes to give two benzoyl radicals, **R**(dot). The polymerization is then **initiated** by the attack of the benzoyl radical on the double bond of the monomer **M** to generate a growing chain radical:

1. Generation of Radicals: Initiator I $\longrightarrow$ 2 R•

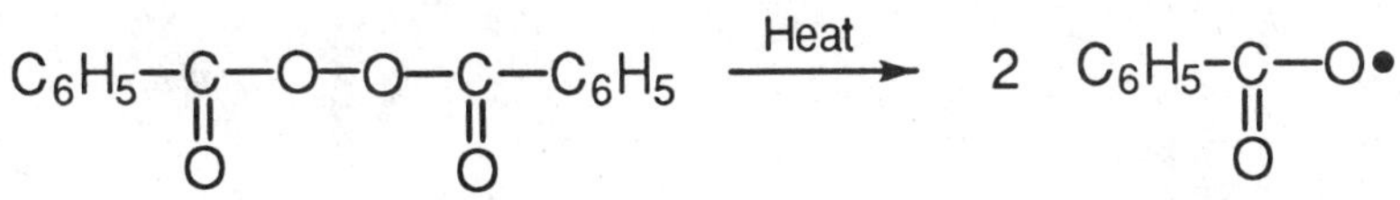

Benzoyl peroxide **Benzoyl Radical**

2. Initiation: R• + M $\longrightarrow$ RM•

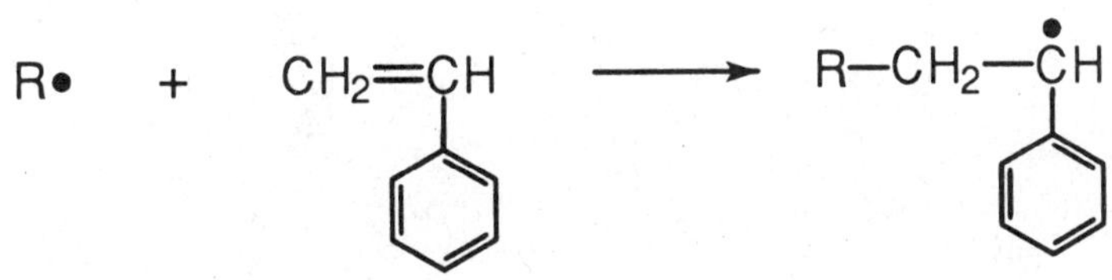

Propagation is the successive addition of monomers to the growing chain radical **R(M)$_n$M**(dot). The reactivity of the growing chain radical, ---MM(dot), does not change upon successive additions of monomer. This process accounts for the long chain-like structure and is the main part of the polymerization, occurring many times to build up high molecular weight polymer molecules. Molecular weights of the order of one or two million are common.

3. Propagation:

$$RM\bullet + M \longrightarrow RMM\bullet$$
$$RMM\bullet + M \longrightarrow RMMM\bullet$$
$$RMMM\bullet + M \longrightarrow RMMMM\bullet$$
$$RMMM\bullet \xrightarrow{M} \xrightarrow[\text{many times}]{M} \xrightarrow{M} \text{----}MMM\bullet$$

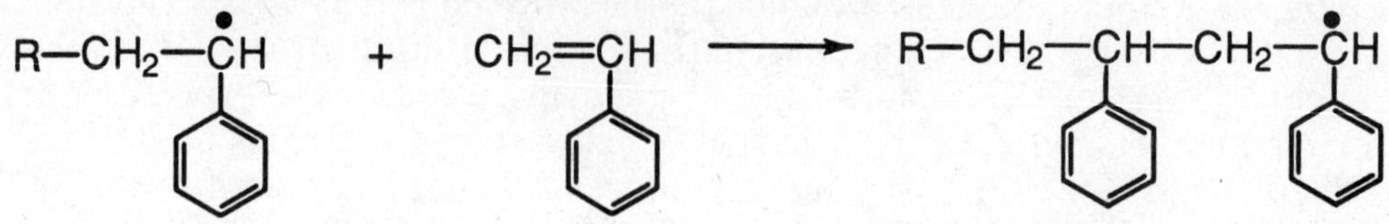

Growing Polymer Chain Radical

$$\text{----}MM\bullet + M \longrightarrow \text{----}MMM\bullet$$

Termination is the destruction of a growing chain radical to give polymer chains that are no longer capable of growing further. Free radicals are extremely reactive and if any two radicals should happen to meet, they will instantly react with each other to "kill" the growing chain and form "dead" polymer. This process is called termination. One such termination reaction is the coupling of two chains at the sites of the two radicals. Other termination processes are possible. Although this polymer chain has terminated and will no longer add monomer, radicals are still being generated to initiate new chain growth. New chains will initiate, grow, and terminate until the monomer, or initiator, has been completely consumed.

4. Termination: $R(M)_x\bullet + R(M)_y\bullet \longrightarrow R(M)_x\text{—}(M)_yR$

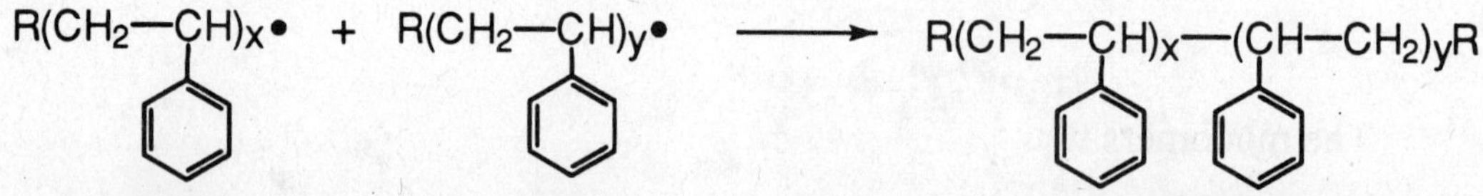

$$\text{-----}MM\bullet + \text{-----}MMM\bullet \longrightarrow \text{-----}MM\text{—}MMM\text{-----}$$

Polymer

The monomers, as they are received from the manufacturer, contain small amounts of inhibitors which are added for the purpose of destroying any stray free radicals that may be generated spontaneously and that would cause premature polymerization. Normally these inhibitors are removed by washing with sodium hydroxide solution, by chromatography, or by distillation. In our experiments, we will simply "overwhelm" the inhibitors by adding a sufficiently large quantity of the initiator. This excess initiator will consume the inhibitor, and when the inhibitor is all consumed, then the polymerization will proceed.

The following monomers and initiator will be used in this experiment.

Styrene (Monomer) $CH_2{=}CH{-}C_6H_5$ bp 146°C

Methyl Methacrylate (Monomer) 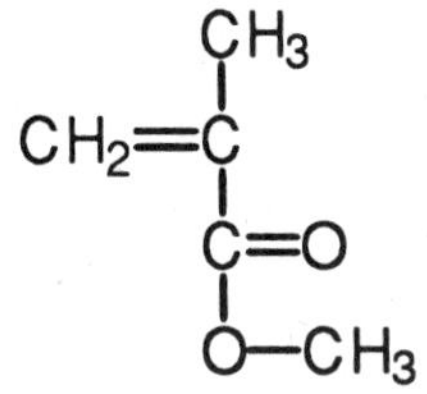 bp 100°C

Benzoyl Peroxide (Initiator) 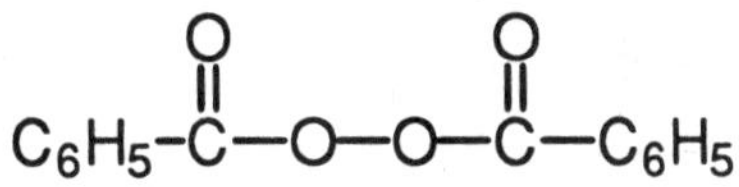

Benzoyl peroxide will be used to form the radicals that initiate the polymerization of styrene and methyl methacrylate. The polymers formed in these processes are named after the starting monomers, usually by adding the prefix "poly" to the name of the monomer. Thus, the polymer from styrene is called polystyrene and the polymer from methyl methacrylate is called poly(methyl methacrylate). Parentheses are often used if the name of the monomer consists of two or more words. Foamed polystyrene is an excellent insulator and is used in drinking cups. Poly(methyl methacrylate) is the same as Plexiglass, or Lucite, and is used for automobile tail lights. Of course, there are many other uses for these polymers.

B. Procedure

> **Note: Benzoyl peroxide can explode if subjected to severe shock. It is also very flammable. Do not use near flames and be careful about sparks. Benzoyl peroxide will decompose explosively with a spark.**

The monomers will be dispensed from dropping bottles using graduated pipets to measure the quantities. Polymerizations will be carried out in 3-inch test tubes. To serve as a constant temperature bath, place a 100-mL beaker half-filled with water on the hot-plate and heat to gentle boiling. Since you will need to add water periodically to the boiling water bath, it is advisable to have a second beaker of boiling water on the hot-plate to use for this purpose, so that the reaction mixtures will not be cooled when water is added. The polymers should be solids, which are difficult, if not impossible, to remove from the test tubes. Do not try to remove residual polymer to save the test tube. Instead, at the completion of the experiment, label the test tubes and turn them in to your instructor.

1. Polymerization of Styrene and Methyl Methacrylate (MMA)

Weigh *ca* 10 mg of benzoyl peroxide into a labeled 3-inch test tube. Then add 0.5 mL of styrene and swirl to dissolve the solid. To a second labeled 3-inch test tube add 0.5 mL of styrene but **do not add any benzoyl peroxide**. This sample will serve as a "control." To identify the tubes, mark them with a pencil on the frosted glass, or **firmly** apply a label. Small rubber bands might also be used for identification purposes.

Weigh *ca* 10 mg of benzoyl peroxide into a third labeled 3-inch test tube and add 0.5 mL of methyl methacrylate. Mix well and mark the tube. Finally add 0.5 mL of methyl methacrylate to a fourth labeled test tube (no benzoyl peroxide initiator) to use as a control. Place all four tubes in the boiling water bath and record the time. Be certain that the benzoyl peroxide is all in solution. If not, swirl the test tubes while heating.

Follow the course of the polymerization by observing the extent to which the liquid increases in viscosity as a function of time compared with the controls. Be alert because the methyl methacrylate polymerizes fairly rapidly. Because of the great increase in viscosity, the final stages of this polymerization can be readily detected.

2. Polymeric Products

After sufficient heating (*ca* 10-15 min. for methyl methacrylate; *ca* an hour for styrene) the systems containing the benzoyl peroxide initiator should yield solid polymers which are very hard. Because the MMA polymerizes rapidly, this product can be worked-up and examined while the styrene is still polymerizing. The PMMA should be brittle enough so that a small amount can be chipped out of the tube. Be careful not to break the test tube. Crush a small amount of the polymer on a watch glass and determine its melting behavior. To save time, heat the sample at about 8-10 °C per min. Follow the directions of your instructor for disposing of the test tubes.

3. Infrared Spectra

For infrared analysis, a film of each polymer is required. With commercial polymers, which have higher molecular weights than the polymers we prepare, it is possible to cast films which can be stripped from the support. Our polymers tend to be too brittle to pick up, so we need to cast films directly on salt plates. In order to get good spectra, it is necessary to prepare films which are neither too thick nor too thin. The following procedure should provide satisfactory films.

Poly(methyl methacrylate) (PMMA): Add 0.2–0.4 mL of toluene to the test tube containing the poly(methyl methacrylate). Heat the test tube in the 100°C water bath for **one minute.** Then, with a micropipet, add **three** drops of the toluene solution to a clean salt plate which is placed on a watch glass. Promptly distribute the solution evenly over the salt plate, using the sealed end of a melting point tube to avoid scratching the salt plate. Allow the plate to stand for at least 5 min., during which time the toluene evaporates. Record the IR spectrum of the film of polymer that has formed on the salt plate.

Polystyrene: Repeat the procedure outlined above by heating the polystyrene in the test tube with 0.2–0.4 mL of toluene for **one minute** at 100°C. Spread **six** microdrops of the solution over a clean salt plate and allow the toluene to evaporate as before to form a hard film. Record the IR spectrum.

Cleaning the Salt Plates: If the films are prepared as described above, it will be possible to remove the films by wiping with a tissue wet with toluene. However, if the films are difficult to remove with cold toluene, a brief immersion of the salt plate in hot toluene may be required. Check with your instructor.

Spectral Comparisons: Compare the IR spectra of the polymers to the spectra of the appropriate monomers. If time permits, you can record the IR spectra of the monomers as thin films between salt plates. Alternatively, your instructor may post these spectra in the laboratory. What structural changes occurred during the polymerization which would account for the differences between the spectra of the polymers and the spectra of the monomers?

Also compare the spectra of your polymers with the spectra obtained from commercial samples of these polymers. These spectra will be posted in the laboratory. Record any differences in the spectra. In particular, what IR absorptions might be attributed to the initiator or its decomposition products?

Dispose of waste chemicals in labeled waste containers provided in the laboratory.

C. Prelab Questions

1. Write the equations for the polymerization of both monomers in your notebook. Identify the **generation of radicals, initiation, propagation,** and **termination** steps.
2. Write a general structure for each polymer.
3. Explain why head-to-tail addition is energetically more favorable than head-to-head addition. (Hint: what new radical results from adding R· to CH_2=CH-X, and why is this new radical formed?)

D. Postlab Questions

1. Why was it necessary to heat these reactants?
2. Compare the results of heating the monomers with and without benzoyl peroxide.
3. Compare the apparent reactivity of styrene and methyl methacrylate toward polymerization.
4. How do you account for the melting behavior of poly(methyl methacrylate)? How does the melting behavior of PMMA compare with that of a low molecular weight crystalline compound?
5. Suggest how these polymers could be made into molded objects.
6. Compare your IR spectra with those of the monomers and commercial polymers. Discuss any differences.

22. The Diels-Alder Reaction

A. Introduction

The [4 + 2] cycloaddition of a 1,3-diene with an alkene to form a cyclohexene adduct is known as the Diels-Alder reaction. The alkene, because of its affinity for 1,3-dienes, is called the "dienophile." The simplest case is the reaction of 1,3-butadiene with ethylene (Figure 1), but this particular reaction requires heat and pressure in order to obtain a good yield of product. If the dienophile has an electron-withdrawing substituent, such as the $-CO_2CH_3$ group, on the carbon-carbon double bond, the reaction may be quite rapid and good yields often can be obtained in just a few minutes. By varying the structure of the diene and the structure of the dienophile, a wide variety of cyclohexene derivatives can be synthesized, making this one of the most useful reactions in organic synthesis.

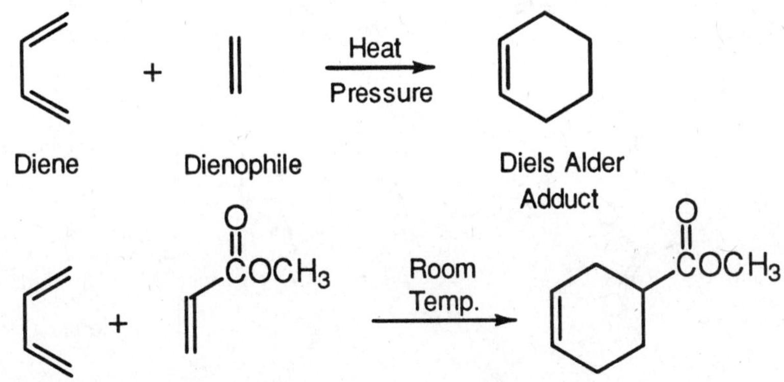

The [4 + 2] cycloaddition (*syn* addition) results from π orbital overlap of the molecular orbitals of the diene and the dienophile. In Figure 1 is shown the interaction of the highest occupied molecular orbital (HOMO) of the dienophile with the lowest unoccupied molecular orbital (LUMO) of the diene.

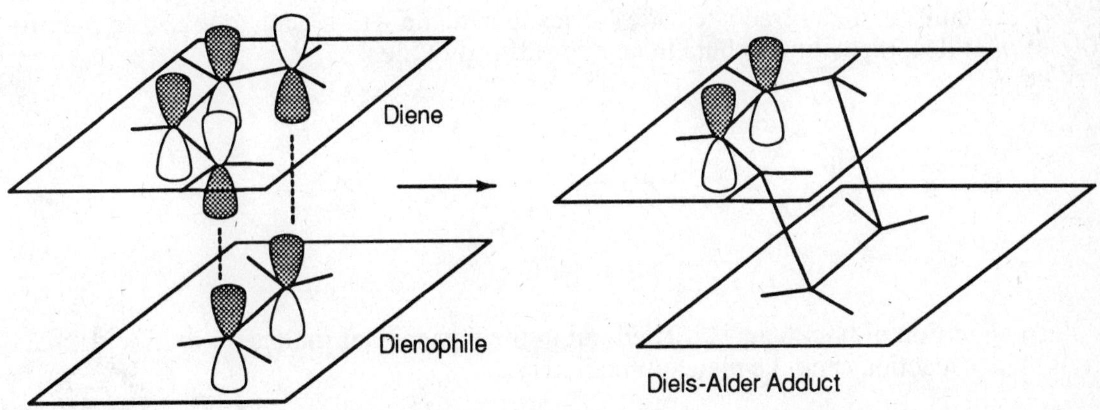

Figure 1. The [4 + 2] cycloaddition of a diene with a dienophile.

The Diels-Alder reaction is described for three diene components: cyclopentadiene, butadiene, and furan. In all three cases, maleic anhydride, an electron withdrawing compound, will serve as the dienophile.

Cyclopentadiene, bp 42°C, is prepared by heating dicyclopentadiene in a distillation apparatus. Under these conditions, dicyclopentadiene undergoes a retro-Diels-Alder reaction to give cyclopentadiene which distills out (Figure 2A). Since dicyclopentadiene is a dimer, one mole gives two moles of cyclopentadiene. For use, cyclopentadiene must be prepared fresh because it dimerizes, *via* the Diels-Alder reaction, back to dicyclopentadiene. Cyclopentadiene is widely used in both organic and organometallic chemistry and has the interesting property of forming sandwich compounds with various metals, such as iron. In the Diels-Alder reaction, cyclopentadiene is a particularly reactive diene and maleic anhydride is a reactive dieneophile. The reaction is over in just a few minutes to give the *endo* isomer (Figure 2B). The alternative *exo* isomer is not formed.

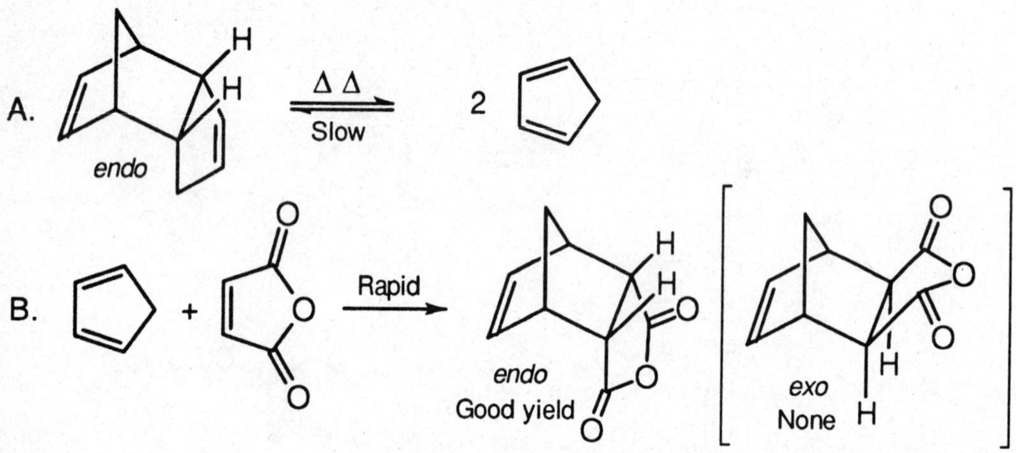

Figure 2. The Diels-Alder reaction. A. Cracking of dicyclopentadiene. B. Reaction of cyclopentadiene with maleic anhydride.

Butadiene is a gas at ambient pressures and is generally supplied as a liquid in tanks under pressure. It is used in great quantities to make synthetic rubber. Since is not practical to provide gas tanks in undergraduate laboratories, butadiene will be generated from 3-sulfolene by thermal decomposition to butadiene and sulfur dioxide.

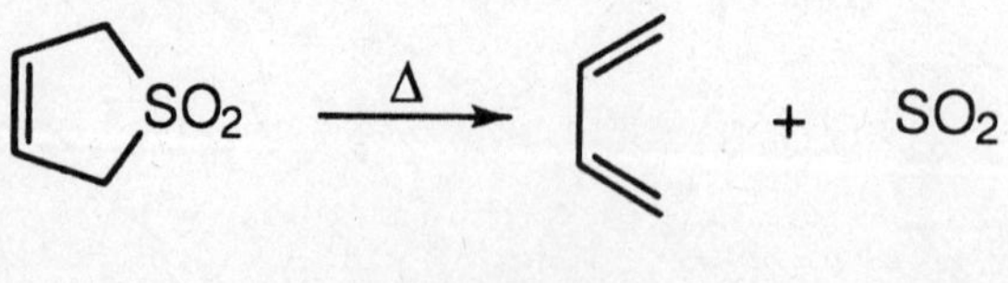

The generation of butadiene is carried out in the presence of maleic anhydride so that the Diels-Alder reaction can take place immediately.

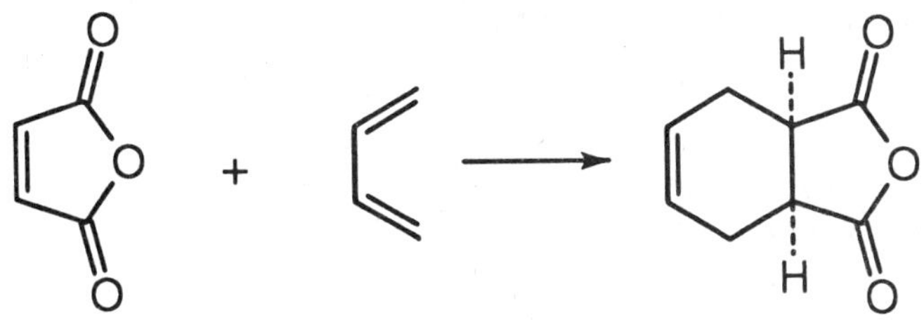

Furan is a liquid, bp 32°C, which is readily available in a pure state. Because the reaction with maleic anhydride is relatively slow at room temperature, the reaction is carried out overnight, or until the next laboratory period. In this case the *exo* stereoisomer results.

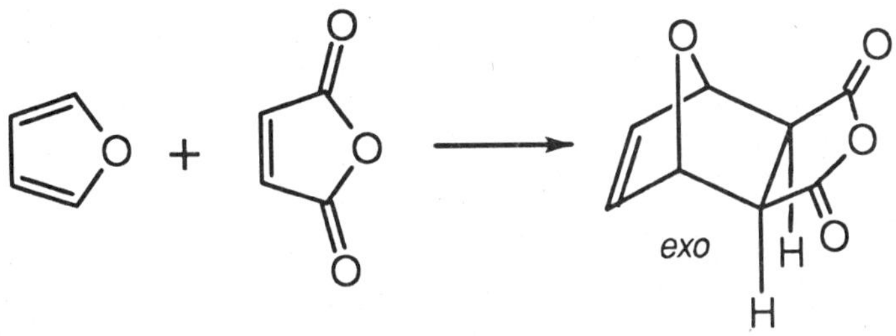

B. Experimental Procedure

In order to provide variety, three Diels-Alder reactions are described. Your instructor may recommend that you do only one or two of them.

Maleic anhydride is prepared commercially by catalytic air oxidation of benzene at 400-500°C. It is generally supplied in large lumps which should be crushed in a polyethylene bag shortly before use. The crushed maleic anhydride (which will be supplied) must be kept in a tightly sealed container because it reacts with moisture in the air, undergoing hydrolysis to maleic acid. Maleic anhydride is the limiting reagent in these reactions, and thus the formation of maleic acid must be avoided. Also, in the high-temperature reaction with sulfolene, the presence of maleic acid can lead to the formation of dark side-products.

1. Cracking of Dicyclopentadiene: Preparation of Cyclopentadiene

Note: Dicyclopentadiene and especially cyclopentadiene have irritating odors. These compounds should be handled only in a hood.

As noted earlier, cyclopentadiene is obtained by the thermal cracking of dicyclopentadiene (Figure 2A). Although there are a variety of ways to carry out this cracking procedure, the following technique is satisfactory if the additional items of equipment needed are available. Each cracking set-up can provide enough cyclopentadiene for several students, so it is not necessary that everyone perform this operation.

The apparatus shown in Figure 3 is set up in the hood. Add about 2 mL of paraffin oil and a spin vane to a 5-mL conical vial set in an aluminum heating block on the hot-plate. (Alternatively, a 10-mL round-bottom flask with 4 mL of paraffin oil may be used.) Attach the Claisen head. Close off one arm of the Claisen head with a septum cap. To the other arm, connect the air condenser and the Hickman still assembly with the cold finger. Close off the side-arm of the Hickman still with a solid cap (or a screw-cap and septum). Handle the assembly very carefully and support it with a clamp. The air condenser is needed, first to condense any dicyclopentadiene that distills at the high temperatures employed, and second to place the Hickman still farther away from the heat of the hot-plate so that the cyclopentadiene (bp 42°C) collecting in the cup does not evaporate before it can be removed. Turn on the stirrer hot-plate to "high" and wait for about 10 min. The temperature required to crack dicyclopentadiene is about 300°C, but do not use a thermometer in the aluminum block unless you have a thermometer that goes to 350°C because the high temperature will break a standard 250°C thermometer.

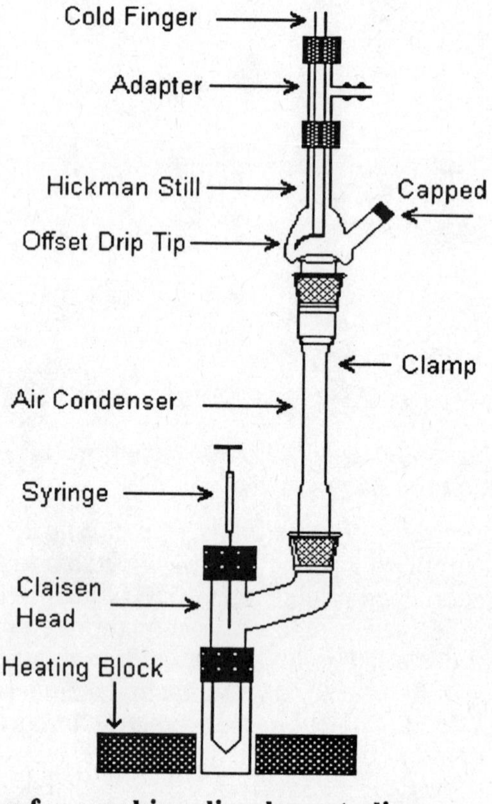

Figure 3. Apparatus for cracking dicyclopentadiene.

BE CAREFUL!! 300°C IS VERY HOT AND YOU WILL BE BADLY BURNED IF YOU TOUCH THE HEATING BLOCK OR HOT-PLATE.

Add ice-water carefully to the cold finger using one of the long micropipets. During the cracking, periodically replace the water with fresh ice-water. Place a blunted needle onto a 1-mL syringe and fill the syringe with dicyclopentadiene. When the hot-plate is up to temperature, add the cyclopentadiene through the septum of the Claisen head at such a rate that the cyclopentadiene distills and collects in the cup of the Hickman still. The cracking is fast if the temperature is high enough and in only a few minutes cyclopentadiene sufficient for several Diels-Alder reactions should collect in the cup.

As the cyclopentadiene collects in the cup of the Hickman still, uncap the side-arm and transfer the **cyclopentadiene** to a small screw-cap vial using a micropipet. Continue to add **dicyclopentadiene** until enough cyclopentadiene has been prepared to supply several students. Additional quantities of dicyclopentadiene can be injected to obtain more cyclopentadiene as it is needed. When the cracking has been completed and all students have been supplied with cyclopentadiene, the system may be dismantled, **but allow the hot-plate and the flask to cool first.** Remove the Hickman still assembly, wash it with acetone, and return it to the instructor. Pour the paraffin oil into the waste jar.

2. Reaction of Cyclopentadiene with Maleic Anhydride

In a 2-mL Craig tube place 0.7 mL of ethyl acetate and 150 mg (1.53 mmol) of maleic anhydride. Stir to dissolve. Add 0.7 mL of hexane, mix well, and then add *ca* 150 µL (*ca* 120 mg, 1.8 mmol) of cyclopentadiene (d = 0.8 g/mL). Note the heat of the reaction. After a few minutes, scratch the walls of the tube in order to induce crystallization. Cool the tube in an ice-bath for a few minutes and collect the crystals by centrifugation. Remove the crystals and allow them to dry to constant weight. The crystals are very pure at this point and recrystallization is not necessary. Calculate the percent yield, determine the melting point, and record the IR spectrum.

3. Reaction of 1,3-Butadiene with Maleic Anhydride

This procedure must be carried out in a hood because of the pungent odor and toxicity of the sulfur dioxide gas evolved.

A two-fold excess of sulfolene will be used because some 1,3-butadiene is apt to escape. Maleic anhydride will be the limiting reagent.

Weigh 100 mg (*ca* 1.0 mmol) of maleic anhydride into a dry 3-mL conical vial. Then add *ca* 250 mg (2.2 mmol) of sulfolene and 0.10 mL of xylene. Add a very small (micro) boiling chip and attach an air condenser.

Place the vial in an aluminum heating block fitted with a thermometer and turn on the heat. As you begin heating, swirl the vial to ensure that the reagents dissolve. Heat rapidly at first and then reduce the heating rate when the temperature reaches 120 or 130°C because there is considerable overshoot. Note the temperature at which vigorous bubbling occurs (SO_2 escaping). If the bubbling becomes too vigorous, remove the vial from the heating block briefly. Heat the block to 180–190°C for 20 minutes. **Note: Monitor the temperature frequently to avoid overheating, which can cause the thermometer to break, as well as decompose the product.** At the end of the heating period, remove the vial from the heating block and allow it to cool to room temperature.

There is often some dark solid material in the product which must be removed by filtration. While the reaction mixture is cooling, prepare a filter column using a micropipet. Insert a very small wad of cotton into the pipet and pack firmly, but not tightly. Add about 5 mm of sand, 10 mm of Celite, then 5 mm of charcoal, and finally about 2 mm of sand. Clamp the column to the stand.

When the reaction mixture has cooled, add 1.0 mL of toluene and filter the solution through the charcoal column. Collect the filtrate in the Craig tube. Rinse the vial with a few drops of toluene and pass this rinse solution through the filter, collecting it with the main filtrate. To the solution in the Craig tube, add hexane dropwise with swirling until the product starts to separate. Then heat the solution carefully on a water, or steam bath, until the solid redissolves. Allow the solution to cool slowly. The solution might need seeding with a small crystal of the product. Scratching the walls of the tube might also induce crystallization. Finally, cool the solution in an ice-bath and collect the product by centrifugation. Air-dry the product, determine the yield and melting point, and record the IR spectrum.

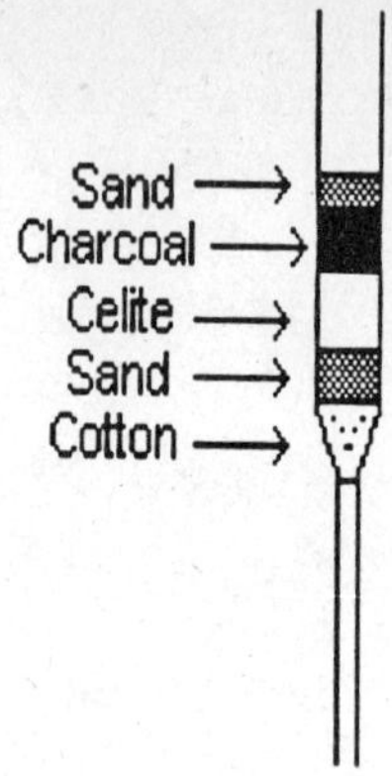

4. Reaction of Furan with Maleic Anhydride.

The reaction between furan and maleic anhydride is slow, and thus it should be set up and allowed to stand until the next laboratory period.

Dissolve 200 mg (*ca* 2.0 mmol) of maleic anhydride in *ca* 0.4 mL of ethyl acetate contained in a 3-mL vial (use marks on vial). Add 200 mg (*ca* 2.4 mmol, 20 microdrops, 0.20 mL) of furan (d = 0.936 g/mL) to the vial and seal the vial with a septum and screw cap. Allow the mixture to stand for at least 18 hours, after which time crystals of the addition product should have formed. Collect the crystals on the Hirsch funnel, wash them in the funnel with a few drops of ethyl acetate, and allow the product to dry. Determine the yield and melting point and record the IR spectrum. The product is not very soluble in carbon tetrachloride and so one of the techniques for determining IR spectra on solids should be used.

Dispose of waste chemicals in labeled waste containers provided in the laboratory.

C. Prelab Questions

1. 1,3-Butadiene is drawn in the *s-cis* (cisoid) conformation in the introduction. Is this the preferred conformation? Explain. Why is the *s-cis* conformation shown?
2. At moderate temperatures, 1,3-butadiene reacts with sulfur dioxide to form an adduct. Give the structure of this adduct.
3. What **kind** of reaction does the "cracking" of dicyclopentadiene represent?

D. Postlab Questions

1. Draw three-dimensional structures for both the *endo* and *exo* adducts formed from 1,3-cyclopentadiene and maleic anhydride. Which stereoisomer is formed? Is this the result of kinetic or thermodynamic control of the reaction? Draw the structures for the adduct of furan and maleic anhydride. Which stereoisomer is formed in this case? Provide a plausible reason for the difference in the stereochemical outcome in these two systems.
2. The relative reactivity of the three dienophiles is: Cyclopentadiene > Butadiene > Furan. How do you account for this order of reactivity?

23. Aromatic Substitution I: Bromination of Aniline, Acetanilide, and Phenol

A. Introduction

Study the material on electrophilic aromatic substitution in your lecture text and notes.

Aromatic compounds are much less reactive than alkenes toward most reagents, yet they do undergo substitution reactions with a variety of **electrophilic** reagents. The specific reaction conditions that bring about substitution depend upon the structure of the aromatic compound and the nature of the electrophilic reagent. However all electrophilic substitutions follow the general mechanism shown below, where E^+ represents any electrophile. The electrophile may be truly ionic such as the NO_2^+ ion, or it may be a strongly polarized reagent due to coordination with a Lewis acid such as $AlCl_3$ or BF_3, or the electrophile may actually become polarized as it reacts with a very reactive aromatic substrate as in the case of bromine reacting with aniline or phenol.

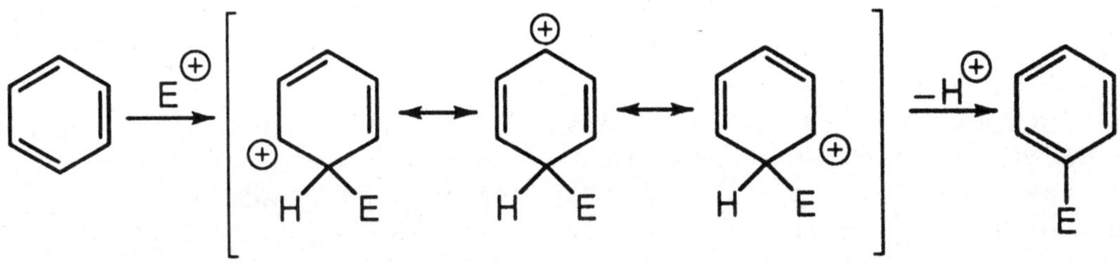

Aniline and phenol are examples of aromatic compounds in which the benzene ring is very strongly activated by electron-donating substituents and both undergo electrophilic aromatic substitution rapidly without the necessity of a Lewis acid catalyst. All strongly activating substituents are *ortho-para* directors, hence the initial bromine substitution is either *ortho* or *para* to the amino group (NH_2) of aniline or the hydroxyl group (OH) of phenol. Because these groups activate the ring so greatly, it is difficult to stop at mono-substitution. With three equivalents of bromine, both compounds undergo substitution in all three *ortho-para* positions.

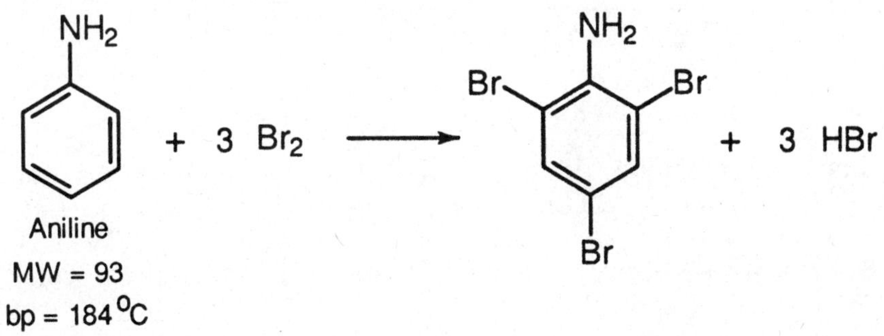

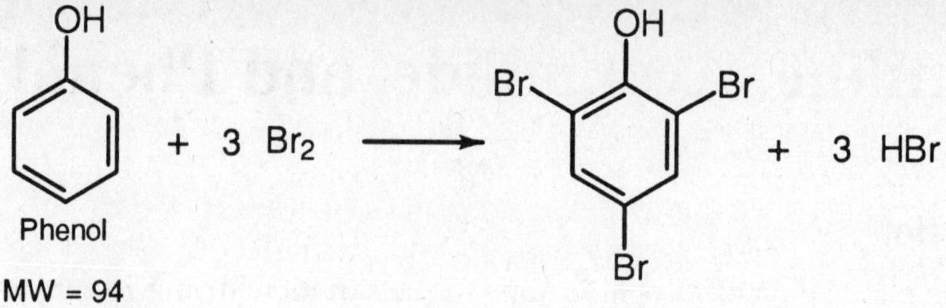

Phenol

MW = 94
bp = 182 °C
mp = 41 °C

Because aniline is an organic base, it reacts with the HBr produced to form the $C_6H_5NH_3^+$ ion.

$$C_6H_5NH_2 + HBr \longrightarrow C_6H_5NH_3^+ + Br^-$$

The $-NH_3^+$ group is a powerful **deactivating** group; thus protonation of the aniline inhibits the reaction. However, if the reaction is performed in acetic acid-water, there is a sufficient amount of free aniline present at equilibrium to effect facile bromination.

If monosubstitution of the ring is desired, the aniline may be converted to acetanilide by reaction with acetic anhydride. This reaction converts the very strongly activating amino group into an acetamido group, $-NHCOCH_3$. The acetamido group, although still an *ortho-para* director, is a much weaker activating group, and thus it is now possible to monosubstitute the ring. Substitution is predominantly *para*, presumably because the ortho positions are sterically hindered by the N-acetamido group.

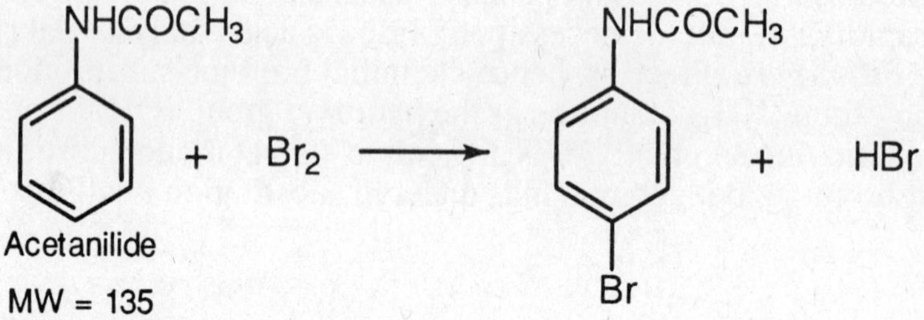

Acetanilide

MW = 135
mp = 114°C

Both aniline and acetanilide will be brominated with a solution of bromine in acetic acid. Bromine (bp 59°C) is very soluble in acetic acid (bp 118°C) and the resulting solutions are quite stable. Acetic acid is a good solvent for many organic substances and it has the advantage in the present case of providing a fairly polar reaction medium which promotes electrophilic aromatic substitution.

The bromination of phenol will be carried out in water using a bromine-acetic acid solution. Phenol itself is fairly soluble in water, but most other phenols are not. If sufficient bromine is used, the product is 2,4,6-tribromophenol, a compound much less soluble that the starting phenol. Formation of this "insoluble" tribromide has been used as a simple way to detect moderately low concentrations of phenol in water (0.001 M or greater).

CAUTION: Bromine is a noxious chemical and very toxic. It must be kept off the skin and must not be inhaled, and the vapors must not be permitted to contact the eyes. Work in a hood. It is advisable to wear gloves and to wear long sleeves when handling bromine or bromine solutions. If bromine, or a bromine solution, is spilled on the skin it must be washed off immediately with a stream of water.

B. Procedure

1. Bromination of Aniline

Weigh 2 microdrops of aniline (*ca* 30 mg) into a 3-inch test tube which is supported in a vial or small Erlenmeyer flask. Add about 0.2 mL of glacial acetic acid and swirl or tap the test tube to dissolve the aniline. When the oil has dissolved, add *ca* 0.4 mL of water and mix well. Then, working in a hood, add 20% bromine in acetic acid dropwise with swirling, or tapping until the yellow-red color persists and there is a definite excess of bromine. Record what happened during the bromination.

Add *ca* 2 mL of water and mix thoroughly. Remove excess bromine by adding a few drops of $NaHSO_3$ solution with thorough mixing.

Collect the solid tribromide on a Hirsch funnel and rinse it well with water. After brief air-drying, transfer the solid to a Craig tube and recrystallize it from 95% ethanol. Use a minimum amount of hot ethanol to dissolve the solid and then set the solution aside to cool, finally using an ice-bath. Collect the product, and after air-drying, record the yield and melting point.

While this product is crystallizing you should start the next experiment.

2. Bromination of Acetanilide

Weigh *ca* 68 mg (0.50 mmol) of acetanilide into a 3-inch test tube and dissolve it in 10–15 microdrops of acetic acid. Add 0.6 mL of a 1.0 M solution of bromine in acetic acid and mix thoroughly. Allow the reaction mixture to stand for about 10 min. At the end of this period there should be an excess of bromine. If not, add more bromine solution dropwise until the bromine color persists. Describe the appearance of the reaction mixture.

Add about 2 mL of water, mix well, and add a few drops of $NaHSO_3$ solution to reduce excess bromine. Again mix thoroughly. The mixture should be white or off-white. Collect the solid on a Hirsch funnel and wash well with cold water (*ca* 1 mL). Air-dry the product briefly, then transfer it to a Craig tube for recrystallization from 95% ethanol. **Use the minimum amount of hot ethanol.** Add the ethanol a few drops at a time while heating (steam or water bath) until the crystals dissolve. Allow the solution to cool to room temperature, and then cool the mixture in an ice-bath. Collect the crystals and wash with a few microdrops of cold ethanol. Air-dry the product and determine the yield and melting point.

3. Bromination of Phenol

Dissolve *ca* 47 mg of phenol in 1–2 mL of water in a 3-inch test tube. Then add dropwise 20% bromine in acetic acid, with thorough mixing, until a permanent bromine color persists. Describe what happens. Add a few drops of aqueous $NaHSO_3$ solution to reduce the excess bromine and mix well. Collect the product on a Hirsch funnel and wash it well with cold water. Air-dry briefly and then recrystallize the solid in a minimum amount of hot ethanol. Collect the product and determine the yield and melting point.

Dispose of waste chemicals in labeled waste containers provided in the laboratory.

C. Prelab Questions

1. Will benzene react with bromine under the conditions used in this experiment? Explain.
2. How will $C_6H_5NH_3^+$ and $C_6H_5N(CH_3)_3^+$ compare in reactivity in electrophilic substitution?
3. Look up the pK_B of aniline in aqueous solution. Discuss what this value means in terms of the base strength of aniline in water. Will aniline appear to be a stronger or weaker base in **acetic acid** as a solvent compared to water as the solvent? Explain.
4. Show why the NH_2 group of aniline and the OH group of phenol are **strongly** activating groups.

D. Postlab Questions

1. Compare the apparent reactivity of aniline, phenol, and acetanilide toward bromine.
2. Show how steric effects can account for the fact that acetanilide gives rise to mainly *para*-substitution.
3. Is acetanilide significantly protonated by the HBr produced? Explain (compare with aniline).
4. Why is the $-NHCOCH_3$ group **less** activating in electrophilic substitution compared to the $-NH_2$ group?

24. Aromatic Substitution II: Nitration of Methyl Benzoate: Relative Reactivity of Aromatic Compounds: A Competitive Nitration

A. Introduction

Nitration is a very important reaction in organic chemistry in terms of both theory and practice. Many of the important concepts concerned with electrophilic substitution originated in studies of nitration. Moreover, the nitro group is particularly valuable as a precursor for aromatic amines, many of which are widely used in commerce.

The first experiment is the nitration of methyl benzoate. When this reaction has been successfully accomplished, a competitive nitration will be carried out between methyl benzoate and acetanilide, the object being to find out which of the two undergoes nitration at the faster rate. This result will provide information as to which group, the -CO_2CH_3 or the -$NHCOCH_3$, has the stronger activating influence on the reaction. These reactions are shown below.

$$HNO_3 + 2\,H_2SO_4 \longrightarrow NO_2^{\oplus} + H_3O^{\oplus} + 2\,HSO_4^{\ominus}$$

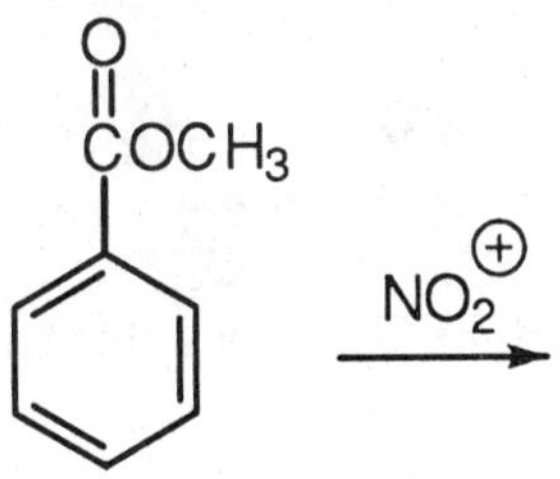
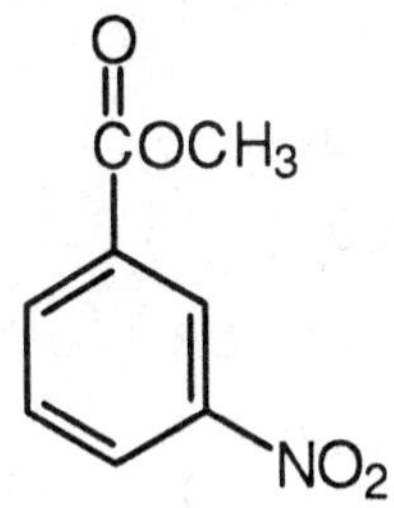

Methyl Benzoate

MW = 136

bp = 200 °C

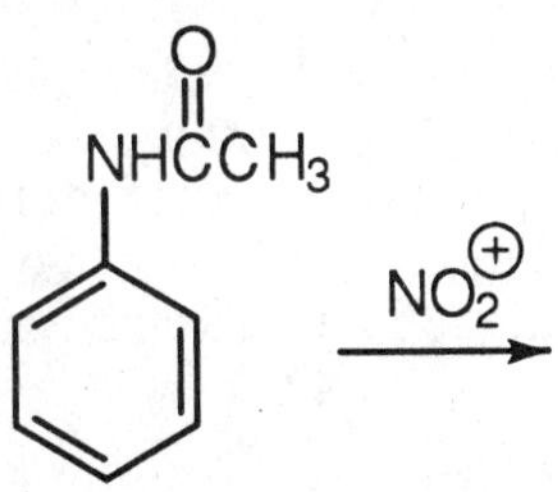
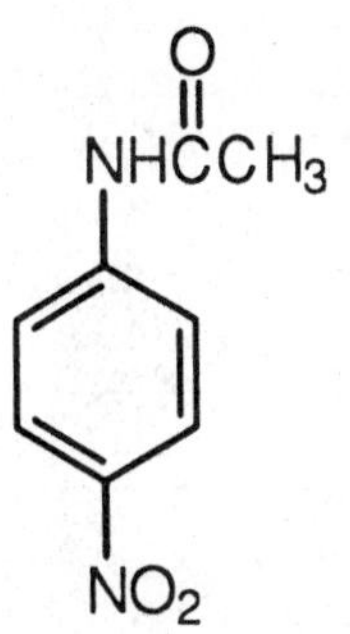

Acetanilide

MW = 135

mp = 114°C

The nitrating reagent is a mixture of nitric acid and sulfuric acid, a combination which generates the nitronium ion, NO_2^+, considered to be the actual nitrating species. The nitronium ion, NO_2^+, is a very reactive electrophile, able to effect nitration not only of activated aromatic rings, but also of many deactivated rings as well. In the present case, the ester group of methyl benzoate, $-CO_2CH_3$, gives rise to a deactivated system which undergoes substitution *meta* to the deactivating group. On the other hand, the acetamido group is an *ortho-para* director (as we found in bromination experiment) and steric hindrance of the ortho position leads to predominantly *para*-substitution.

A small amount of acetic acid is used as an inert co-solvent in order to ensure identical conditions in both experiments. Although the nitration of methyl benzoate alone can be carried out without co-solvent, the added solvent power of acetic acid is needed in the competitive nitration which follows.

CAUTION: Concentrated nitric acid and concentrated sulfuric acid are very corrosive and must be handled with care. If you get either on your skin, immediately wash off the acid with large amounts of running water. To transfer these reagents use ONLY GLASS DROPPERS OR GLASS MICROPIPETS.

B. Procedure

1. Nitration of Methyl Benzoate

Weigh *ca* 100 mg (0.74 mmol; 8 microdrops) of methyl benzoate into a 5-mL conical vial containing a spin vane. Add *ca* 10 microdrops (0.10 mL) of glacial acetic acid and 20 microdrops (0.20 mL) of concd. sulfuric acid.

Half-fill a 100-mL beaker with water cooled to *ca* 10°C with a little ice and place the vial containing the methyl benzoate in the water bath. Center the assembly on the stirrer and stir fairly rapidly, but avoid splashing. The water bath will absorb the heat evolved during the nitration without getting the reaction mixture so cold that it becomes too viscous to stir.

Using micropipets, prepare a "nitrating mixture" by placing 10 microdrops (0.10 mL) of glacial acetic acid, 20 microdrops (0.20 mL) of concd. sulfuric acid and 8 microdrops (0.080 mL) of concd. nitric acid in a 3-mL conical vial. Swirl to ensure complete mixing.

Using a glass micropipet, add the nitrating mixture dropwise over a period of 2 or 3 minutes, keeping the mixture well stirred. When all of the nitrating mixture has been added, remove the water bath and let the solution stir at room temperature for 15 min. Then add a few small pieces of ice. When the ice has melted, adjust the volume to 3 or 4 mL with additional water, if necessary. Record the change in appearance of the reaction mixture.

The product should be a solid (which may be a bit sticky) and should settle to the bottom of the reaction vial, or stick to the spin vane. (Occasionally some of the product will float on the surface.) Because it may be difficult to collect the solid directly by filtration, we will remove the aqueous layer, leaving the product (and spin vane) in the vial. Using a micropipet, carefully draw off the aqueous layer without removing any of the solid. After most of the water has been removed, while holding the vial in one hand, gently tip and rotate the vial so that the liquid flows away from the solid, enabling you to remove almost all of the liquid while leaving the solid in the vial.

Wash the product by adding 2 or 3 mL of cold water. Swirl to ensure good mixing, allow the solid to settle, and again carefully remove the water layer with a micropipet.

Repeat this washing one more time and remove as much water as possible. Then, while holding the vial over the steam bath, dissolve the solid in the minimum amount of hot ethanol (*ca* 1 mL). Using a micropipet, quickly transfer the hot solution to a Craig tube, leaving the spin vane in the vial. Rinse the vial with a few drops of hot ethanol and add this wash to the Craig tube. Reheat if necessary to give a clear solution. Then let the solution cool to allow crystallization to proceed. It may be helpful to generate a few seed crystals to initiate crystallization. If you used an excessive amount of ethanol, you will have to concentrate the solution, or else add water to the hot ethanol solution until it is nearly to the "turbid point." Finally cool the mixture in an ice-bath. Collect the crystals, allow them to dry, determine the yield and melting point, and record the IR spectrum.

2. Competitive Nitration of Acetanilide and Methyl Benzoate

In this experiment, the relative reactivities of methyl benzoate and acetanilide will be determined by performing a competitive nitration employing essentially the same conditions as described above. An equimolar mixture of the methyl benzoate and acetanilide will be treated with a quantity of nitric acid which is **slightly less** than the amount needed to react with only one of the competing compounds. If the compounds have similar reactivities, comparable amounts of two products will result. But if there is a substantial difference in reactivity, only the more reactive compound will be nitrated, giving rise to one major product. Consequently this experimental approach can give a direct indication of relative reactivity.

Because it is essential that a deficiency of nitric acid be used, we will weigh the nitric acid first and then take appropriate amounts of the competing reactants.

Using a glass micropipet, weigh 2 microdrops (*ca* 45 mg) of concd. nitric acid into a 3-mL conical vial. The concd. acid is 70–71% nitric acid and thus 45 mg of concd. acid is equivalent to *ca* 32 mg (0.50 mmol) of HNO_3. Then add 10 microdrops (0.10 mL) of glacial acetic acid and 20 microdrops (0.20 mL) of concd. sulfuric acid. Mix thoroughly.

The quantities of the acetanilide and methyl benzoate must be adjusted to ensure that there is a deficiency of nitric acid for the nitration reaction. The following amounts are based on 45 mg of concd. nitric acid. Weigh 74 mg (0.55 mmol) of acetanilide and 75 mg (0.55 mmol) of methyl benzoate into a second 5-mL conical vial. Then add 10 microdrops (0.10 mL) of glacial acetic acid and 20 microdrops (0.20 mL) of concd. sulfuric acid. Mix well. You should have a clear solution (probably light amber in color).

Half-fill a 100-mL beaker with water cooled to ca 10°C with a little ice. Add a spin vane to the conical vial and place the vial in the beaker of cool water. Center the assembly on the stirrer and stir fairly rapidly, but avoid splashing.

Add the nitric acid solution dropwise with a micropipet over a period of 3 or 4 minutes, keeping the reaction mixture well-stirred at all times. When all of the nitric acid solution has been added, remove the water bath and stir the mixture at room temperature for *ca* 5 min. Then add a few small pieces of ice. When the ice has melted, adjust the volume to 3 or 4 mL with additional water, if necessary. Record the appearance of the reaction mixture. What does the odor of the reaction mixture indicate to you?

Collect the solid by filtration on the Hirsch funnel and wash it thoroughly with several small portions of cold water. Transfer the product as quantitatively as possible to a Craig tube (scrape the product away from the spin vane) and recrystallize it using the minimum quantity of hot ethanol. Allow the solution to cool slowly and then chill it in an ice-bath. Collect the crystals and allow them to dry. Calculate the yield based upon nitric acid as the limiting reagent and determine the melting point.

From the melting point data determine which product was obtained, methyl *m*-nitrobenzoate , or *p*-nitroacetanilide. What can you conclude about the relative reactivities of methyl benzoate and acetanilide?

Dispose of waste chemicals in labeled waste containers provided in the laboratory.

C. Prelab Questions

1. (a) Explain why the -CO_2CH_3 group is deactivating.

   (b) Show why this deactivating group leads to *meta* substitution.

2. Why are concd. nitric acid and concd. sulfuric acid used rather than dilute solutions?

3. Under the conditions we use for nitration of methyl benzoate, is dinitration a concern? Explain.

D. Postlab Questions

1. Explain how NMR spectroscopy should be useful in proving that methyl benzoate undergoes *meta* nitration and acetanilide undergoes *para* nitration.

2. In the competitive nitration, what method(s) of analysis might be useful to determine whether or not any of the nitration product from the less reactive substrate was formed?

3. Suppose nitrobenzene and methyl benzoate were compared in a competitive nitration. What result would you expect to find? Explain.

25. The Grignard Reaction: The Synthesis of Triphenylmethanol

NOTE: It is essential that you prepare equipment for this experiment during the previous laboratory period. Clean the glassware that you will use and allow it to air-dry in your desk. The Grignard reaction is run under anhydrous conditions and absolute dryness is essential. Weigh the magnesium into the 10-mL round-bottom flask.

A. Introduction

Review the material in your lecture text and notes on the preparation, properties and uses of Grignard reagents in organic synthesis. A wide variety of organic chlorides, bromides and iodides react with magnesium to form Grignard reagents. In this experiment, we prepare phenylmagnesium bromide, C_6H_5MgBr, from bromobenzene and then carry out the reaction of this Grignard reagent with methyl benzoate to form triphenylmethanol, $(C_6H_5)_3COH$.

The path of the Grignard reaction between phenylmagnesium bromide with methyl benzoate to give triphenylmethanol is shown in Figure 1.

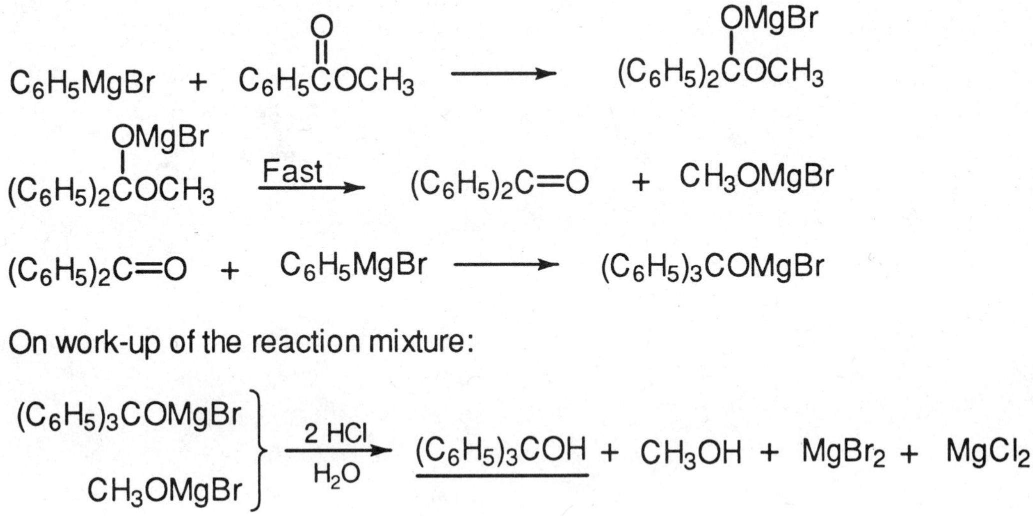

Figure 1. The path of the reaction of a Grignard reagent with an ester.

The initial reaction is the addition of phenylmagnesium bromide to the carbonyl group of methyl benzoate. The intermediate adduct rapidly eliminates CH_3OMgBr to give benzophenone (diphenyl ketone), $(C_6H_5)_2CO$. Since ketones are very reactive toward Grignard reagents, the benzophenone rapidly adds a second equivalent of phenylmagnesium bromide to give the alcohol, in the form of its magnesium salt. Acidic hydrolysis of the magnesium salt with dilute hydrochloric acid converts this salt into triphenylmethanol and also hydrolyzes the CH_3OMgBr to methanol, CH_3OH. Note that it takes **two** moles of the Grignard reagent to react with **one** mole of the ester to give triphenylmethanol.

Benzophenone, $(C_6H_5)_2CO$, can also be used to react with C_6H_5MgBr to form triphenylmethanol. In this case the stoichiometry is one mole of the ketone to one mole of the Grignard reagent.

Success of a Grignard reaction depends upon the total exclusion of water. Water presents two problems: Traces of water will completely inhibit the formation of the Grignard reagent; i.e., the reaction will not "start." Or, once the Grignard reagent is formed, if water enters the system, the water will very rapidly destroy the Grignard reagent by hydrolysis.

$$C_6H_5MgBr + H_2O \longrightarrow C_6H_6 + BrMgOH$$

Thus, it is essential that the equipment be absolutely dry and that all reagents be as nearly anhydrous as possible.

The diethyl ether is particularly critical. It is more than just a "good" solvent. Ethers in general, and diethyl ether in particular, promote the formation of Grignard reagents by forming Lewis acid-base complexes with the magnesium component of the Grignard reagent. While diethyl ether is not very soluble in water, it readily absorbs a moderate amount of water. Consequently the ether must be specially dried. Commercially available anhydrous ether will be provided in the laboratory, and care must be taken to ensure that it does not pick up moisture from the air.

A Grignard reagent reacts instantaneously with oxygen to form oxidized products: first a peroxide and then a salt of an alcohol derived from the Grignard reagent itself.

$$C_6H_5MgBr + O_2 \longrightarrow C_6H_5OOMgBr \xrightarrow{C_6H_5MgBr} 2\ C_6H_5OMgBr$$

In order to avoid oxidation, we will follow the standard research procedure of replacing the air by nitrogen, N_2, which provides an inert atmosphere that not only excludes oxygen, but also prevents moisture in the air from entering the reaction system.

A side product in this experiment is biphenyl (also called diphenyl) C_6H_5-C_6H_5, formed by the coupling of two phenyl radicals at the surface of the magnesium.

$$2\ C_6H_5Br + Mg \longrightarrow C_6H_5{-}C_6H_5 + MgBr_2$$

The biphenyl is formed directly in a reaction between magnesium and bromobenzene, rather than from a reaction of the phenylmagnesium bromide with bromobenzene. Very high concentrations of bromobenzene often increase biphenyl formation. That fact has to be balanced against the fact that if the concentration of bromobenzene is too low, the reaction may stop and be difficult to restart. For that reason, we will employ a moderately high concentration of bromobenzene to ensure that the reaction goes rapidly.

The first step, formation of the Grignard reagent, is usually the most troublesome one in the sequence. The reaction between magnesium and bromobenzene is heterogeneous and heterogeneous reactions are much more subject to inhibition than are homogeneous reactions. Thus, highly purified magnesium (Grignard magnesium) should be used. The bromobenzene and diethyl ether must be pure and anhydrous. Even if all of these conditions are met, it is often difficult to get the reaction to start. One problem is that the surface of the magnesium is coated with a film, generally magnesium oxide, MgO, which inhibits the reaction with the bromobenzene. Crushing or breaking the magnesium with a dry metal spatula immediately before starting the reaction "activates" it.

A method for starting Grignard reactions, which we will use, is to add 1,2-dibromoethane as an initiator. It is more reactive than most other common halides and it leads to no undesired side products. The reaction generates ethylene (ethene) and magnesium bromide.

$$BrCH_2CH_2Br \quad + \quad Mg \longrightarrow CH_2{=}CH_2 \quad + \quad MgBr_2$$

The ethylene escapes as a gas and the $MgBr_2$ remains in solution. The reaction etches the surface of the magnesium, creating active sites which make the magnesium much more reactive toward bromobenzene. This enhanced reactivity is fleeting, so the bromobenzene must be added as soon as it is apparent that the 1,2-dibromoethane is reacting. Therefore, it is advisable to add a little of the bromobenzene to the 1,2-dibromoethylene.

Another way to start a Grignard reaction (which can be tried if the above method fails) is to crush, or break apart, the magnesium with a spatula or a glass rod, while the magnesium is immersed in the ether-bromobenzene mixture. This exposes a fresh surface of magnesium to the action of the bromobenzene. Once started, the mixture turns cloudy, heat is evolved, and the ether boils vigorously. On large scale preparations, but not on microscale, cooling baths must be employed. However, on any scale it is important that the reaction not be allowed to stop. Once reaction has stopped, it may be very difficult to restart.

This experiment introduces techniques used by research workers dealing with air-sensitive materials. Be sure you understand the purpose of each step of the procedure.

B. Experimental Procedures

The equipment for carrying out the Grignard reaction consists of a 10-mL round-bottom flask to which is connected a Claisen head, a water-cooled condenser, and a bubbler to monitor the flow of nitrogen (Figure 2).

The flask (with the magnesium), the Claisen head and the condenser should be placed in an oven at 110°C at the start of the period and heated for at least 15 min.

The nitrogen bubbler is used to regulate the flow of nitrogen and provides a pressure relief valve.

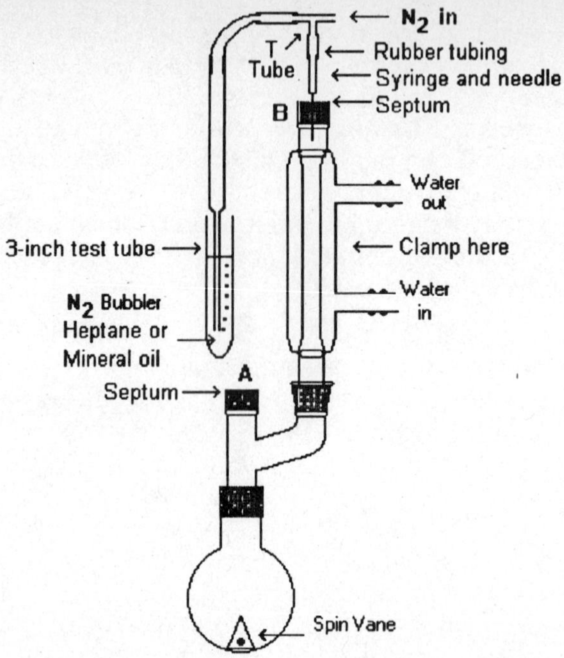

Figure 2. Apparatus for the Grignard Reaction.

Weigh *ca* 53 mg (48.6 mg = 2.00 mmol) of Grignard-grade magnesium into a 10-mL round-bottom flask and then add a spin vane. **(This step should by done during the previous laboratory period.)** Use tweezers to handle the magnesium. **Note: In this procedure, magnesium is the limiting reagent.**

While the equipment is drying, or the system is being flushed with nitrogen (see below), prepare the following solutions in small screw-cap vials. Keep the vials tightly capped in order to prevent moisture from entering. The quantities are based upon 53.0 mg of magnesium (2.18 mmol). Correcting for the magnesium which reacts with 1,2-dibromoethane, leaves *ca* 2.00 mmol of Mg (the limiting reagent) for reaction with the bromobenzene. The quantities of the reagents should be adjusted for the actual amount of magnesium weighed out.

Vial 1. 1,2-Dibromoethane, three microdrops (*ca* 0.18 mmol) and 0.50 mL of anhydrous ether and about 10 drops of the solution prepared in vial 2.

Vial 2. Bromobenzene, 392 mg (2.5 mmol), and 1.5 mL of anhydrous ether.

Vial 3. Methyl benzoate, 136 mg (1.00 mmol) and 1.0 mL of anhydrous ether.

After the apparatus has been dried, and while it is still warm, it should by assembled using screw-caps and O-rings as shown in Figure 2. But first, crush the magnesium repeatedly, using a metal spatula to aid in getting the reaction started. By means of a clamp attached to the condenser, center the flask over the stirrer hot-plate, but *ca* 2 cm above it. It is important that septum A does not leak, in order to prevent ether from evaporating.

Your instructor will give you directions for using the nitrogen system.

To flush the system with nitrogen, turn on the nitrogen and adjust the flow to give a fast stream of bubbles escaping from the bubbler. If you have a leak in the system, you may not get bubbles (check the septa and tighten the lock nuts). Then loosen the locknut at A, enabling the nitrogen to escape from this point (bubbling will cease). Turn on the stirrer and adjust it to give a fairly fast stirring rate; turbulence will aid in flushing the flask. Allow the system to purge for at least 5 min. During this period, attach the water lines to the condenser and pass a slow stream of water through it. Also, complete preparing the solutions in Vials 1, 2, and 3. At this point have your system checked by your instructor. When you are ready to start the reaction, tighten the locknut at A, decrease the nitrogen flow to give a slow bubble rate (1 per sec.) and turn off the stirrer.

Pick up the ether solution of 1,2-dibromoethane (Vial 1) in a **dry 1-mL syringe fitted with a dry hypodermic needle** and add the solution to the reaction vial through the septum at **A.** The reaction should start, but if, after a few minutes, it does not start, remove the cap and septum at **A** and insert a **dry** microspatula with the pointed end down. Working around the spin vane, crush the magnesium chips repeatedly **under the liquid.** The solution should turn cloudy and then as bubbling starts (ethylene escaping) the solution should become clear. When the reaction is clearly underway, remove the spatula and replace the cap and septum. (While the cap was off, nitrogen escaped at A, keeping the system protected from the air). Turn on the stirrer, lowering the assembly if necessary, and start adding the solution of bromobenzene in ether by injecting it at **A** using your 1-mL syringe. Add about 0.2 mL of the solution initially. A vigorous reaction should ensue, causing the ether to boil. Add the rest of the bromobenzene continuously over a period of a 4–5 min. When the rate of boiling slows, turn on the heater of the hot-plate, lower the flask, and heat to reflux. Continue stirring and refluxing until the magnesium has all reacted (or the reaction stops).

At this point, the solution should be clear, probably light amber, containing the Grignard reagent. (There may be fragments of the magnesium remaining.) Then, with continued stirring, add the solution of methyl benzoate, over a minute or two, through septum **A,** using the 1-mL syringe. When the addition is complete, heat the mixture under reflux for 30 minutes. It is important not to boil off the ether during the reflux period. If the ether is lost, more anhydrous ether should be added through the septum at **A.** Record any color changes that occur.

At the end of the heating, turn off the nitrogen, cool the mixture in an ice-bath, remove the cap at A, and slowly add 1 mL of water. Then add 3 M HCl dropwise until all of the solids dissolve. Caution: excess magnesium will react with the evolution of hydrogen. Remove the ice-bath, disconnect the round-bottom flask, and dismantle the remainder of the equipment. There should be a clear ether layer, probably colored, floating on an acidic aqueous layer. Check with pH indicator paper to be sure that the aqueous solution is acidic; if it is not, add more acid. **Note:** If necessary, the experiment can be stopped at this or any subsequent point and completed during the next period.

The following extraction procedure should be carried out with ordinary, not anhydrous, ether. Using a micropipet, transfer the solutions to a 5-mL reaction vial. Rinse the flask with a small amount of ether, which then should be added to the vial. Then use tweezers to transfer the spin vane to the vial. Add sufficient ether to bring the **ether layer** to about 2 mL. Center the vial over the magnetic stirrer (no heat), then stir the contents rapidly, but without splashing for *ca* 1 min. Turn off the stirrer and carefully remove and discard the **lower** aqueous layer with a pipet. Add 1–2 mL of water and stir again for *ca* 1 min. Remove the **lower** layer and repeat the washing procedure one more time. Add *ca* 0.5 mL of saturated aqueous sodium chloride, stir, and remove the lower layer as before.

Prepare a drying tube by packing about 3 cm of anhydrous sodium sulfate into a micropipet that has been plugged with a small wad of cotton. Transfer the ether solution to the drying tube using a micropipet. **Caution: Flush the ether in and out of the pipet a few times to saturate the atmosphere in the pipet to avoid leaking during the transfer.** Be sure that you are not transferring any droplets of water. Collect the effluent in a 5-mL conical vial. Rinse the drying column with a small amount of ether and collect the rinse in the same vial.

Add a micro boiling stone and **carefully** evaporate most of the ether on the steam bath, or hot-plate set to a very low temperature. It is helpful to use a stream of nitrogen to aid evaporation. Transfer the residual oil to a Craig tube and wash out the vial with a small amount of ether to complete the transfer.

Using a stream of nitrogen, carefully evaporate most of the ether. Then add about 1 mL of hexane (or ligroin); a solid should form. Warm the mixture carefully, but avoid bumping. Depending on how much product you have, the solid may not all dissolve, but warming for a few minutes will "digest" it, allowing impurities to dissolve. Cool in an ice-bath and isolate the product by centrifugation, but save the mother liquor. Determine the yield and melting point of the product and record the IR spectrum. If the melting point is not sharp, the triphenylmethanol can be recrystallized from hexane or ligroin.

If there is time available, evaporate the mother liquor to isolate the by-products, mainly biphenyl. Biphenyl is reported to melt at 72°C. Record the weight and melting point.

Dispose of waste chemicals in labeled waste containers provided in the laboratory.

C. Prelab Questions

1. Write reactions for the treatment of phenylmagnesium bromide with: water, D_2O, methanol, oxygen.
2. Write the structural formula of the phenyl radical. How might it be formed? How could it give rise to biphenyl?
3. Give the structural formula of the Lewis acid-base complex which consists of one mole of phenylmagnesium bromide and two moles of diethyl ether.
4. Write an equation for the reaction between the excess magnesium and hydrochloric acid.

D. Postlab Questions

1. Devise a Grignard synthesis to make: (a) (*para*-$(CH_3C_6H_4)_2C(OH)C_6H_5$ (b) $(C_6H_5)_2C(OH)$(*para*-$CH_3C_6H_4$)
2. How could you show experimentally that $(C_6H_5)_2CO$ is or is not more reactive than $C_6H_5CO_2CH_3$ toward C_6H_5MgBr?
3. In washing the ether solution, why is saturated sodium chloride used rather than pure water?
4. Treatment of triphenylmethanol with concd. sulfuric acid is known to give rise to a stable carbocation. **Explain.**

26. Chromic Acid Oxidation of *para*-Nitrotoluene to *para*-Nitrobenzoic Acid

A. Introduction

Chromium in its highest oxidation state, Cr(VI), provides several very useful oxidizing agents. Chromium trioxide (CrO_3), sodium dichromate ($Na_2Cr_2O_7$), and potassium dichromate ($K_2Cr_2O_7$) are relatively inexpensive compounds which may be used as sources of Cr(VI). When used as an oxidant for organic substrates in acidic solutions, Cr(VI) is reduced to Cr(III), a three-electron change.

$$CrO_3 + 6\,H^+ + 3\,e^- \longrightarrow Cr^{3+} + 3\,H_2O$$

Oxidations with Cr(VI) are striking because the yellow-orange color of Cr(VI) is replaced by the blue-green color of Cr(III). The actual Cr(VI) species present depends upon the solvent and the acidity. In basic solutions it is the yellow chromate ion (CrO_4^{2-}) while in moderately acidic solutions the $HCrO_4^-$ ion and the yellow-orange $Cr_2O_7^{2-}$ ion are in equilibrium. In strongly acidic solutions H_2CrO_4 is the main species present. At high concentrations of sulfuric acid, chromium trioxide (also called chromic anhydride) separates as orange-red crystals.

Several Cr(VI) complexes have found use in organic chemistry. Pyridine-chromium trioxide (C_5H_5N-CrO_3) and pyridinium chlorochromate, PCC ($C_5H_5NH^+$, $ClCrO_3^-$) are two widely used oxidants. The latter is useful because, unlike other chromium oxidizing agents, it oxidizes primary alcohols to aldehydes without the subsequent oxidation of the aldehydes to carboxylic acids.

In this experiment we use Cr(VI) to oxidize *para*-nitrotoluene to *para*-nitrobenzoic acid in acetic acid solution. Sulfuric acid is added to $Na_2Cr_2O_7$ to generate H_2CrO_4 which acts as a powerful oxidizing agent. This oxidation is very fast in acetic acid solutions and care must be taken to ensure that it does not get out of hand.

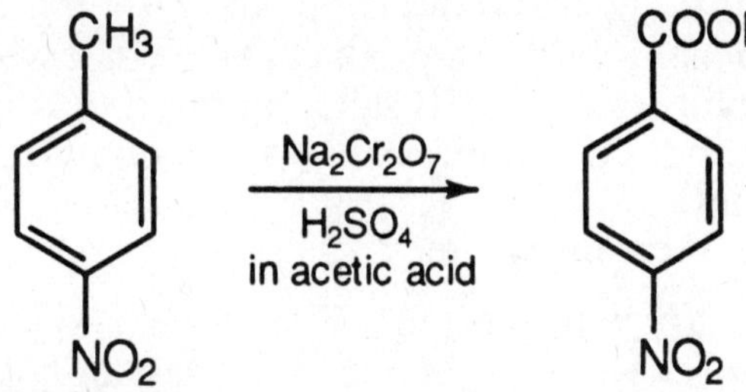

MW = 137
mp = 54 °C

The mechanism of oxidation is not established with certainty, but it appears that the first step may involve H-atom transfer from a benzylic C-H to a Cr(VI) species. A radical, if formed, would be immediately oxidized to a carbocation which would then be captured by some anion. Once the group is "functionalized", subsequent oxidation would be very fast (as it is with alcohols).

B. Experimental Procedures

Be sure your safety goggles are fitted tightly.

Note: The dust of Cr(VI) salts should not be inhaled or ingested because Cr(VI) species are toxic and are also believed to be carcinogenic. All Cr residues (solutions) are to be collected in the waste bottles provided in the laboratory.

Weigh 137 mg (1.00 mmol) of *p*-nitrotoluene into a 50-mL Erlenmeyer flask. Add 400 mg (1.34 mmol) of $Na_2Cr_2O_7 \cdot 2H_2O$ and then 2.0 mL of glacial acetic acid. Heat the flask on a steam cone or hot-plate, at *ca* 100°C, with swirling to effect solution.

Add 0.7 mL of concd. H_2SO_4 **(caution)** to a 3-inch test tube supported in a test tube rack or a 50-mL Erlenmeyer flask.

Allow the *p*-nitrotoluene solution to cool briefly and then, using a micropipet, add 3 or 4 drops of sulfuric acid and swirl the flask carefully to mix the reactants thoroughly, but avoid splashing. Add the remainder of the concentrated H_2SO_4 **VERY SLOWLY (DROP-WISE)** to the hot acetic acid solution over a period of 3 or 4 min. **Do not add the sulfuric acid too rapidly or the reaction may become violent and "take off" out of the flask.** Record the color change. At some point, solids will separate, but continue to swirl the reaction mixture.

After the addition of sulfuric acid is complete, heat the flask for another 5 min. with occasional swirling. Then add *ca* 0.2 mL of ethanol dropwise with swirling. The ethanol will reduce excess Cr(VI) to Cr(III).

Next, add water in small portions until the flask is *ca* three-fourths full and continue to heat with occasional swirling for *ca* 10 minutes. This process of "digestion" improves the crystallinity and purity of the solid product which is not very soluble in the hot solution.

Cool the flask, allow it to stand at room temperature for a few minutes, and collect the solid product on a Hirsch funnel. Wash the product thoroughly with cold water on the Hirsch funnel to remove all of the blue-green Cr(III) salts. Turn off the vacuum and add a small amount of water to wash the sides of the funnel and to cover the solid. Then turn on the vacuum again to draw off the wash water. Repeat this process until the product is nearly colorless (perhaps off-white or cream-colored).

Air-dry the product briefly, transfer it to a Craig tube, and recrystallize it from 95% ethanol; the product should dissolve in about one mL or less of hot ethanol. The solution should be free of suspended insoluble material, but if it is not, filter the hot solution through a cotton plug packed in a micropipet and return the hot filtrate to the Craig tube.

Allow the solution to cool slowly to room temperature and then place it in an ice-bath. Collect the crystals by centrifugation, allow them to dry, and determine the yield and melting point. Because the melting point is so high, you can save time by heating rapidly until the temperature approaches 200°C, then cutting the heating rate back to raise the temperature more slowly. Because the acid is not soluble in CCl_4 or $CHCl_3$, the IR spectrum must be obtained on a "mull" of the solid.

Dispose of waste chemicals in labeled waste containers provided in the laboratory.

C. Prelab Questions

1. Write the balanced half reaction for $Cr_2O_7^{2-}$ going to Cr^{3+} in acidic solution.
2. Ethanol is added to the reaction mixture to reduce excess Cr(VI) to Cr(III). What products can be formed by oxidizing ethanol? What will happen to these compounds in the work-up?
3. Suggest a means for synthesizing *para*-nitrotoluene which should be commercially feasible.
4. Which system is more strongly acidic: 0.7 mL of concd. H_2SO_4 in 2 mL of glacial acetic acid, or 0.7 mL of H_2SO_4 in 2 mL of water? Explain.

D. Postlab Questions

1. Since our oxidizing conditions involve hot sulfuric acid solutions, should we expect sulfonation to be a side reaction in the oxidation we performed? Explain.
2. Suppose you had standard 0.1000 **M** NaOH, an indicator, such as phenolphthalein, an analytical balance, and a micro-buret available. Explain how you could determine the (apparent) purity of the *para*-nitrobenzoic acid you synthesized.
3. What products would be formed from the Cr(VI) oxidation of the following compounds using the procedure outlined in this experiment?

(a) *meta*-xylene

(b) *para*-methylbenzyl alcohol

(c)

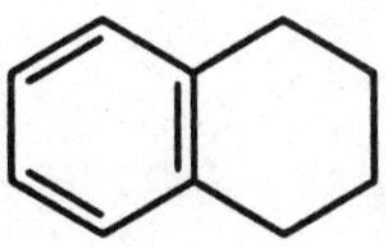

(d) 2,4,6-trinitrotoluene

(e) *para-t*-butyltoluene

(f) styrene

27. Oxidation of Alcohols by a Chromate-Impregnated Resin: Following a Reaction by TLC

A. Introduction

In this experiment, 9-hydroxyfluorene is oxidized by chromic acid to the ketone, 9-fluorenone. Alternatively, under the same conditions, benzoin is oxidized to the yellow α-diketone, benzil (Figure 2). A primary purpose of the experiment is to learn how to follow a reaction by thin-layer chromatography.

The oxidizing agent is chromium trioxide that has been impregnated into an ion exchange resin (Amberlyst A-26). Amberlyst A-26 is a cross-linked insoluble form of polystyrene with quaternary ammonium groups attached to the phenyl rings of the polystyrene. When the resin is treated with an aqueous solution of chromic acid, H_2CrO_4 ($CrO_3 + H_2O$), the quaternary ammonium groups of the resin form salts with the chromate ion (CrO_4^{2-}). The alcohol to be oxidized is dissolved in toluene and the solution is heated under reflux with the chromate impregnated resin. The alcohol diffuses into the resin, the oxidation takes place, and the product ketone diffuses out. The advantage of using a solid phase oxidizing agent is that when the reaction is over, the inorganic products of the oxidation, such as salts of chromium, are easily separated from the reaction mixture by drawing off the solution from the resin. The reaction mixture is simply drawn off from the resin using a micropipet. The oxidations of 9-hydroxyfluorenene and benzoin are shown in Figure 1.

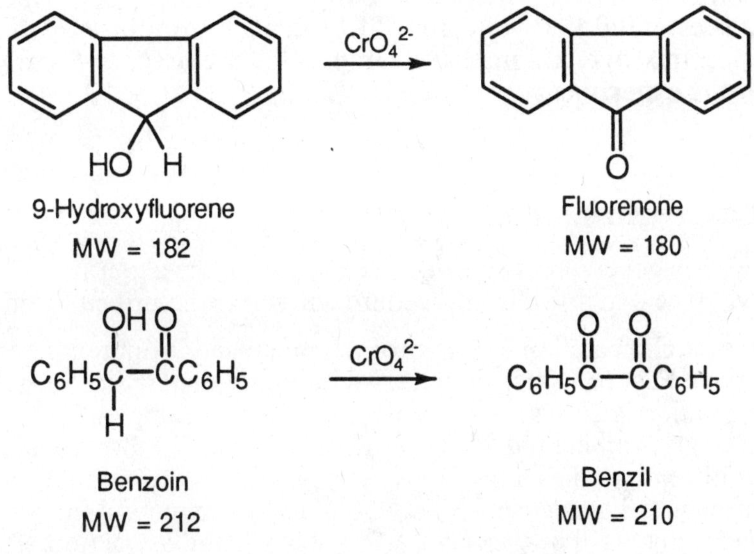

Figure 1. Oxidation of 9-hydroxyfluorene and benzoin.

A major objective in this experiment is to learn how to follow the extent of the reaction while it is in progress. For every reaction, a decision must be made as to the length of time to allow in order to ensure that the reaction proceeds to completion. Often this is done by guesswork based on experience. However, if we measure the extent of reaction at various times during the course of the reaction, this information will allow a determination of the time of completion. This knowledge will prevent a premature termination of the reaction before completion, and will save time because the reaction need not be allowed to proceed longer than is necessary. If we find that the reaction does not go to completion, this result might indicate that there was not enough of some reagent. In this event, more reagent could be added and the reaction continued.

One of the best and simplest methods for following a reaction while it is proceeding is thin-layer chromatography (TLC). The starting material, 9-hydroxyfluorene, is easily separated from the product, 9-fluorenone, by TLC. Similarly the product benzil is easily separated from benzoin by TLC. In each case the R_f values for reactant and product are quite different because the starting alcohol is fairly polar, as are most alcohols, whereas the product ketone is of lower polarity. Thus, by running TLC analyses at various stages of the reaction and by observing the relative sizes of the spots for reactant and product, the degree of conversion of reactant to product at various times can be roughly ascertained.

In this procedure you periodically remove a sample of the reaction mixture and then run the TLC immediately. A sample taken before the reaction has started should show only one spot, that due to the starting material. As the reaction proceeds under refluxing toluene, another TLC is run on a new sample. In the new TLC, the spot for the product ketone, should appear together with a spot for the alcohol. As the reaction continues, and more TLCs are run, the spot for the product should get larger, while the spot for the starting material should get smaller. When the TLC shows only a spot for the product, with the spot for the starting material having completely disappeared, the reaction is complete and may be worked up.

Note: You are following the progress of the reaction to determine when it is complete. Thus, it is essential that you run a TLC on each sample as soon as it is removed from the reaction mixture. Do not save samples to run later; to do so you would miss a major point of the experiment.

B. Procedure

1. Thin-Layer Chromatography

Review the general procedure for TLC given in the experiment "Thin-Layer Chromatography." The details of this procedure will not be described again here.

For the TLC separation of 9-hydroxyfluorene and 9-fluorenone (or benzoin from benzil) 2 x 6 cm Eastman silica gel plates with fluorescent indicator will be used. The developing solution is 10% acetone in hexane. To take samples of the reaction mixture, a thin section of glass tubing (*ca* 25 cm x 4 mm OD) that fits through the bore of the water condenser will be needed. To remove a sample of the reaction mixture, pass the tube **carefully** through the condenser (Figure 2), until it is immersed in the reaction mixture. Some of the solution will be drawn into the tube by capillary action. Remove the tube and touch a **capillary pipet** to the solution caught in the end of the tube. The solution will be drawn into the capillary by capillary action. Return the excess in the tube to the flask. Spot the plate with this solution and, about a cm away, using a **clean** capillary pipet, spot the plate with a methylene chloride solution of the starting material. Place the plate in the developing chamber and allow the developing solution to rise 2 or 3 cm. Remove the plate, mark the solvent front, allow the plate to dry, and examine it under the UV lamp (short wave length). Mark the spots in the usual way.

To reuse a capillary pipet, clean it by dipping the end in toluene and then drawing the toluene out by touching the tip of the pipet to a piece of filter paper. Repeat the procedure 2 or 3 times.

2. Oxidation of 9-Hydroxyfluorene (or Benzoin)

Preparation of chromated resin: Suspend Amberlyst A-26 ion exchange resin in a 15% aqueous solution of chromic acid. Stir for about 30 min and collect the resin by suction filtration. Wash the resin several times with water and then with acetone. Allow the resin to dry and it is then ready for use. There will be some loss of weight due to water being removed from the resin. The aqueous chromic acid may be saved for use at a later date. The chromated resin may be reactivated after use.

To use as a standard for TLC analysis, place a few milligrams of starting material in a 3-inch test tube, or a small vial, and dissolve it with a few drops of methylene chloride. Place a capillary pipet in the test tube to use in spotting the starting material onto the TLC plates. The reaction is carried out in a 10-mL round bottom flask with a water cooled reflux condenser as shown in Figure 2.

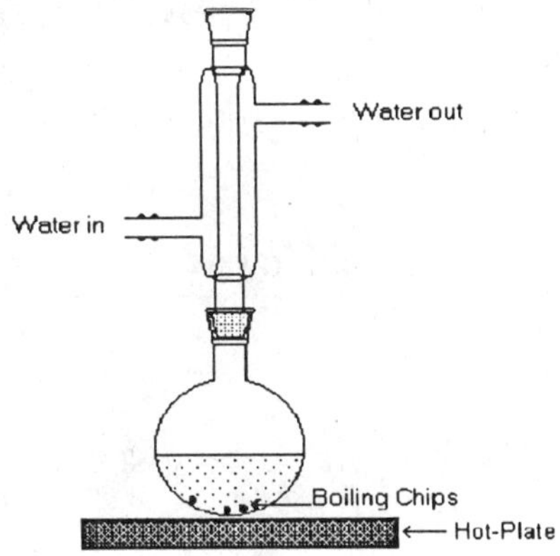

Figure 2. Heating a reaction under reflux.

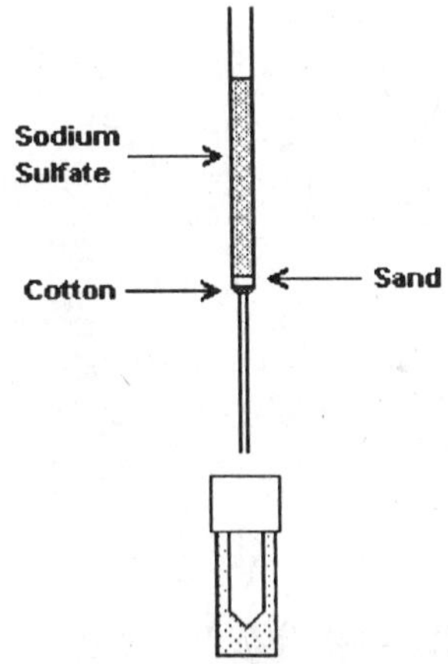

Figure 3. Na_2SO_4 drying tube.

Place 100 mg (0.55 mmol) of 9-hydroxyfluorene (or 100 mg, 0.47 mmol, of benzoin) and 3.5 mL of toluene in the flask together with 1.0 g of the chromated resin and one or two boiling chips. Attach a water-cooled reflux condenser and heat to reflux. Five minutes after refluxing commences, take a sample of the reaction mixture for TLC analysis. While the reaction is in progress, spot a fresh TLC plate with a sample of the starting material. About 0.5 cm away place a spot of the sample from the reaction mixture. Run this TLC. Continue to heat under reflux and after 10 minutes, sample the reaction mixture again and run the TLC together with the standard as before. Note the time that each sample is taken **using pencil** at the top of the TLC plate. Continue in this fashion and sample the reaction mixture every 10 minutes. The interval does not have to be exactly 10 minutes, but be sure to mark the time when you actually take the sample.

When the spot for the starting material no longer appears on the TLC plate, or is very faint, the conversion of reactant to product is complete. The reaction will probably take 20–40 min.

3. Isolation of the Product

When the reaction is judged to be complete, allow the reaction mixture to cool and then, using a micropipet, remove the liquid from the resin. An easy way to do this is to place the tip of the pipet on the bottom of the flask, withdraw the liquid carefully so as not to draw up any of the resin, and place it in a 5-mL vial. Alternatively, a filter pipet may be used. If done properly, the particles of resin will not be drawn up with the liquid. Wash the resin two times with 0.5-mL portions of methylene chloride, draw off the methylene chloride with the pipet in the same manner, and add the washings to the reaction mixture in the vial.

Prepare a sodium sulfate drying tube as shown in Figure 3. Pass the solution of the product through the column, catching the effluent in a small vial or evaporator. Evaporate the solvent using mild heat from a hot plate under a slight stream of nitrogen. When the volume is very small, remove the vial from the hot plate and stir with a spatula, while maintaining a stream of nitrogen. When the solid appears dry and powdery, remove it and obtain a crude weight and percent yield. Recrystallize 50–60 mg from about 1–2 mL of hexane using the Craig tube and determine the melting point and the percent recovery on recrystallization. Measure the R_f values of reactant and product from two or three of the TLC plates. The values found for each compound should agree closely. Record the infrared spectrum of the recrystallized product.

Place your "spent" chromated resin in the bottles provided for this material. The resin can be reactivated by your instructor.

Dispose of waste chemicals in labeled waste containers provided in the laboratory.

C. Prelab Questions

1. Write the structure of the initial oxidation products of the following alcohols: 1-butanol, 2-butanol, cyclohexanol. Which of the three would be oxidized to an acid?
2. How will you know when the oxidation is complete? Give specific details.
3. In running a TLC, why is it absolutely essential that the level of the developing liquid be below the initial spot on the TLC plate?
4. Why is it necessary to measure the distance that the solvent moves on the TLC plate?

D. Postlab Questions

1. Write the half-reaction for the reduction of the chromic acid (H_2CrO_4) in acidic solution.
2. From the relative sizes of the TLC spots, estimate the degree of conversion of the reaction when each sample was taken. Estimate the time at which the reaction was 50% complete.
3. Assuming that the chromium goes to chromium hydroxide, $Cr(OH)_3$, write a balanced equation for the oxidation of 2-propanol to acetone by CrO_4^{-2}.
4. Cite the evidence you have obtained which establishes that the oxidation actually occurred.

28. The Haloform Reaction. Oxidation of Aryl Methyl Ketones to Aromatic Carboxylic Acids

A. Introduction

Many aryl methyl ketones are readily available through the Friedel-Crafts acetylation of aromatic compounds. For the case of acetophenone:

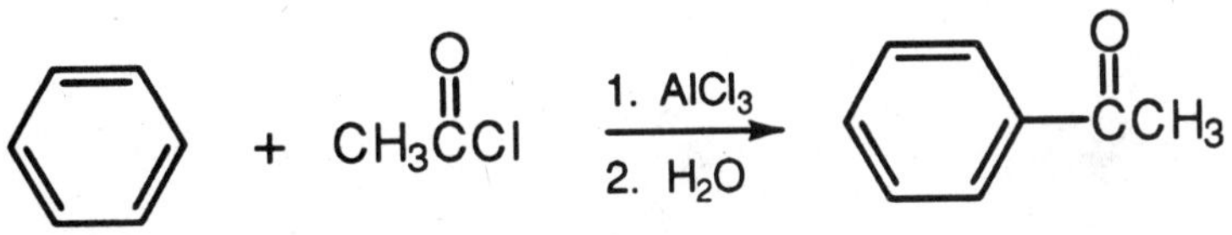

These aryl methyl ketones can be oxidized with sodium hypochlorite (NaOCl) to give aryl carboxylic acids:

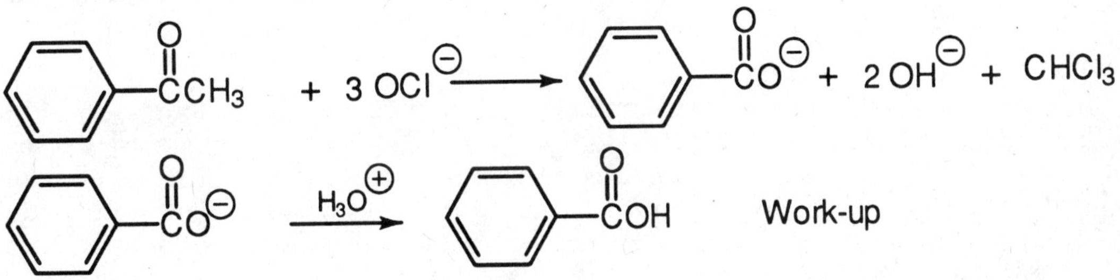

For small-scale preparations, common "home bleach" is a suitable source of sodium hypochlorite. It is prepared by passing chlorine into aqueous sodium hydroxide. Most commercial bleaches contain *ca* 5% by weight of NaOCl, giving *ca* 0.7 M NaOCl.

$$Cl_2 + 2\,OH^{\ominus} \longrightarrow OCl^{\ominus} + Cl^{\ominus} + H_2O$$

A solution of NaOCl contains a low (equilibrium) concentration of Cl_2 which is the actual chlorinating agent. The ketone reacts with hydroxide ion to generate an enolate ion which reacts rapidly with Cl_2 to give a chloroketone.

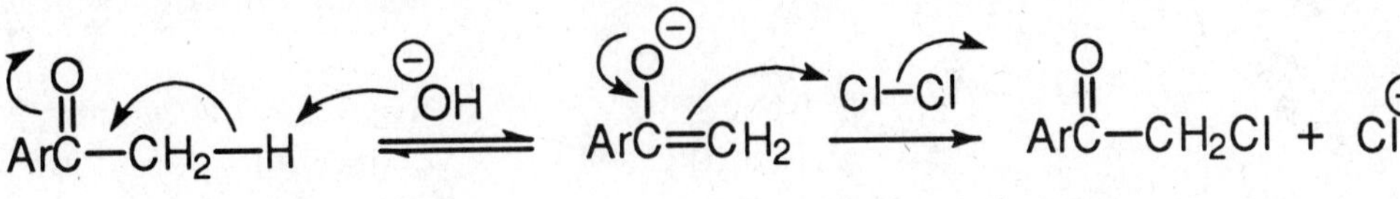

The enolization-chlorination sequence is repeated twice to give a trichloromethyl ketone.

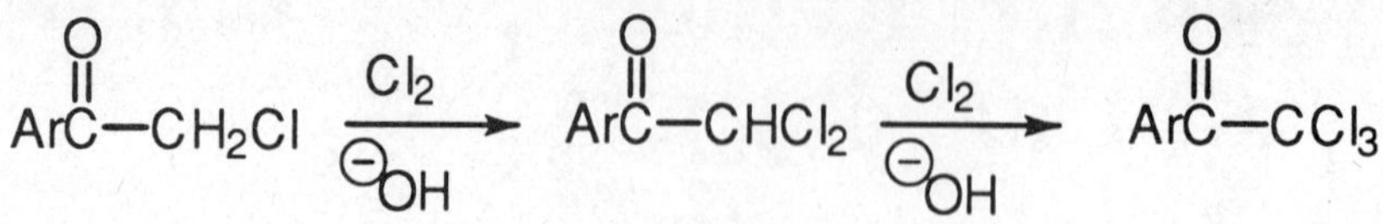

Finally, reaction of the trichloromethyl ketone with hydroxide ion gives an intermediate which cleaves by expelling the trichloromethyl anion to form the arylcarboxylic acid. (The $^{-}$:CCl_3 anion is stabilized by the electronegative Cl atoms.) A subsequent fast acid-base reaction gives the carboxylate anion and chloroform.

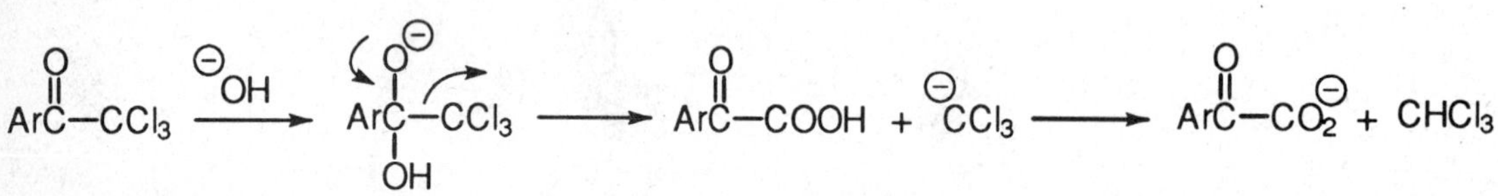

Bromine or iodine can be used in place of chlorine. This oxidation is called the **haloform reaction** because chloroform is formed from chlorine, bromoform from bromine, and iodoform from iodine.

An excess of NaOCl is used to ensure that the reaction goes to completion. In addition, a small amount of 1,2-dimethoxyethane, $CH_3OCH_2CH_2OCH_3$, is added to increase the solubility of the aryl methyl ketone in the aqueous system. This solvent, the dimethyl ether of ethylene glycol (sometimes called "glyme" or "monoglyme"), is miscible with water and is a good solvent for organic compounds. When highly insoluble aryl methyl ketones are oxidized, this cosolvent greatly accelerates the rate of reaction.

B. Procedure

1. Oxidation of an Aryl Methyl Ketone

Each student will receive *ca* 0.1 g of an unknown aryl methyl ketone to be oxidized by the following procedure. The molecular weight will be in the range of 130–170. You are to isolate the aromatic carboxylic acid and determine the yield and melting point. (You can calculate a yield without knowing the structure because the molecular weights of $ArCOCH_3$ and ArCOOH are almost identical.) The infrared spectrum will confirm the presence of the carboxylic acid group. The procedure is based on an assumed molecular weight of 130 to ensure that an excess of NaOCl is used in all cases.

Preheat the hot-plate to about 120°C so it will be ready when it is required to heat the reaction mixture. Weigh *ca* 100 mg of the unknown aryl methyl ketone into a 50-mL Erlenmeyer flask containing a spin vane. Then add 0.5 mL of 1,2-dimethoxyethane followed by 0.4 mL of 10% NaOH and 4.0 mL of 5% NaOCl.

Place the flask on the heated hot-plate and stir the mixture at a moderate rate. Initially the mixture will be turbid because the ketone is not very soluble. Continue heating and stirring for *ca* 30 min. By this time there should be a clear solution, indicating that the reaction is complete.

Test the solution with starch-iodide paper to be sure that the NaOCl is in excess. First, cut a strip of starch iodide paper into six or eight pieces. Moisten a piece of the test paper with dilute acetic acid and then add a microdrop of the reaction mixture to the acidified test paper; a melting point tube is useful as a micro stirring rod (use the sealed end). A deep blue color indicates excess $Cl_2(OCl^-)$. If a negative test is obtained, add more NaOCl in small increments until the solution tests positive.

Allow the solution to cool slightly and add aqueous $NaHSO_3$ to reduce the excess Cl_2 (OCl^-) to chloride, testing with the starch iodide paper to follow the reaction. When a negative test has been obtained, add 6 **M** HCl dropwise with stirring until the solution is strongly acidic, as indicated by pH paper or Congo Red paper (which turns blue at pH values below *ca* 3).

Heat the mixture to near-boiling for a few minutes to "digest" the precipitate and give better crystallinity. Then allow the solution to cool, first to room temperature and then in an ice-bath. Collect the product on a Hirsch funnel and wash it well with several small portions of ice-water to remove inorganic salts. Air-dry the aromatic carboxylic acid (or dry it in an oven at 70°C for 10–15 min.), determine the yield and melting point, and record the infrared spectrum. (Most of these acids can by recrystallized from ethanol, but this should not be necessary if the aryl methyl ketone was pure.)

2. The Iodoform Test for Methyl Ketones

When the haloform oxidation is carried out with iodine rather than chlorine, iodoform, CHI_3, a yellow solid, mp 120°C is formed. Formation of iodoform indicates that a compound is a methyl ketone (or an alcohol that can be oxidized to a methyl ketone). Iodine (I_2) is only very slightly soluble in water (*ca* 0.001 **M**), but it is readily soluble in aqueous iodide solution due to the reversible formation of the red-brown triiodide ion, I_3^-:

$$I_2 + I^- \longrightarrow I_3^-$$

Instructor preparation: The I_2/KI reagent used in this experiment is prepared by dissolving 10 g of I_2 and 20 g of KI in 100 mL of water.

Dissolve 2 or 3 microdrops of acetophenone (or an unknown) in 0.5 mL of glyme and add 1 mL of 10% NaOH. Then add I_2/KI reagent dropwise until a definite dark color of iodine persists. If no iodoform separates, warm the mixture in hot water (60°C) for a few minutes. If the color fades, add more of the iodine reagent. Cool and remove excess iodine by the addition of a few drops of aq. NaOH and *ca* 1 mL of water. At this point you should have a yellow precipitate. Collect the yellow solid, wash it well with cold water, allow it to dry, and determine the melting point.

Although in modern research laboratories IR and NMR spectroscopy have largely replaced many traditional chemical tests, the iodoform reaction is still useful as a confirmatory test.

Dispose of waste chemicals in labeled waste containers provided in the laboratory.

C. Prelab Questions

1. What is the oxidation state of chlorine in hypochlorite?
2. 1-Phenylethanol gives a positive iodoform test. Write out the steps to explain the reactions which occur.
3. What simple chemical test(s) could you do to establish that the product you obtain is a carboxylic acid and not something else?

D. Postlab Questions

1. Treatment of *ortho*-xylene with CH_3COCl and $AlCl_3$ followed by hydrolysis gives virtually a single product (one isomer). Treatment of this product with NaOCl in base followed by acidification gives a crystalline solid. Write equations for these reactions.
2. Phenyl ethyl ketone, $CH_3CH_2COC_6H_5$, fails to give a positive iodoform test. Does this result mean that this compound does not react at all with the I_2/KI reagent? Explain.

29. Synthesis of Isopentyl Acetate by Fischer Esterification

A. Introduction

When a carboxylic acid is heated with an alcohol in the presence of a strong acid, an ester is formed, a process called Fischer esterification. The general reaction for the process is shown in Figure 1A. In this experiment we will prepare isopentyl acetate from acetic acid and isopentyl alcohol (Figure 1B).

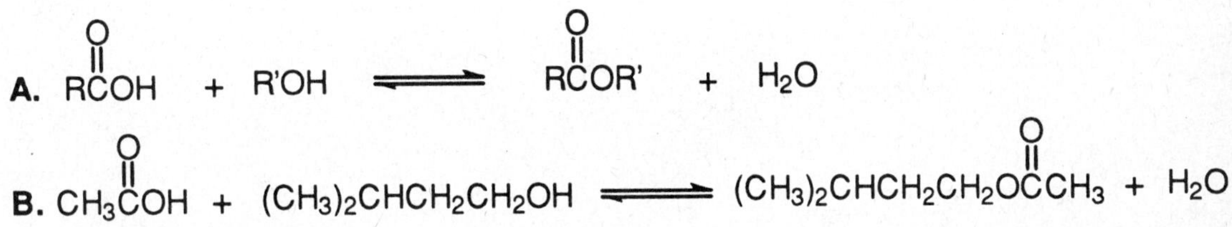

Figure 1. The Fischer esterification.

The mechanism of the Fischer esterification is outlined in Figure 2. The major role of the acidic catalyst is to "activate" the carbonyl group of the carboxylic acid toward addition of the alcohol, but it also promotes the dehydration of the intermediate ester hydrate $RC(OH)_2OR'$. All of the steps are reversible as shown.

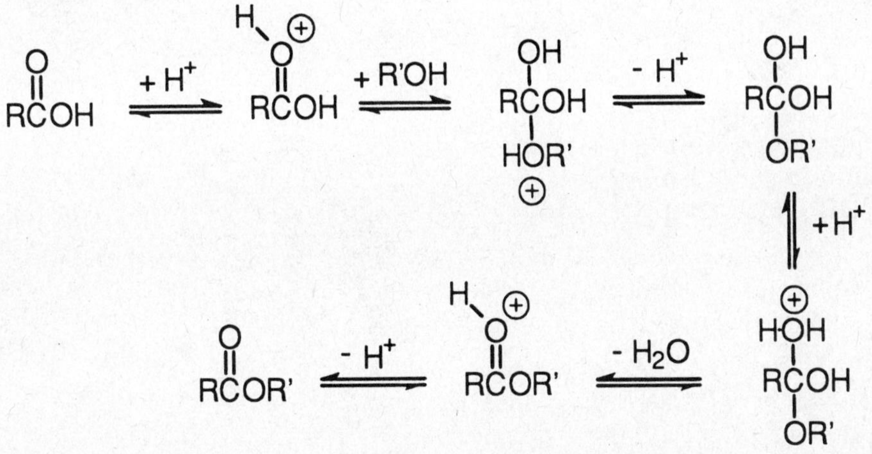

Figure 2. The mechanism of the Fischer esterification.

The Fischer esterification is an equilibrium process. The position of the equilibrium and the rate at which equilibrium is attained are dependent upon the structures of the carboxylic acid and the alcohol. The rate of equilibration is also dependent upon the temperature and upon the nature and concentration of the acid catalyst. In practice, it is possible to shift the equilibrium toward the desired ester by utilizing the Le Chatelier principle. If one of the products of the reaction is removed, the equilibrium is shifted to the right. For example, in esterification, removal of the water shifts the equilibrium to the right and thus increases the yield of the product ester. Alternatively, a large excess of one of the reactants, usually the less expensive one, also shifts the reaction to the right and increases the yield of the ester product. We will utilize the second approach by incorporating into the reaction a large excess of the less expensive reactant, acetic acid (about 4:1 molar ratio of acetic acid to isopentyl alcohol) to increase the yield of isopentyl acetate. The excess acetic acid is easily removed in the work-up of the reaction mixture. The ester will be purified by distillation using a Hickman still and subsequently analyzed for purity by gas chromatography.

Many other routes to esters are available, e.g. by the reaction of alcohols with carboxylic acid anhydrides or acid chlorides. These reagents can lead to virtually quantitative formation of esters. However, the Fischer route remains extremely important for both "theoretical" reasons (*viz.*, to study the mechanism of the reaction) and for the fact that many useful esters are easily prepared from readily-available reactants.

As noted above we will prepare isopentyl acetate from acetic acid and isopentyl alcohol:

$$CH_3\overset{\overset{O}{\|}}{C}OH + (CH_3)_2CHCH_2CH_2OH \longrightarrow CH_3\overset{\overset{O}{\|}}{C}OCH_2CH_2CH(CH_3)_2 + H_2O$$

MW	60.1	88.2	130.2
bp°C	118	132	142
d^{20} g/mL	1.049	0.809	0.876
n_D^{20}	1.3720	1.4075	1.4003

B. Procedure

Place 0.800 mL (647 mg, 7.3 mmol) of isopentyl alcohol and 1.6 mL (1.7 g, 28 mmol) of glacial acetic acid in a 5-mL conical reaction vial containing a spin vane. Then, **using a micropipet,** add 5 microdrops of concd. H_2SO_4.

Fit the vial with an air condenser, add a spin vane, and heat to reflux with stirring on the stirrer hot-plate. Set the hot-plate to high until the heat block reaches *ca* 100°C, then reduce the setting to maintain a block temperature of *ca* 160°C. Heat at reflux for 45 min.

Remove the vial with the air condenser still attached, and after cooling for a short time in air, cool the vial briefly in tap water and then cool it in ice water. Disconnect the air condenser and add cold water until the vial is nearly full. Stir rapidly with the spin vane for about a minute, but avoid splashing. Stop stirring and allow the layers to separate; an organic layer containing the ester floating on an aqueous layer. With a micropipet, remove and discard the **lower** layer.

The next operation is to dissolve the ester in ether and to remove the acetic acid and traces of sulfuric acid by extraction. First add *ca* 1.5 mL of ether and nearly fill the vial with cold water. Stir again for a minute as before, allow the layers to separate, and remove the lower aqueous layer. If necessary, more ether may be added to keep the volume of the ether layer at *ca* 1.5 mL. Then add *ca* 1 mL of 5% $NaHCO_3$, stir, and remove and discard the lower layer. In the same manner, wash the ether layer with *ca* 1 mL of saturated aqueous NaCl. After removing as much of the lower layer as possible, pass the ether solution of the ester through a column of anhydrous sodium sulfate prepared in a micropipet. Take care to be certain that no water droplets are transferred with the ether solution. The column is made up in a micropipet containing in order, a cotton plug, 1–2 mm of sand, and 5–6 cm of anhydrous sodium sulfate (Figure 3). Collect the effluent in a 3-mL conical vial. Wash the column with *ca* 1 mL of ether and collect it in the same vial.

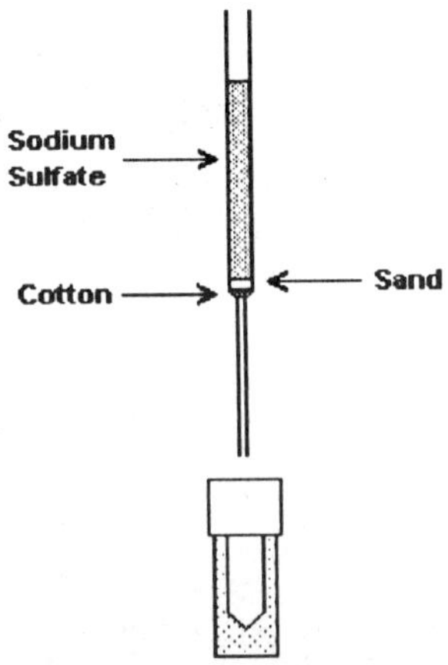

Figure 3. Na_2SO_4 drying tube.

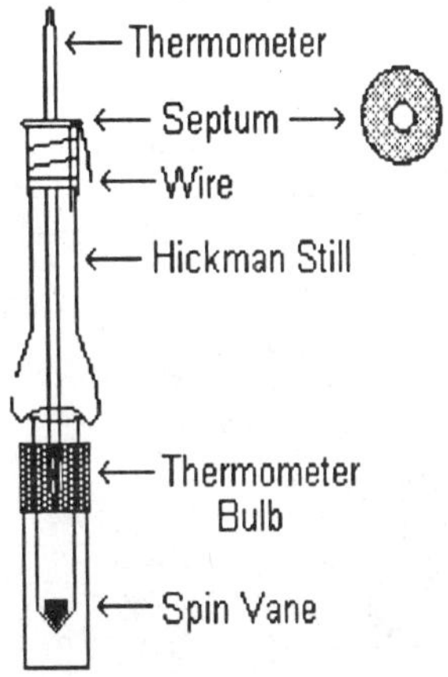

Figure 4. Distillation apparatus.

To evaporate the ether, add a clean dry spin vane to the vial, and place the vial in the aluminum block preheated to *ca* 50-65°C. Promptly begin stirring at a moderate speed. It is important to keep the spin vane spinning to avoid excessive frothing. Gradually raise the temperature of the heating block, causing the ether to boil off. When the block temperature reaches about 100°C, hold the temperature at about this value for a few minutes to ensure that all of the ether has been removed.

Next we distill the ester that remains after evaporation of the ether. The equipment for the distillation is shown in Figure 4. A thermometer is placed in the Hickman still and adjusted so that the top of the thermometer bulb is about even with the top of the lock nut. One way to suspend the thermometer is to insert the thermometer through a septum that has a hole in it that is just large enough to grip the thermometer. **Do NOT attach a lock nut to this septum.** Then attach the Hickman still to the vial using an O-ring and a screw cap. To ensure that the system is not sealed, a small wire hook may be placed under the septum holding the thermometer.

Continue to raise the block temperature. Eventually the ester will begin to boil and will condense in the recessed region of the Hickman still. A block temperature in the range of 200-215°C (well above the boiling point of the ester) will be required.

CAREFUL: Remove the thermometer from the block when the temperature is sufficient for distillation to proceed, otherwise you may break the thermometer.

Observe the cup of the Hickman still and record the temperature of the distillate as the ester collects. You should get a fairly narrow boiling range. Continue heating until no more ester collects and the head temperature falls. Remove the thermometer (collect any ester adhering to the bulb by touching the latter to the receiver) and turn off the heater and stirrer. Using a micropipet, transfer the distillate to a small vial and weigh it to determine your isolated yield.

Assay the purity of the ester by gas chromatography (4 ft, 1/4 in column, Carbowax, 100–110°C). Record the infrared spectrum of your product. Only a small drop of the sample between salt plates is needed. What is the most characteristic absorption for this compound? Determine the refractive index of the ester.

Dispose of waste chemicals in labeled waste containers provided in the laboratory.

C. Prelab Questions

1. Look at the mechanism outlined in Figure 2. What do the reverse steps accomplish?
2. What is the purpose of the $NaHCO_3$ wash? Why might CO_2 be evolved?
3. What data do you have that indicates that the ester should float on water?
4. What are the most likely impurities in the distilled ester? Explain.

D. Postlab Questions

1. Compare the odors of the reactants with the odor of the ester.
2. What additional technique could be used to characterize the ester you made?
3. What would happen if you heated your ester with NaOH solution? What would you expect to observe?
4. Fischer esterification is not useful for most tertiary alcohols. Explain.
5. What would happen if the ester we made were to be heated with a large volume of ethanol containing some sulfuric acid?

30. Acetylation: Aspirin and N-Acetylanthranilic Acid

A. Synthesis of Acetylsalicylic Acid: Aspirin

1. Introduction

a. Discovery of Aspirin

Aspirin is one of the most-widely used analgesic, antipyretic, and anti-inflamatory drugs. The discovery of the effects of aspirin by Rev. Mr. Edward Stone, of Chipping Norton, England was based upon the myth that the causes and cures of diseases were somehow related. The Rev. Stone was aware of the bitter-tasting bark of the willow tree, which grows in damp places, and he reasoned that perhaps the bark of the willow might relieve symptoms of diseases occurring in damp locations. He made a preparation of willow bark and found that, indeed, the potion did relieve the symptoms of various diseases. After five years of using the concoction, Mr. Stone wrote a letter describing his results, which was read before the Royal Society in London and was printed in the Philosophical Transactions of the Society, dated June 2, 1763. After a century, the active ingredient of the willow bark had been extracted and found to be an acid that irritated the mouth. The acid was named salicylic acid, from *Salix*, the Latin name of the willow tree. Later it was found that the acetylated form was much less irritating, without losing the therapeutic properties. The name "aspirin" was the exclusive right of the German firm of Bayer, who marketed the synthetic drug throughout the world. They lost their rights to the name at the time of World War I.

b. Acetylation of Salicylic Acid

Aspirin is prepared by the acid-catalyzed acetylation of salicylic acid by acetic anhydride. Protonation of a carbonyl group of the acetic anhydride activates it toward attack by the phenolic group of salicylic acid. Displacement of acetate and loss of a proton gives the acetylated product. The reaction goes readily under mild heating and the product is easily isolated as large white crystals.

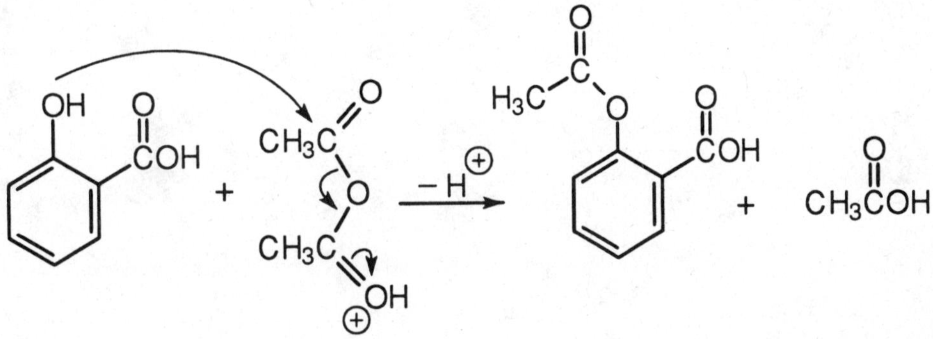

2. Procedure

Place 150 mg (1.09 mmol) of salicylic acid, 0.3 mL (0.3 g, 3 mmol) of acetic anhydride (d = 1.08 g/mL) and two microdrops of 85% phosphoric acid (H_3PO_4) in a 5-mL vial fitted with a spin vane. Attach an air condenser to the vial using an O-ring and a lock nut. Heat the vial for 5 min. with stirring in a 100 mL beaker half-filled with boiling water. Remove the vial and add 0.2 mL of water to the hot solution and stir for one or two minutes. You may see a little fizzing as the water hydrolyzes the excess acetic anhydride to acetic acid. Then add 2 mL of water and cool the mixture in an ice bath for 5 min. Scratching the walls of the vial should initiate crystallization. Collect the product on a Hirsch funnel, allow it to dry in air for about 10 minutes and then recrystallize it from toluene-hexane. First, dissolve the product in *ca* 1.5 mL of hot toluene contained in 5-mL vial. Add hexane dropwise until it appears that the product starts to crystallize (0.5–1.5 mL). Allow the solution to cool to room temperature and then cool it in an ice bath for about 5 min. Collect the crystals on the Hirsch funnel and wash them on the funnel with 1 or 2 mL of cold hexane. Allow the crystals to dry, determine the yield and melting point, and record the IR spectrum.

Test the solubility of salicylic acid and aspirin in both water and sodium bicarbonate. Test both with 2% $FeCl_3$ solution. Account for the difference in behavior.

B. N-Acetylanthranilic Acid: A Triboluminescent Material

1. Introduction

a. Triboluminscence

Triboluminescence is the emission of light from a substance as a result of grinding, crushing, or tearing it apart. The light is usually emitted in flashes and is thought to be due to piezoelectric discharges. The compound N-acetylanthranilic acid is a convenient compound to synthesize in order that this property of triboluminescence can be observed.

b. N-Acetylation

Acetylation of amines is easily accomplished by heating with acetic anhydride. The reaction is generally fast and is complete after only a few minutes. In the case of anthranilic acid, with excess acetic anhydride, 2-methylbenzisoxazinone is formed. This compound, on heating briefly with water, is hydrolyzed to the amide. The crystals of N-acetylanthranilic acid form readily and are easily isolated by filtration.

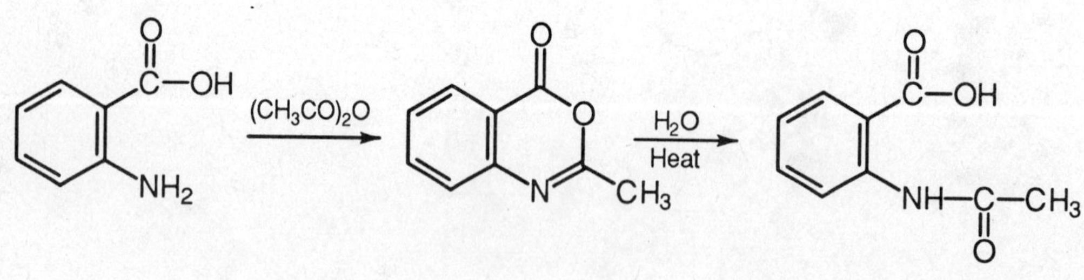

2. Procedure

Place 0.9 mL of acetic anhydride in a 5-mL vial fitted with a spin vane. While stirring add 300 mg (2.2 mmol) of anthranilic acid. Connect the water-cooled reflux condenser and heat to reflux. (Note: On heating, the reaction mixture may seem to solidify, but on continued heating all the material will go into solution.) Continue to heat with stirring and brief shaking until all of the solids have dissolved. Heat under reflux for 5 minutes, allow the reaction mixture to cool briefly, then add 0.3 mL of water through the condenser. Heat again to boiling and allow the reaction mixture to cool to room temperature undisturbed. The object is to obtain large-sized crystals to aid in observing triboluminescence. Cool in ice and filter on the Hirsch filter. Rinse the remaining crystals out of the flask and onto the filter using a little ice-cold methanol. Allow the crystals to dry and determine the melting point. Heating briefly in a warm oven will accelerate drying.

In order to observe triboluminescence, it is necessary that the crystals be thoroughly dry and that the observation be made in a dark room. Any stray light will obscure the triboluminescent effect. Place a few crystals on a slide or watch glass. Enter the dark room and allow your eyes to become accustomed to the dark for about a minute. Then, in the dark, gently grind the crystals using another slide or watch glass. The triboluminescent effect will be observed as sharp flashes of light.

Reference: Erikson, J., *J. Chem. Educ.* **1972,** *49*, 688.

Dispose of waste chemicals in labeled waste containers provided in the laboratory.

C. Prelab Questions

1. Draw the structure of the methyl ester of salicylic acid, the main constituent of oil of wintergreen. How could you synthesize this compound?
2. Write the reaction for the hydrolysis of acetic anhydride.
3. What does "piezoelectric" mean (often termed the piezoelectric effect).

D. Postlab Questions

1. What evidence do you have that indicates that salicylic acid was actually acetylated?
2. What reaction of acetylsalicylic acid should occur under the strongly acidic conditions of the stomach?
3. What is the purpose of providing aspirin tablets that are coated?
4. Formulate a reaction mechanism to show how, with excess acetic anhydride, 2-methylbenzisoxazinone might be formed. (Hint: acetylation of the amino group should occur first.) How many moles of acetic anhydride are consumed in this reaction?
5. Show how 2-methylbenzisoxazinone is hydrolyzed to N-acetylanthranilic acid. Why is this compound readily hydrolyzed?

31. Preparation of a Liquid Crystal: The Enol Acetate of 4-Cholestene-3-one

A. Introduction

1. Liquid Crystals

In a crystalline compound the molecules are arranged in a specific order with respect to one another. On melting, the molecules of the compound generally assume a random order and the melt is said to be isotropic. By isotropic is meant that a measurement, such as the refractive index, is the same when viewed in any direction. On the other hand, certain compounds melt to form what are known as liquid crystals, where the molecules of the melt retain a certain degree of order. In these cases the melts are anisotropic, which means that a measurement, such as the refractive index, will vary depending upon the direction of the measurement. The liquid crystal state is a transition state between the rigid order of the crystalline solid and the random order of the isotropic phase. At some higher temperature the molecular order is lost and the anisotropic liquid crystal is converted to the randomly-ordered isotropic phase.

The structure of the molecule determines whether or not the compound shows liquid crystal behavior. Rod-shaped molecules are particularly susceptible to this condition. Liquid crystals are classified according to the arrangement that the molecules assume in the liquid crystal state. **Nematic** order occurs when the molecules, shown as rods, line up roughly parallel to each other, but without any evidence of layers. **Smectic** order arises when the molecules not only line up parallel, as in the nematic type, but there is also the appearance of layers (Figure 1).

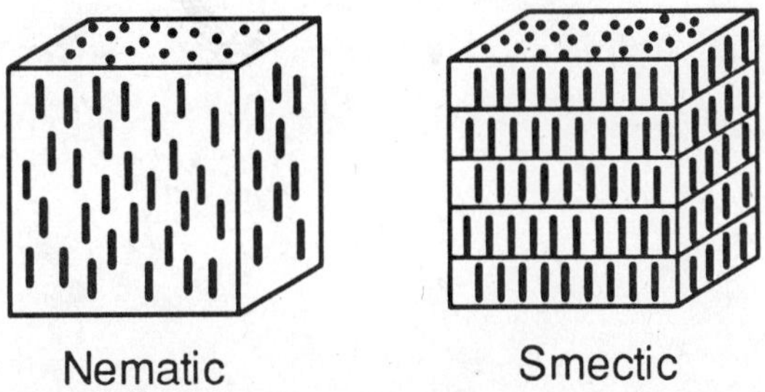

Figure 1. Molecular arrangements in nematic and smectic liquid crystals.

For **cholesteric** liquid crystals, it is necessary that the compounds possess chirality. The molecules not only line up and form layers, but the layers assume a helical arrangement with respect to each other (Figure 2). The arrows illustrate the changing orientation of the molecules of one layer with respect to those of the layer above it. In the example, the molecules of each succeeding plane are shown lying at a 45° angle with respect to the plane above it. Other angles are possible depending on the structure of the compound. On going from top to bottom the rotation illustrated is in the counter-clockwise direction and so we say that it is left-handed. The term **cholesteric,** comes from the name cholesterol because many cholesterol derivatives exhibit this type of liquid crystal behavior.

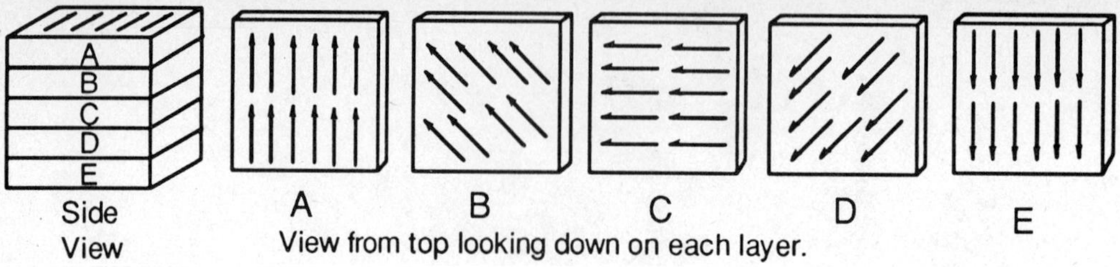

Figure 2. Orientation of molecules in a cholesteric liquid crystal.

In this experiment, a derivative of chloresterol will be prepared and its liquid crystal properties examined qualitatively. The liquid crystal behavior may be observed by melting a small sample of the compound between two glass slides. The thicknesses of the cholesteric layers are of the order of the visible wavelength of light and therefore reflected light will undergo interference effects, which result in a profusion of vivid iridescent colors, even though the compound itself is colorless. If one slide is moved over the other one, the colors will change as the layers are disturbed.

2. The Enol Ester of 4-Cholestene-3-one

The overall reaction to form the enol acetate is shown below.

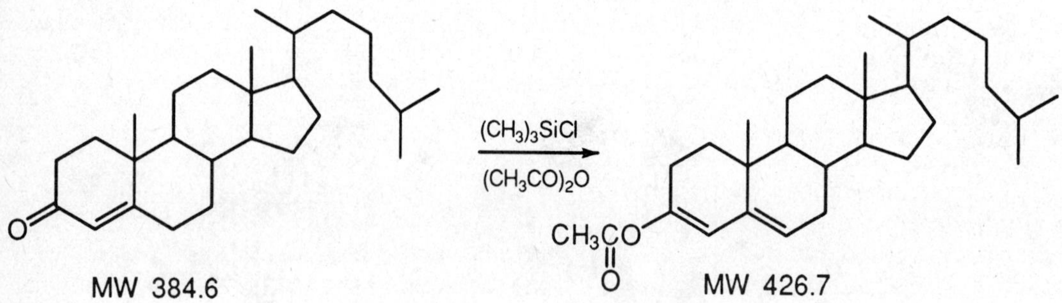

The reaction path for the formation of the enol acetate is:

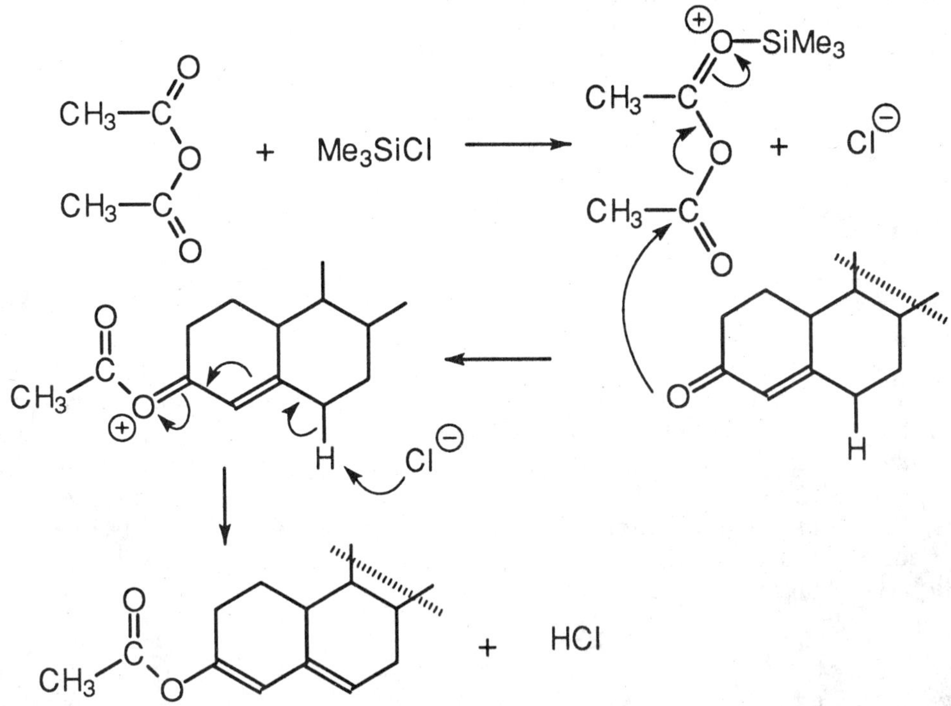

B. Procedure

A. Synthesis of the Enol Acetate

Only fresh acetic anhydride and chlorotrimethylsilane should be used, because both readily hydrolyze. Place 150 mg (0.39 mmol) of 4-cholestene-3-one, 1.0 mL (1.1 mmol) of acetic anhydride and 0.4 mL (3.2 mmol) of chlorotrimethylsilane in a 5-mL vial. Attach a water-cooled condenser and a drying tube and heat under reflux for about 45 min.

The reaction should be followed by TLC and worked up when the spot of starting material disappears or becomes very small. Sample the reaction mixture every 10 min. and run the TLC using Eastman TLC plates with fluorescent indicator. (A sample may be removed through the condenser by means of a small glass tube as is described in the experiment on the oxidation of 9-hydroxyfluorene.) The developing solution is 1:1 methylene chloride:hexane. Make sure that the spot on the TLC plate is thoroughly dry before placing the plate in the developing chamber. Acetic anhydride is slow to evaporate and if there is any left on the plate, it will distort the chromatogram. Holding the plate in the air over the hot-plate will help to dry the spot. Examine the developed plates with a UV lamp. Discontinue the reaction when the spot for the starting material becomes very small or disappears. It should not be necessary to run the reaction longer than *ca* 45 min.

At the end of the reflux period, allow the reaction mixture to cool to room temperature and then cool it in an ice bath. Scratch the walls of the vial to induce crystallization. When it appears that crystallization is complete (about 10 minutes in the ice bath) collect the product on the Hirsch funnel. Wash the product with a few mL of ice-cold methanol using the methanol to help rinse out any crystals remaining in the vial. A small amount of the product may be recrystallized from methanol using the Craig tube, but the product is fairly insoluble. Determine the melting point of the recrystallized sample. Also record the IR spectra of the product and of the starting material.

B. Liquid Crystal Behavior

On melting, the product forms a liquid crystal. Observe the colors that appear during the determination of the melting point. Then place a small sample of the compound between glass slides and place the sandwich on a warm hot-plate, hot enough to melt the sample, but not too hot. When the sample has melted, remove the slides and observe the colors, both by reflected light and by transmitted light. Record your observations. Slide one of the glass slides over the other by a few millimeters and note the changes that occur as the cholesteric layers are disturbed. Place the slide sandwich between two crossed polarizers and examine the effect by transmitted light. (Two pairs of Polaroid sunglasses that are positioned at a 90° angle will work well. Polarizing sheets may be obtained from Edmonds Scientific Co., Barrington, NJ.) What do you observe? Explain the effect of the liquid crystal on polarized light.

Heat the sample again on the hot plate to a higher temperature until the colors disappear. Why have the colors disappeared? In this state, observe the sample again between crossed polarizers. What do you observe? Allow the sample to cool. What do you observe on cooling?

Dispose of waste chemicals in labeled waste containers provided in the laboratory.

C. Prelab Questions

1. Draw the structure of the enol acetate of 4-cholestene-3-one and circle each of the chiral carbon atoms. How many stereoisomers are possible?
2. What is plane polarized light? What use is made of it in studying chiral molecules?
3. What organic "side product" is produced in this reaction? What will happen to it in the work-up?
4. Could we prepare an enol acetate by the Fischer esterification method? Explain.

D. Postlab Questions

1. Compare the IR spectra of the starting material and the product. What evidence can you cite that indicates that the reaction has occurred as written?
2. Why are enol acetates more easily hydrolyzed by sodium hydroxide than simple esters? Write equations for the reaction of the enol acetate with aqueous sodium hydroxide and show how the final product is formed. Use curved arrows to show the electron flow.
3. Liquid crystals are widely used in commerce today. Name some instruments (devices) that use liquid crystals.

32. The Claisen-Schmidt Reaction: Benzalacetone and Dibenzalacetone

A. Introduction

The Claisen-Schmidt reaction is an **aldol-type** condensation that typically involves the reaction of an aromatic aldehyde with a methyl ketone in the presence of a strong base such as hydroxide or alkoxide. The initially formed β-hydroxy ketone dehydrates under the reaction conditions to form an α,β-unsaturated ketone. The overall reaction and the mechanism of the reaction are shown in Figure 1.

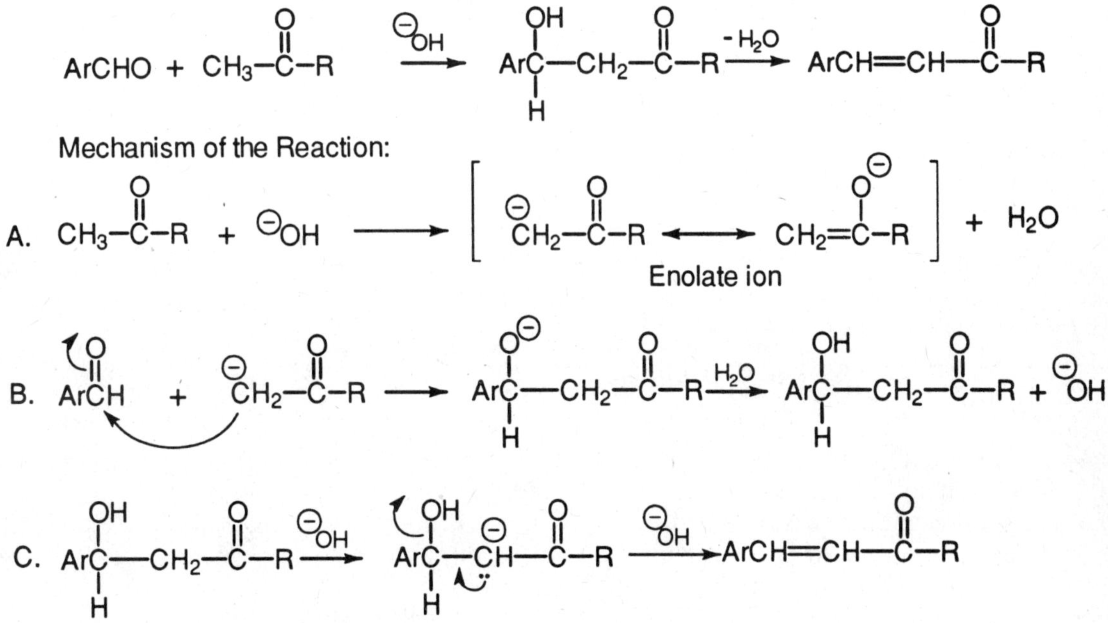

Figure 1. The Claisen-Schmidt reaction, an Aldol-type condensation.

Since the aromatic aldehyde has no α-hydrogen atoms, it cannot form an enolate ion and thus cannot undergo an aldol condensation by itself. In contrast, the methyl ketone has an alpha hydrogen and can form an enolate ion, $^-$:CH_2COR (Figure **1,A**). This enolate ion then reacts with the aldehyde (Figure **1,B**) rather than with another ketone molecule because the aldehyde carbonyl is far more reactive than the ketone carbonyl. The reaction between the aldehyde and the enolate ion is the rate-determining step; i.e., it controls the overall reaction rate. The subsequent dehydration (Figure **1,C**)is favored because the C=C of the α,β-unsaturated ketone that is formed is also conjugated with the aromatic ring, giving a stable extended conjugated system.

This experiment features an aldol-type reaction between benzaldehyde and acetone. The product obtained from this reaction depends upon the ratio of benzaldehyde and acetone in the reaction mixture. The condensation of one molecule of benzaldehyde with one molecule of acetone gives benzalacetone, $C_6H_5CH{=}CHCOCH_3$ (4-phenyl-3-butene-2-one) also called "benzylidene acetone" (Figure 2A). With a large excess of acetone, it is possible to obtain benzalacetone in good yield.

If the reaction is carried out using a ratio of two moles of benzaldehyde for each mole of acetone, the initially formed benzalacetone undergoes a subsequent condensation-dehydration with benzaldehyde to form dibenzalacetone (1,5-diphenyl-1,4-pentadien-3-one). The second condensation is similar to the first and takes place with the methyl group of the benzalacetone (Figure **2,B**). The overall reaction to form dibenzalacetone is shown in Figure **2,C**.

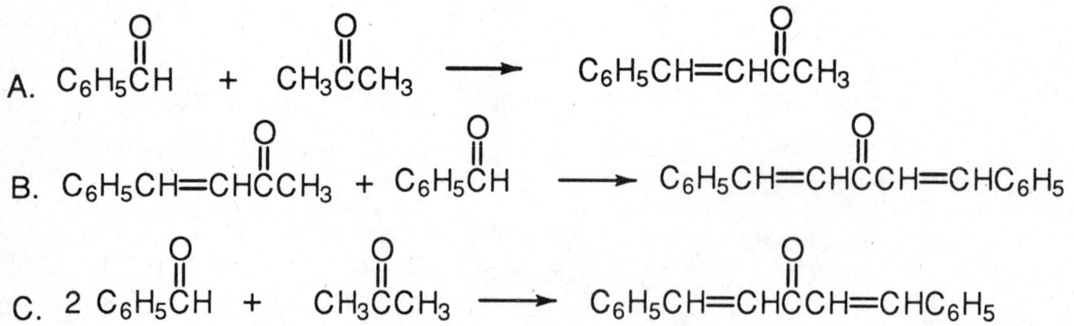

Figure 2. The Claisen-Schmidt reaction between benzaldehyde and acetone.

In practice it is easy to isolate and purify dibenzalacetone since it is a fairly high-melting solid, insoluble in the reaction medium. On the other hand, isolation of benzalacetone on a microscale level is more difficult because it is a low-melting solid (p 42°C) which tends to form an oil. With larger quantities, distillation under reduced pressure affords material which subsequently crystallizes (and can then be recrystallized from petroleum ether).

We will prepare both condensation products. The dibenzalacetone will be isolated as a solid and characterized by its melting point. The benzalacetone, obtained as an oil, will be converted to a crystalline derivative, the semicarbazone, and will be characterized in this form.

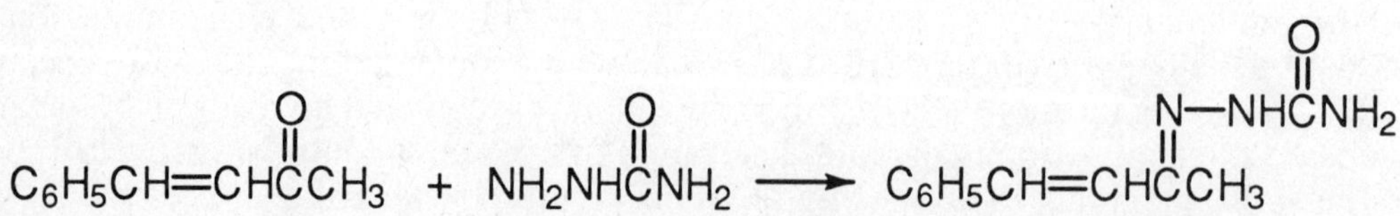

Two experimental procedures are given below. To save time, plan to do them both together rather than completing one before starting the other.

B. Experimental Procedure

1. Dibenzalacetone

Note: Benzaldehyde is slowly air-oxidized to benzoic acid. While benzoic acid does not inhibit the condensation reactions, its presence makes uncertain the quantity of benzaldehyde taken. Thus it is advisable to use freshly-distilled benzaldehyde.

It is important to use the stoichiometric amounts of benzaldehyde and acetone. Benzaldehyde is a relatively high-boiling liquid, but acetone is quite volatile (bp 56°C) and is readily lost by evaporation, a problem which becomes quite significant on the microscale level. Consequently, we will weigh the acetone directly into an ethanol solution, thereby greatly decreasing the vapor pressure of the acetone.

Add 1.0 mL of 95% ethanol to a 4-dram screw-cap vial. Cap the vial, sealing it tightly, and weigh it accurately. Then add **six microdrops** of acetone to the vial, promptly reseal the vial, swirl to effect a homogeneous solution, and reweigh to determine the weight of the acetone added. The quantity of benzaldehyde specified below is based on a **58 mg sample** (1.00 mmol) of acetone.

In a second sample vial, weigh out 212 mg (2.00 mmol) of benzaldehyde. It is important that the millimoles of benzaldehyde be exactly twice the millimoles of acetone actually weighed. Adjust the quantity based on the millimoles acetone actually used. A **slight** excess of benzaldehyde (3 or 4%) is permissible, but do not have a deficiency. Then, with a micropipet, quantitatively transfer the ethanol/acetone solution to the vial containing the benzaldehyde.

Obtain 0.5 mL of aqueous 10% NaOH (*ca* 2.8 **M**) in a 3-inch test tube. Add the NaOH solution in small portions over a period of 2 or 3 minutes to the acetone solution of benzaldehyde. Between additions, cap the vial and swirl it vigorously to ensure good mixing. The reaction is exothermic. If necessary, cool the vial by dipping it in water at room temperature (have a beaker with water in it ready). After all of the NaOH solution has been added, cap the vial tightly and shake it periodically over a period of 20–30 min. Record changes in color, etc. Dibenzalacetone should separate as a solid. Thorough mixing is needed to ensure that all of the sparingly soluble intermediate, benzalacetone, is able to undergo further reaction with benzaldehyde to produce dibenzalacetone.

After the reaction is complete, collect the product on a Hirsch funnel. Break up any lumps and then wash the solid repeatedly with cold water to remove all traces of NaOH which may be occluded in the solid. Draw off as much water as possible, then dissolve the product in a slight excess, *ca* 2 mL, of hot ethanol. Allow the solution to cool slowly. The product may "oil out" if the solution cools too rapidly; seeding is helpful. Finally, cool the mixture in an ice-bath and collect the crystals of dibenzalacetone. Determine the yield and melting point after air drying. Record the IR spectrum.

2. Benzalacetone -- Isolated as the Semicarbazone

> Note: Benzalacetone has been reported to be a skin irritant. Avoid getting it on your hands. If you do, wash it off with soap and water without delay.

Weigh *ca* 80 mg (0.75 mmol) of benzaldehyde into a sample vial. Then add 0.5 mL of ethanol and 0.5 mL (400 mg, 8.6 mmol) of acetone (a large excess). Swirl to dissolve and add 0.25 mL of 10% aq. NaOH. Cap the vial tightly and shake it periodically for 10–15 min. Record any color changes. Does any solid separate?

After the reaction is over, add 0.5 mL of methylene chloride and then nearly fill the vial with water. Cap the vial tightly and gently invert and shake it for 1–2 min. (Note: If a reaction vial is used, rather than a sample vial, the extraction may be performed by stirring with a spin vane rather than by capping and shaking.) Always seal the vial tightly to avoid leakage onto your hands. Allow the layers to separate. Swirling and tapping aid in giving a cleaner separation. With a micropipet, carefully remove and discard the **upper aqueous layer.** Perform the extraction with water four more times to ensure the complete removal of acetone.

After the last extraction, remove most of the water, add a **micro** boiling chip, and cautiously evaporate the methylene chloride over a steam bath. **Be careful to avoid uncontrolled foaming.** Heat the residual oil "strongly" on a steam bath for a few minutes to ensure that no trace of acetone remains. You will probably have the product as an oil, floating on a drop of water.

Dissolve the residue in *ca* 1.0 mL of ethanol. Add 1.0 mL of water, then 100 mg (0.9 mmol) of semicarbazide hydrochloride, $NH_2CONHNH_2 \cdot HCl$, and 150 mg (1.8 mmol) of anhydrous sodium acetate (or 250 mg of $CH_3CO_2Na \cdot 3H_2O$). Swirl to dissolve the reactants and then heat the solution below boiling for at least 10 min. Transfer the hot solution (micropipet) to a Craig tube and allow it to cool slowly. When crystallization is complete, collect the solid by centrifugation. Dissolve the solid in hot ethanol (*ca* 1 mL) and allow the solution to cool slowly. Beautiful crystalline needles may result. Collect the crystals, dry them, and determine the yield and melting point.

Dispose of waste chemicals in labeled waste containers provided in the laboratory.

C. Prelab Questions

1. Why are aldehydes more reactive than ketones toward nucleophiles?
2. Is NaOH a true catalyst in this condensation? Explain.
3. In the synthesis of benzalacetone we use a large excess of acetone. Why? Is the excess of acetone apt to lead to an aldol condensation of acetone with itself? Explain.
4. Why remove all acetone from benzalacetone before the next reaction with semicarbazide?
5. Write the products expected from the base-catalyzed condensation of diethyl ketone with benzaldehyde. Write the mechanism of this reaction.

D. Postlab Questions

1. Give a mechanism for the reaction of a ketone with semicarbazide to form a semicarbazone.
2. Why is sodium acetate added in the preparation of the semicarbazone of benzalacetone?
3. Write a structure for the product formed by condensing benzaldehyde with acetophenone.
4. Write stereostructures for all possible stereoisomers of benzalacetone and dibenzalacetone. In each system, only one stereoisomer is the actual product. Which stereoisomer is formed? Explain.

33. Aryldiazonium Ion Reactions: Formation, Displacement, Coupling

A. Introduction

1. Formation and Displacement Reactions of Diazonium Salts

The diazotization of primary aromatic amines with nitrous acid gives aryldiazonium ions which are moderately stable in water at 0°C. The relative stability of aryldiazonium ions (compared to alkyldiazonium ions) reflects a stabilizing interaction of the π-system of the aromatic ring with the π-system of the N_2^+ group. These aryldiazonium ions (named as arenediazonium salts) are intermediates in the synthesis of a wide variety of compounds. For example, reaction with potassium iodide, cuprous chloride, cuprous bromide, cuprous cyanide, and water give aryl iodides, chlorides, bromides, cyanides, and phenols, respectively (Figure 1).

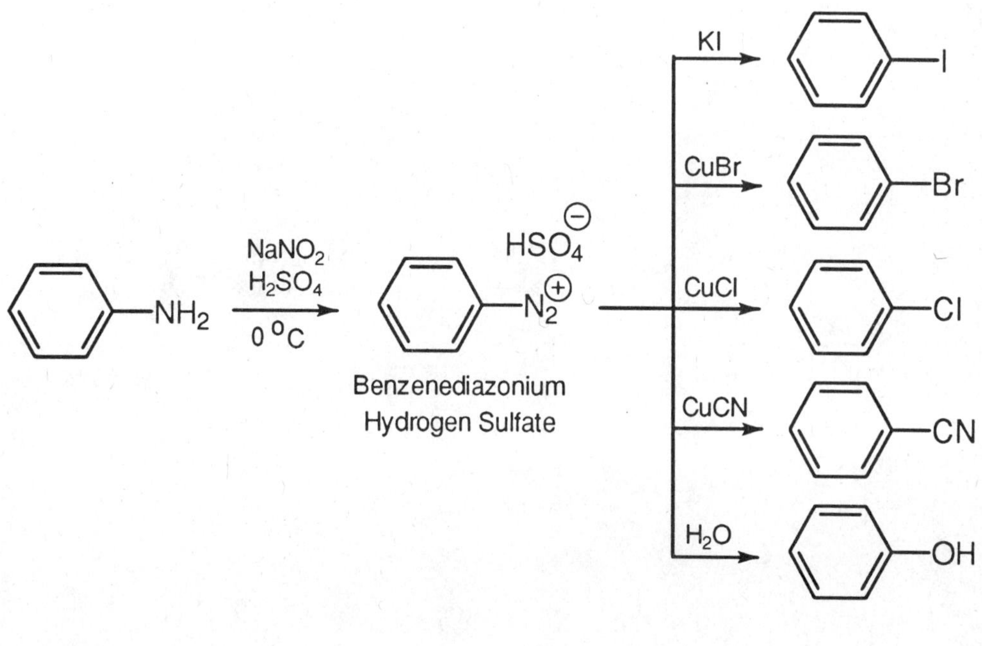

Figure 1. Preparation and reactions of benzenediazonium hydrogen sulfate.

In contrast, alkyldiazonium ions are quite unstable and rapidly decompose by loss of nitrogen to form carbocations, which then react in several ways to give a variety of products. For this reason, alkyldiazonium ions have limited utility in organic synthesis.

In all of these reactions of aryldiazonium ions, the driving force is the loss of N_2, a very stable molecule. When cuprous salts are used, the reactions are believed to involve formation of an aryl free radical, with the copper serving as a redox catalyst:

$$ArN_2^+ + CuX_2^- \xrightarrow[-\,N_2]{\text{electron transfer}} \left[Ar\bullet + CuX_2\right] \longrightarrow ArX + CuX$$

The noncatalyzed reactions appear to be S_N1-type substitutions involving highly reactive aryl carbocations which are captured by nucleophiles such as I^- and H_2O:

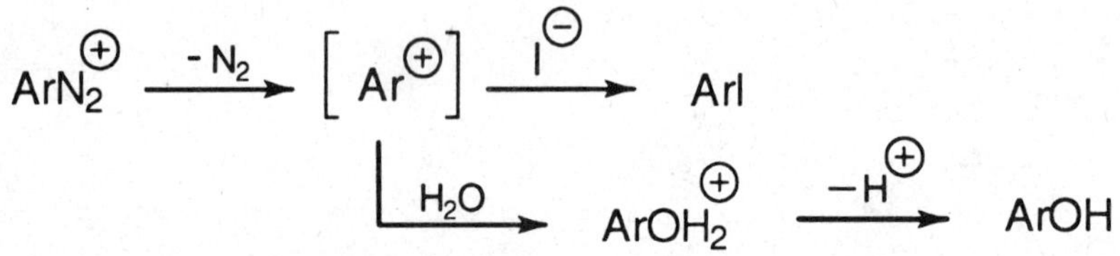

In this experiment, two of the reactions shown in Figure 1 will be carried out using anthranilic acid as the primary aromatic amine. The amine group will be diazotized by reaction with nitrous acid and the diazo group will be replaced with iodide and hydroxide groups through reaction with potassium iodide and water, respectively (Figure 2).

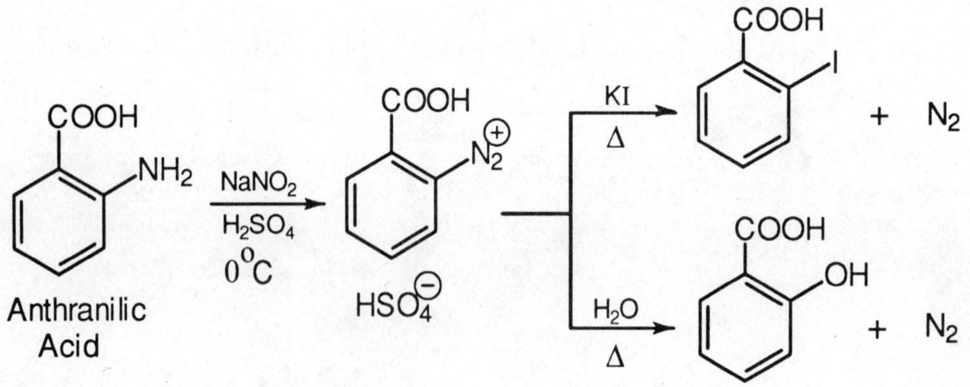

Figure 2. Reaction of the anthranilic acid with KI and H_2O.

2. Coupling of Aryldiazonium Salts to form Dyes

One of the most important commercial uses of aryldiazonium salts is the synthesis of dyes through a coupling reaction with activated aromatic compounds (dye couplers). This reaction is the basis of the azo dye industry and many azo dyes are commercially produced through the coupling reaction of diazonium salts with various compounds. By this means, dyes for fabrics, plastics, and a host of other commodities are produced.

In this experiment the dye methyl red will be synthesized by the coupling of the diazonium salt of anthranilic acid with N,N-dimethylaniline (Figure 3).

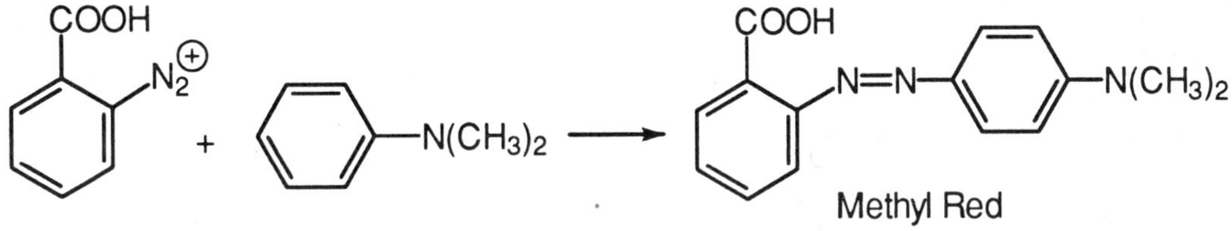

Figure 3. The synthesis of methyl red by reaction of the diazonium salt of anthranilic acid and N,N-dimethylaniline.

Diazotization: The chemistry of forming diazonium salts is somewhat involved. Nitrous acid, a weak acid (pK *ca* 4), is unstable and on standing, or warming, decomposes to form N_2O and water. Therefore, nitrous acid must be prepared *in situ* from the reaction of sodium nitrite with a mineral acid such as hydrochloric or sulfuric acid. The HNO_2 is generated in the presence of the amine so that the reaction to form the diazonium salt can take place immediately. In solution, the nitrous acid appears to be in equilibrium with a small amount of N_2O_3, considered to be the active nitrosating agent. The N_2O_3 reacts with the amino group to form an intermediate N-nitroso compound, ArNHNO, which undergoes subsequent rapid reaction in the acidic solution to generate the diazonium ion ArN_2^+.

$$ArNH_2 \xrightarrow{N_2O_3} ArNHNO \xrightarrow{+H^{\oplus}} ArN_2^{\oplus}$$

The diazotization and coupling reactions must be carried out under carefully controlled conditions of temperature and pH. The temperature should be close to 0°C to suppress hydrolysis of the diazonium salt. During the diazotization the pH must be adjusted carefully so that the equilibrium:

$$ArNH_2 + H^{\oplus} \rightleftharpoons ArNH_3^{\oplus}$$

favors the amine salt form and keeps the concentration of the free amine low in order to prevent the diazonium ion, ArN_2^+, from reacting with the free amine to form a side product, ArNHN=NAr. However, it is the free amine that reacts with nitrous acid, and so the pH must be high enough to provide a sufficient concentration of free amine to undergo the diazotization reaction.

Coupling: The coupling reaction is an electrophilic aromatic substitution in which a weak electrophile, the diazonium ion, reacts with an aromatic ring which is strongly activated by a powerful activating group such as $-N(CH_3)_2$ or -OH. Coupling occurs generally at the *para* position due to steric hindrance at the *ortho* sites.

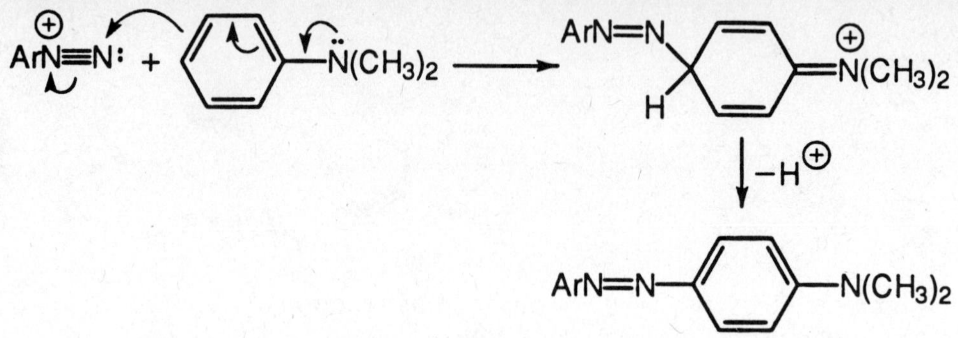

When the N,N-dimethylaniline is added to the acidic solution of the diazonium salt of anthranilic acid, very little coupling occurs because the solution is so acidic that the amine exists primarily as the salt, the amine hydrosulfate. Since the amine salt is unreactive, it is necessary to raise the pH in order to increase the concentration of free amine so that coupling can take place. The pH will be increased by adding sodium acetate, which reacts with the mineral acid to form acetic acid, a weak acid, raising the pH to the point where a sufficient amount of free amine is liberated, enabling it to undergo the coupling reaction with the diazonium ion to form the azo dye. In the case of methyl red, although the coupling reaction is fast, the formation of crystals of the azo dye occurs slowly, so prolonged standing is necessary to achieve a good yield of product.

Methyl red, like phenolphthalein, is a useful indicator in acid-base titrations. A good acid-base indicator is one that undergoes a distinct color change over a narrow pH range. For phenolphthalein, the pH range over which the indicator changes from colorless to red is 8.2-10.0. In the case of methyl red the pH range for the color change is 4.2-6.2. Thus methyl red should be used in titrations that have end points in the pH 4.2-6.2 range.

Under basic conditions (high pH) methyl red is colored due to the azo chromophore in conjugation with the aromatic rings. Under acidic conditions (low pH) methyl red becomes protonated on one of the azo nitrogens (Figure 4). The protonation is accompanied by an electronic shift to produce a quinoid-type structure which has a color different from that of the azo form of the non-protonated species; hence methyl red changes color with changes in acidity. As a part of this experiment, the color of methyl red under acidic and basic conditions will be observed.

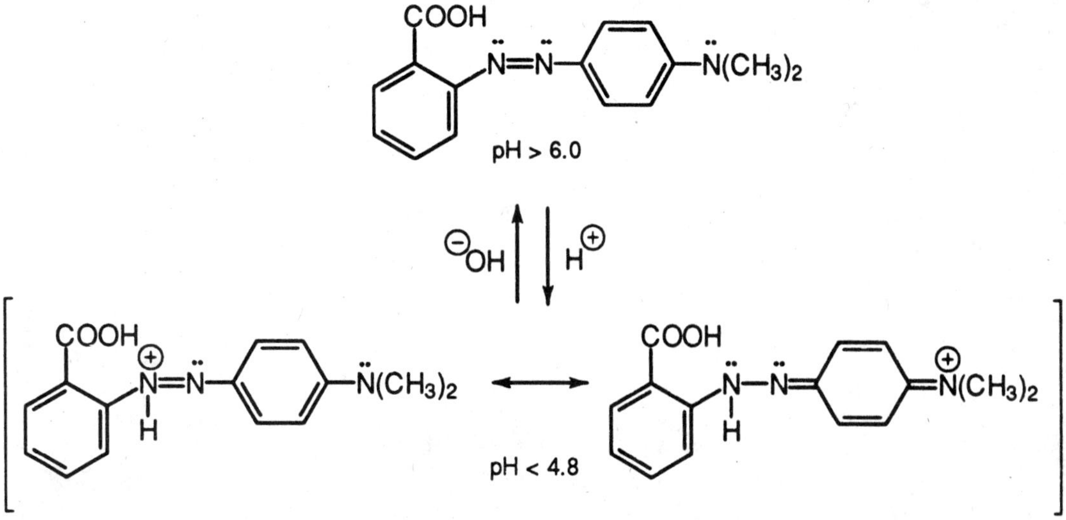

Figure 4. Structural changes of methyl red as a function of pH.

B. Procedure

1. Preparation of the Diazonium Salt of Anthranilic Acid

Place 400 mg (2.9 mmol) of anthranilic acid in a small Erlenmeyer flask and add 1.5 mL of 1.0 M (10%) H_2SO_4 and 4.5 mL of water. Warm until the anthranilic acid is dissolved and then cool in an ice bath for at least 5 min. The sulfate salt generally precipitates at this point. Prepare a solution of sodium nitrite ($NaNO_2$) by dissolving 250 mg (3.6 mmol) in 3 mL of water. Cool this solution in an ice bath for 5 min. When both of the solutions have been cooled below 5°C, gradually add the sodium nitrite solution to the anthranilic acid mixture using a micropipet. Swirl the flask continuously during this addition. The solid should slowly dissolve, leaving a clear solution.

In the latter stages of the addition, test the solution occasionally for excess sodium nitrite by removing a small drop on the end of a stirring rod and touching it to a small piece of starch-iodide paper. A deep blue color will indicate the presence of excess nitrite. If, after all of the sodium nitrite has been added, the solution still tests negative, add a little more sodium nitrite until a positive test is obtained. When the solid has dissolved completely and the solution finally gives a positive test for sodium nitrite with starch-iodide paper, the diazotization is complete. (Avoid adding excess $NaNO_2$.)

At this point, the volume of the reaction mixture should be *ca* 9.0 mL. Divide the solution into three equal portions as follows. Using a pipet, transfer 3.0 mL of the reaction mixture to a 5-mL conical vial that has been previously cooled in an ice-bath. (Use the graduations on the vial.) Divide the remainder equally between two 6-inch test tubes and cool both in ice. Each portion will now contain about 1 mmol of the diazotized anthranilic acid. The portion in the conical vial will be used to synthesize methyl red and the portions in the test tubes will be used for reaction with potassium iodide to prepare *o*-iodobenzoic acid and for hydrolysis to prepare salicylic acid. Store all solutions in ice baths to prevent premature decomposition of the diazonium salt.

NOTE: The preparation of methyl red should be started first because of the prolonged standing time required for the crystalline product to form. The hydrolysis and the reaction with potassium iodide should be run simultaneously as soon as the methyl red reaction is underway.

2. Reaction with Dimethylaniline: Synthesis of Methyl Red, an Azo Dye

Place a spin vane in the vial containing the 3-mL portion of the diazotized anthranilic acid and commence stirring, while cooling with an ice-bath. Add 0.15 mL (d = 0.956, *ca* 1.2 mmol) of N,N-dimethylaniline to the reaction mixture and stir for about 5 min. Note the color change.

While the reaction mixture is stirring, dissolve 150 mg (1.1 mmol) of sodium acetate trihydrate (or 90 mg of anhydrous sodium acetate) in 0.5 mL of water contained in a small test tube. Cool the solution in an ice-bath and add it to the "coupling mixture" in the vial. The sodium acetate raises the pH, increasing the concentration of dimethylaniline, the active species that couples with the diazonium salt. The solution should turn deep red on addition of sodium acetate. If it does not turn red, the solution is too acidic. In this event, add 10% sodium hydroxide dropwise until the solution turns deep red.

Stir the reaction mixture for 5 min in the ice bath, remove the spin vane with forceps, and then allow the mixture to stand at room temperature. During the first half hour, swirl the vial occasionally and then let it stand for at least an hour longer, or until the next laboratory period. The mixture will thicken as the methyl red crystallizes. Stir the mixture occasionally with a stirring rod or spatula. The crystals of methyl red form rather slowly, so the mixture must stand for a time to ensure a good yield of product. Collect the methyl red on a Hirsch funnel and wash it with 3–4 mL of cold water. The wash water can be added to the vial first in order to help transfer more of the methyl red. Determine the yield and melting point.

Dissolve a crystal or two of methyl red in a few drops of alcohol. Add some water and then add a few drops of dilute HCl and observe the color change. Add a few drops of dilute NaOH and again observe the color. Repeat the addition of acid and base while observing the color changes under the different pH conditions. What structural features of methyl red are responsible for the different color in acid and base?

3. Reaction of the Diazonium Salt with Potassium Iodide: Synthesis of *o*-Iodobenzoic Acid

This reaction and the following one should be run at the same time.

A hood is required in order to contain the vapors of iodine which may be evolved in the reaction. To the diazonium salt in one of the test tubes add a solution of 165 mg (1.0 mmol) of potassium iodide dissolved in 1 mL of water. Then heat the solution in a boiling water bath for 10 min. Note the result. Cool the solution to room temperature and add a few drops of dilute sodium bisulfite, $NaHSO_3$, to reduce iodine to iodide. The mixture may still be yellow-amber due to impurities which are not reduced. Then cool the mixture in an ice bath. Collect the crystals of *o*-iodobenzoic acid on a Hirsch funnel and wash well with cold water.

Dissolve the crystals in 1 mL of methanol in a 5-mL conical vial and filter the solution through a micropipet containing the following: a small plug of cotton, about 5 mm of sand, 1 cm of decolorizing carbon, and topped off by 5 mm of sand. The carbon removes colored impurities from the solution. Collect the effluent in a 5-mL vial. When the top of the solution in the filter tube reaches the top of the sand, add *ca* 0.5 mL of methanol and continue to collect the filtrate. This methanol is necessary to wash the column free of all product.

Add water dropwise to the filtrate until a cloudy solution persists. This should not take more than about 1–1.5 mL of water. Reheat to dissolve the crystals and then allow the solution to cool slowly. If crystallization does not commence, seed the solution with a few crystals of the product or scratch the walls of the flask. Then cool the mixture in an ice bath for about 15 min., with occasional stirring. Collect the crystals by filtration using the Hirsch funnel and wash with a little cold water. Determine the yield and melting point and record the IR spectrum. Compare the IR spectrum of the iodide with that of the starting material and note any significant differences.

Since the melting point of *o*-iodobenzoic acid is close to that of salicylic acid, run a mixed melting point with salicylic acid to show that the iodide is not salicylic acid that might have been formed by hydrolysis.

Dissolve a few crystals of the product in a small amount of methanol and test with a few drops of $FeCl_3$ solution. (See the comparison below.)

4. Hydrolysis of the Diazonium Salt: Synthesis of Salicylic Acid

Heat the other test tube containing the diazonium salt solution in a boiling water bath for 10 min. Note what happens. Cool the mixture to room temperature and then in an ice bath. Collect the solid by filtration using the Hirsch funnel and wash it well with ice water. Use some of the wash water to help remove the product from the test tube.

Dissolve the crystals in 1 mL of methanol in a 5-mL conical vial and filter the solution through a micropipet containing the following: a small plug of cotton, about 5 mm of sand, 1 cm of decolorizing carbon, and topped off by 5 mm of sand. Collect the effluent in 5-mL vial. When the top of the solution in the filter tube reaches the top of the sand, add *ca* 0.5 mL of methanol and continue to collect the filtrate.

Add an equal quantity of water (*ca* 1-1.5 mL) to the filtrate and add some seed crystals or scratch the walls of the flask. Then cool the mixture in an ice bath for about 15 min., with occasional stirring. Collect the crystals by filtration using the Hirsch funnel and wash them with a little cold water. Determine the yield and melting point and record the IR spectrum. Compare the IR spectrum of the salicylic acid with that of the starting material and note any significant differences.

Dissolve a few crystals of the product in a small amount of methanol and test with a few drops of $FeCl_3$ solution. Describe what happens. Compare the results with the $FeCl_3$ test of *o*-iodobenzoic acid. Explain.

Dispose of waste chemicals in labeled waste containers provided in the laboratory.

C. Prelab Questions

1. Complete the following reactions which are involved in this experiment.

$$C_6H_5NH(CH_3)_2^+ \quad + \quad CH_3CO_2^- \quad \rightleftharpoons$$

$$C_6H_5NH(CH_3)_2^+ \quad + \quad OH^- \quad \rightleftharpoons$$

   Which equilibrium lies farther to the right? Explain.

2. In diazo coupling reactions we must avoid making the solution too basic because a diazonium ion reacts as follows:

$$ArN_2^+ \quad + \quad OH^- \longrightarrow ArN{=}NOH \xrightarrow{OH^-} ArN{=}NO^-$$

   Predict what happens when a solution of $ArN{=}NO^-$ is acidified.

3. Why is it necessary to ensure that the solution is not too acidic in the coupling reaction leading to methyl red?

D. Postlab Questions

1. Methyl orange is made by diazotizing the sodium salt of sulfanilic acid (*p*-aminobenzenesulfonic acid) followed by coupling with N,N-dimethylaniline. Write equations for these reactions, giving the structure of methyl orange.

2. Explain how you could distinguish between N,N-dimethylaniline and 2,4-dimethylaniline by chemical means. What would you do and what would you see?

3. Would methyl red be a suitable indicator for the titration of a typical carboxylic acid, RCO_2H, with standard 0.1000 M aq. NaOH (e.g., to determine the equivalent weight of RCOOH)? Explain. (The color change for methyl red occurs between pH 4.2 (red) and 6.2 (yellow). Hint: What is the pH at the end point in the titration of RCO_2H with NaOH?

34. Dyes Produced by Diazo Coupling and Diazo Imaging

A. Dyes by Diazo Coupling

1. Introduction

We learned in the chapter on diazotization that most aryl diazonium salts are relatively unstable and are generally prepared in solution at 0°C. On the other hand, the fluoroborate salts of certain diazonium compounds are quite stable and can be isolated in the dry state. One such diazonium salt is 4-diazo-N,N-diethylaniline fluoroborate, which has sufficient stability to be sold commercially as a solid.

In the first part of this experiment, 4-diazo-N,N-diethylaniline fluoroborate will be allowed to react with a variety of dye couplers in order to test the effectiveness of the coupling reactions and to observe the colors of the dyes produced. The test involves dissolving the diazonium salt together with the dye coupler and an organic acid. The acidic conditions prevent coupling, but on neutralizing the acid with ammonia, coupling occurs to produce the dye. Activated aromatic compounds and compounds containing active methylene groups couple with diazonium salts to give dyes (Figure 1).

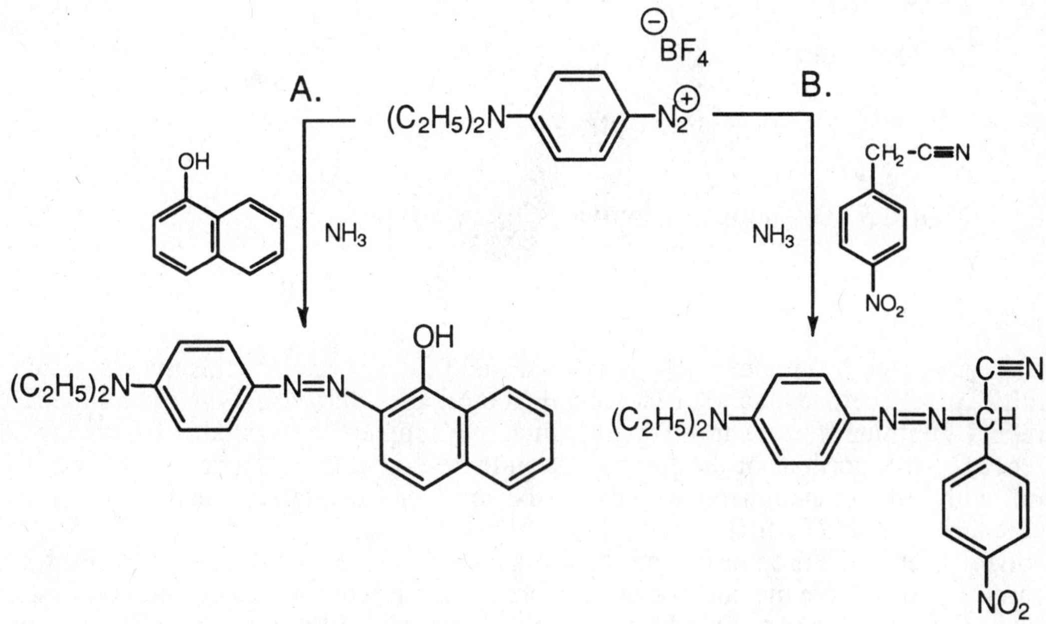

Figure 1. The reaction of 4-diazo-N,N-diethylaniline fluoroborate with A, α-naphthol; and B, 4-nitrophenylacetonitrile.

Additional dye couplers of the aromatic type and the active methylene type are shown in Figures 2 and 3 respectively. Coupling occurs at the sites marked by the arrows.

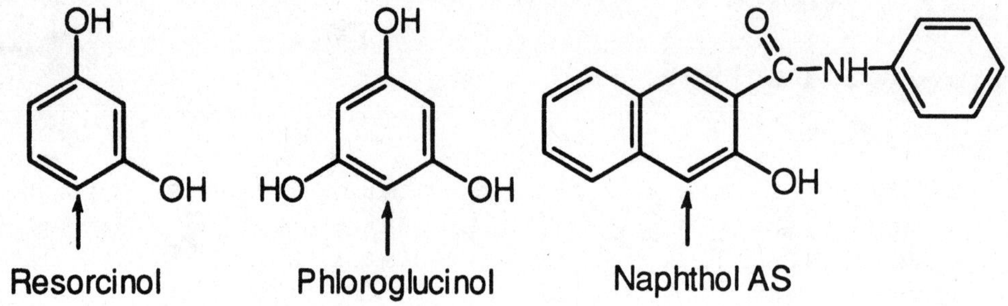

Figure 2. Dye couplers; activated aromatic type.

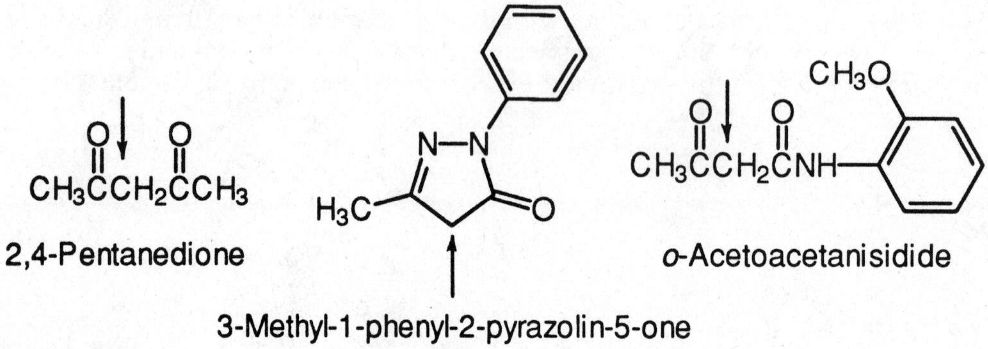

Figure 3. Dye couplers; active methylene type.

2. Procedure

Dissolve 100 mg of *p*-toluenesulfonic acid (other organic acids may be used) in 25 mL of a 1:1 acetone:ethanol, followed by 100 mg of 4-diazo-N,N-diethylaniline fluoroborate. Determine from your instructor which dye couplers are available for testing. Transfer *ca* 0.5-mL portions of the diazo solution to several glass test tubes and dilute to *ca* 1 mL with 1:1 acetone-ethanol. To each tube add 1 or 2 crystals of the different dye couplers available (a different coupler to each tube) (see Figures 1, 2 and 3). Stir to dissolve. Then add a drop or two of ammonia (*ca* 3 M) to one of the test tube and watch carefully to observe the color of the dye produced. Record the color that you first see when the ammonia is added. Continue to watch carefully for a few minutes and record any color changes that you observe. Repeat this procedure for each of the test tubes. Save the remaining diazonium solution for use in the next section on diazo imaging.

B. Diazo Imaging

1. Introduction

Diazo imaging was invented in 1923 by Koegel, a German monk, in order to avoid the need for hand copying large volumes of script. The process was developed under the trade name of the Ozalid process, but this method of copying has been largely superceded by the Xerox process. Diazo imaging is based on the fact that diazonium salts are sensitive to light and decompose to form colorless compounds. Advantage is taken of this property to form an image which consists of an azo dye.

The process depends upon the high stability of the fluoroborate salts of diazonium compounds, and the pH dependency of the coupling reaction; basic conditions are required for coupling to occur.

A solution is made up to contain the diazonium fluoroborate salt, a dye coupler, and an organic acid. The organic acid prevents coupling at this stage of the process. Paper is then coated with the diazo solution and dried. A transparent mask with an opaque image is placed in firm contact with the impregnated paper and exposed to light. Under the light, the diazonium salt decomposes to form colorless products.

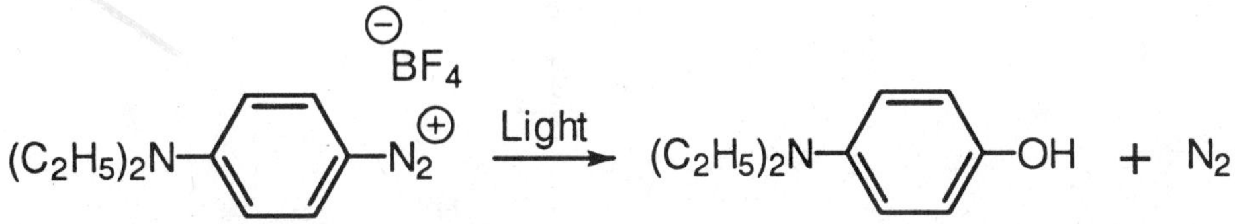

Under the opaque portions of the mask the diazonium salt is shielded from the light and remains unaffected. After exposure, the paper is treated with the vapors of ammonia which neutralizes the organic acid to produce a basic environment. Under these conditions the unexposed, and undecomposed, diazonium salt couples with the dye coupler to form the dye. The result is a copy (azo dye) of the image on the mask. A schematic drawing of the imaging experiment is shown in Figure 4.

Many different couplers are available (Figures 1, 2, and 3), and each produces its own individual color.

2. Procedure

To perform the imaging experiment the following materials will be needed: a 4" x 5" sheet of good quality well-sized writing paper (Crane's Kid Finish is good), a mask, which can be an image photo copied onto a transparent plastic sheet, a 1 or 2-inch foam brush (obtainable inexpensively at paint stores) to coat the paper, a 4" x 5" sheet of glass, or Plexiglass, to hold the mask in contact with the paper during exposure to light, and a flood lamp. A 1500-mL beaker containing a little concentrated ammonia and covered with a watch glass should be placed in the hood for community use.

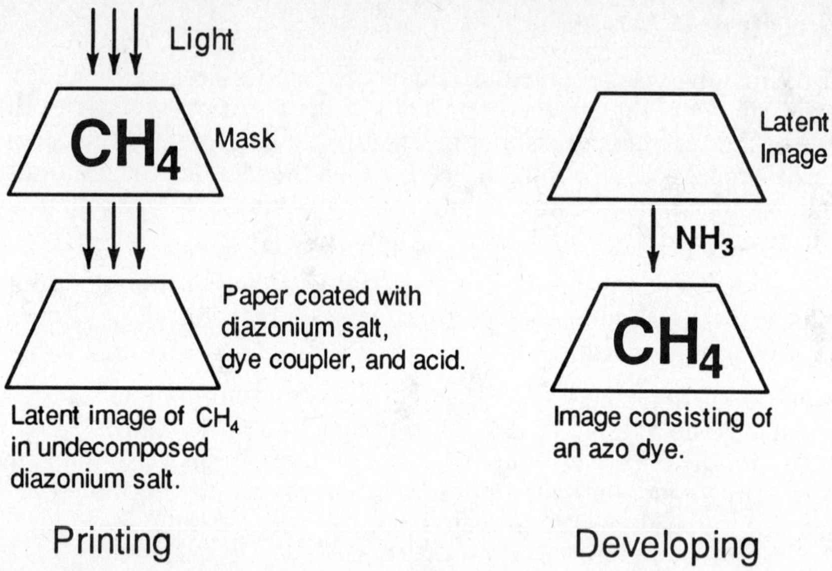

Figure 4. The diazo imaging experiment.

Dissolve *ca* 200 mg of either resorcinol or phloroglucinol in the acetone-ethanol-diazonium salt solution remaining from Part A. With the foam brush, coat the paper evenly with the diazo-coupler solution. Allow the paper to dry for a few minutes. Place the paper, coated side up, on a smooth surface, cover the paper with the plastic mask, and place the glass, or plexiglass, plate on top to give good contact between mask and paper. Illuminate the sandwich for about 5 min. with a flood light positioned 6–8 inches above the surface. In the areas where paper is exposed to the light, the diazonium salt is decomposed into colorless materials. You can tell when the exposure is about right, because the yellow color imparted by the diazonium salt becomes bleached. In those areas where the light is shielded by the mask, the diazonium salt is not decomposed and remains capable of coupling with the dye coupler.

Remove the paper and you should see a very faint yellow outline of the image on the mask. Suspend the paper in the 1500-mL beaker containing the ammonia vapor. The image should appear within a few seconds. If the paper acquires the overall color of the dye, the illumination was not long enough. Try again with a longer exposure. The exposure will depend on the intensity of the light and the distance from the paper. The ideal exposure gives a dark image on a white, or light-colored, background.

Since the 20-mL of diazo solution is sufficient to coat several sheets of paper, students should make up solutions and share the solutions with others. Couplers other than resorcinol or phloroglucinol may also work well. Students could experiment with some of the other couplers and record the effectiveness of each in diazo imaging. Do all of the couplers work well for diazo imaging? If not, why not?

Dispose of waste chemicals in labeled waste containers provided in the laboratory.

C. Prelab Questions

1. Draw resonance structures of the anion produced in the reaction of α-naphthol with a base. Show how this anion couples with the diazonium salt.
2. Draw resonance structures of the anion produced in the reaction of 4-nitrophenylacetonitrile with a base. Show how this anion couples with the diazonium salt.

D. Postlab Questions

1. Draw the structures of the dyes produced in the coupling reactions.
2. A chromophore (which is a structural feature associated with color in a molecule) is often a long series of conjugated double bonds. Circle the chromophores in the structures of the dyes produced.

35. Imaging Dyes in Color Photography

A. Introduction to the Chemistry of Photography

> Each student should bring to the laboratory a small conventional black-and-white photographic print, which will be converted to a dye-colored print.

In the experiment on azo dyes, we found many colored compounds — dyes — which can be used for many purposes such as dying fabrics, paints, plastics, for printing, etc. Similar types of colored compounds can be used in color photography. In order to gain some insight into the processes involved, a short overview of conventional black-and-white photography is presented and followed by a brief look at the chemistry involved in color photography.

Conventional photography utilizes the light sensitivity of silver bromide and other silver halides. During the brief exposure of the film in the camera to light, an invisible latent image is formed in the silver bromide deposited as microcrystals in the film. The latent image consists of silver atoms produced by the initial reduction of silver ions to silver metal by exposure to light. It is believed that as few as four or five silver atoms form a "speck" of silver on a crystal of silver bromide and that this amount of silver is sufficient to catalyze the reduction of the entire crystal of silver bromide to free silver by a chemical reducing agent, such as hydroquinone. After development, the image is composed of varying densities of silver, resulting from the overall distribution of the silver specks in the original latent image. The amplification factor on going from the latent image to the developed image is of the order of 10^9. The final step is to dissolve and wash away the remaining unreduced silver bromide that, if left in place, will darken in light and destroy the image. This step involves washing with a solution of sodium thiosulfate (hypo), a substance which dissolves silver bromide. An abbreviated reaction sequence is shown in Figure 1.

1. Formation of the Latent Image by Light

$$4\ AgBr \xrightarrow{\text{Light}} Ag_4 + 2\ Br_2$$

2. Chemical Development

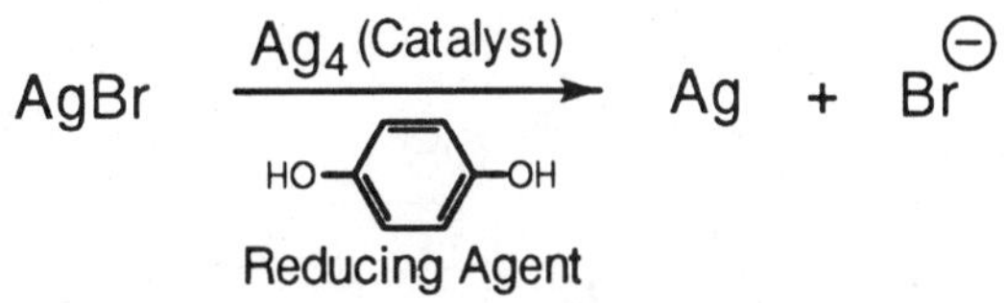

3. Fixation: Dissolving the Excess AgBr

$$AgBr + 3\ S_2O_3^{2-} \longrightarrow Ag(S_2O_3)_3^{5-} + Br^-$$

Figure 1. Chemical reactions in black-and-white photography.

Both black-and-white photography and color photography utilize the light sensitivity of silver bromide to form a latent image and finally the image. In B&W photography, the final image is composed of particles of silver metal. In color photography dye couplers have been incorporated into the emulsion of the film, or paper and a special developing agent, a diamine derivative, is used to reduce the light activated silver bromide to free silver, to give an image in silver. The developing agent, which has now been oxidized by the silver, reacts with the dye coupler to form the dye. The overall process is shown below.

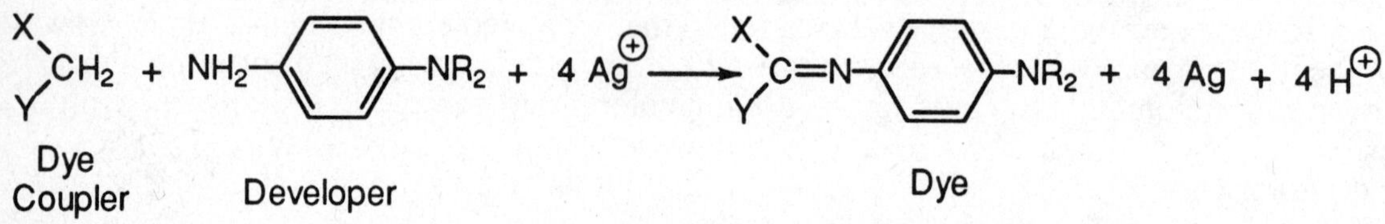

The first step in the process is the oxidation of the diamine color developer by AgBr to form a radical cation, a step which is catalyzed by the silver specks of the latent image produced during the exposure to light. Two molecules of the radical anion disproportionate to form a molecule of the starting diamine and a molecule of a diimine.

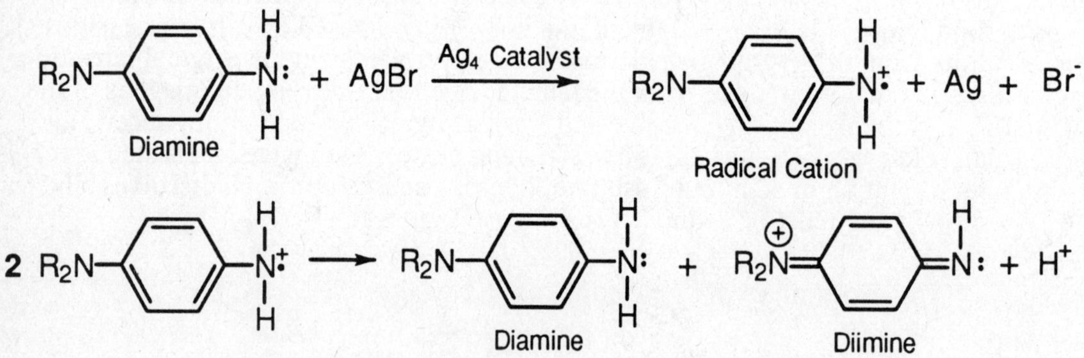

In the basic media the dye coupler is converted to the anion.

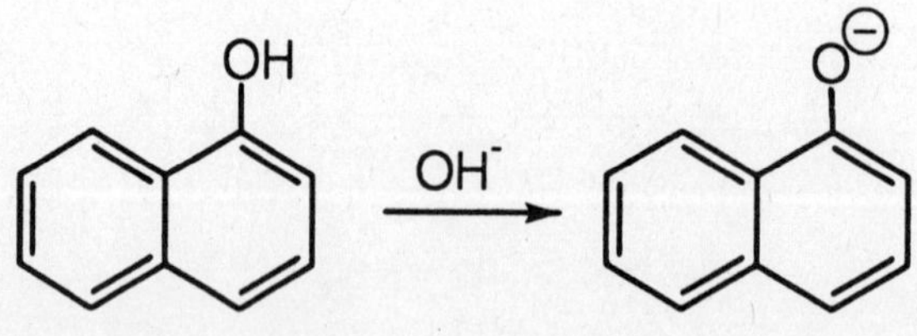

The diimine couples with the anion of the dye coupler to give an intermediate, which then undergoes oxidation to the dye.

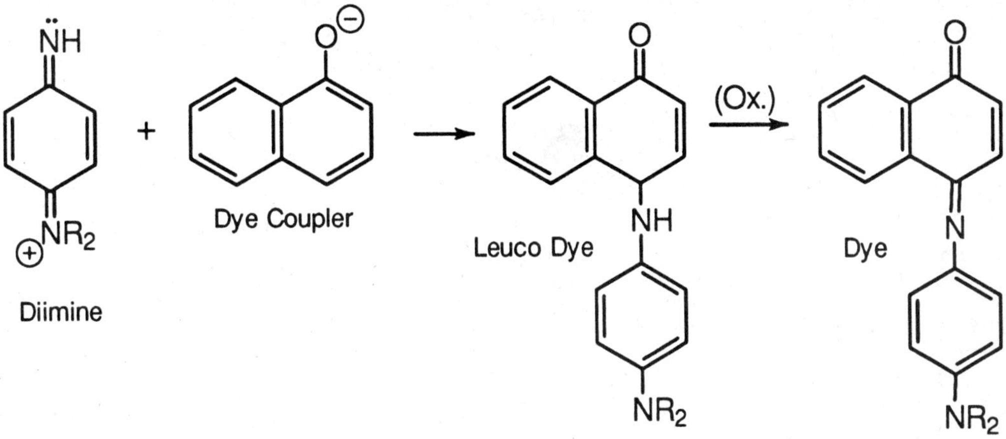

The dye is formed at the same site as the silver bromide crystal that was reduced, so that there is now an image consisting of dye molecules together with an image of silver. The next step is to remove the silver image by oxidizing it back to silver bromide (bleaching) and dissolving it with sodium thiosulfate. In the final image, only the dye remains.

In order to achieve an image in full color, three emulsion layers are utilized, each sensitized to respond to a particular primary color: blue, red, and green. The dye coupler in each of the layers is designed to produce a color complimentary to the color to which the layer is sensitive. Thus, the dye coupler in the blue-sensitized layer produces a yellow dye, the coupler in the red-sensitized layer produces a cyan dye, and the coupler in the green-sensitized layer produces a magenta dye. The combination of the three image layers in yellow, cyan, and magenta give the final image in full negative color. On printing, these colors are reversed to give a final image in full positive color.

Several dye couplers are shown in Figure 2.

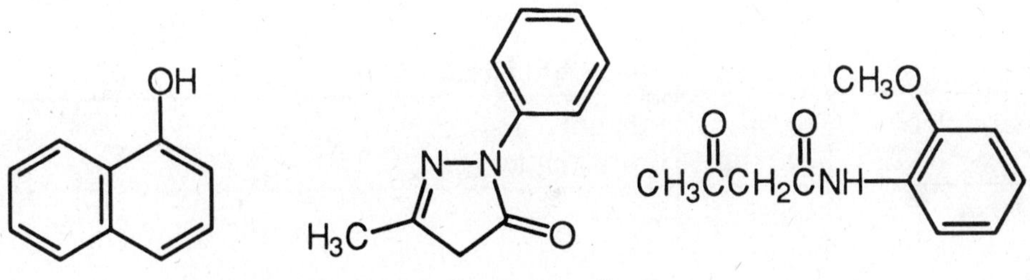

Figure 2. Dye couplers.

B. Testing Dye Couplers

This procedure involves testing several dye couplers for coupling activity with oxidized developer and determining the color of the dye produced. N,N-Dimethyl-1,4-

phenylenediamine will be oxidized by ammonium peroxydisulfate , $(NH_4)_2S_2O_8$, in the presence of the dye coupler. The diimine, produced on oxidation, couples in an alkaline medium with the dye coupler to give the dye. Compare the colors of the dyes with the colors produced from the same dye couplers examined in the experiment on diazo coupling.

The following solutions should be made and stored in dropping bottles in the laboratory.

Color Developer:	100 mg Sodium Sulfite 300 mg N,N-Dimethyl-1,4-phenylenediamine Dihydrochloride 30 mL Distilled Water
Oxidizing Solution:	1% Ammonium Peroxydisulfate
Alkaline Medium:	15% Sodium Carbonate in 3:1 Water:ethanol

Add *ca* 1 mL of the alkaline medium to each of three test tubes. (Alternatively, a well-plate may be used.) Add a crystal or two of each dye coupler (one to each well or test tube) and stir to dissolve. Add a drop or two of the color developer solution followed by a few drops of the peroxydisulfate oxidizing solution. Record the colors of the dyes produced.

C. Converting a Black-and-White Photograph to a Single Color Image

A black-and-white photograph, preferably one made on resin-coated paper, will be converted to a dye-colored photograph in either yellow, magenta, or cyan.

The following solutions are required.

Bleach	500 mL 25 g 10 g	Water Potassium Ferricyanide Potassium Bromide
Color Developer	100 mL 300 mg 500 mg 200 mg 2 g 100 mg	Water Sodium Bisulfite N,N-Dimethyl-1,4-phenylenediamine Dihydrochloride Calgon Sodium Carbonate Potassium Bromide
Dye Coupler	10 mL 100 mg	Ethanol Dye Coupler (Figure 2)
Fixer	500 mL 60 g 5 g	Water Sodium Thiosulfate Sodium Sulfite

The bleach and fixer should be put out in plastic or enameled trays for common use. The color developer and the dye coupler solutions should be available in bottles for preparing the color developer-dye coupler solution described below.

The silver image is first converted to AgBr by soaking in the bleach solution ($K_3Fe(CN)_6$-KBr) until the image is a pale tan color. The following reaction takes place.

$$Ag + K_3Fe(CN)_6 \xrightarrow{KBr} AgBr + K_4Fe(CN)_6$$

Wash the photo in running water for 1 min. and then expose the image for about a minute to a strong light bulb held 6–8 inches from the photograph. The light reduces a few atoms of silver to "specks" of free silver on the crystals of silver bromide. These "specks" will catalyze the reduction of the silver using the color developer.

Select one of the dye couplers (Figure 2), and make up a color developer-dye coupler solution as follows:

Color Developer	100 mL
Dye Coupler Solution	10 mL

Once this solution is prepared, it should be used by several students.

Place the color developer-dye coupler solution in a shallow tray or petri-dish and redevelop the image in this solution for about five minutes, or longer if necessary. The AgBr image is redeveloped (reduced back to silver) by the color developer in the presence of a dye coupler. During reduction of the silver bromide, the amino group of the diamine color developer is oxidized to the diimine, and the diimine reacts with the dye coupler to form the dye, see reactions above.

When the color of the image appears to be completely formed, remove the print and wash it for 1 min. in running water. Convert the silver image back to AgBr by soaking the print in the bleach solution until the silver image has disappeared (*ca* 2 min.) and only the dye image remains.

Place the print in the fixer for about 2 min. to dissolve the AgBr. Wash the print for 5 min. in running water and allow it to dry. The final image is composed of the dye produced in the reaction of the diimine with the dye coupler during redevelopment of the image by the color developer. The silver has all been removed. The following table shows the various steps in the process.

Steps in the Conversion of a B&W Photograph to Color

1.	Bleach	2 min.	Ag oxidized to AgBr
2.	Wash	1 min.	
3.	Expose to strong light	1 min.	Active sites produced on AgBr crystals
4.	Redevelop in Color Dev.	5 min.	AgBr reduced to Ag, oxidized developer couples with dye coupler to form dye.
5.	Wash	1 min.	
6.	Bleach	2 min.	Ag oxidized to AgBr
7.	Fixer	2 min.	AgBr dissolved
8.	Wash	5 min.	

Dispose of waste chemicals in labeled waste containers provided in the laboratory.

D. Prelab Questions

1. Why is a basic environment necessary for coupling to take place? Consider the structure of the color developer as it would exist in both acid and base.
2. Would you expect phenyl ethyl ketone to be as good a coupling agent as *o*-acetoacetanisidide? Explain in chemical terms.

E. Postlab Questions

1. Draw the structures of each dye produced.
2. Identify the chromophore of each dye by drawing a circle about it on your drawings on the report page.
3. Why is it important to use microcrystals of AgBr in photographic film? What quality in the finished photograph is affected by the size of the AgBr crystals?

36. Qualitative Organic Analysis. Part 1: Preliminary Measurements and Classification Tests

A. Introduction

The identification of "unknown" compounds is a common challenge in organic chemistry. In basic research, a compound may be a totally new material, never before isolated. Then the researcher must establish the structure of the substance unequivocally by a combination of spectroscopic and chemical methods. Infrared and nuclear magnetic resonance spectroscopy along with mass spectrometry are modern techniques regarded as indispensable by most chemists. But chemical methods, which were the only methods available until a half century ago, served to establish the structures of a very large number of compounds. These methods are still important and, moreover, they illustrate how functional group properties can be used to solve structural problems.

In this course we are not at the point where we can elucidate the structure of "new" molecules. But we are at a stage where we can identify many compounds which have been previously characterized. We will restrict our attention to a few classes of functional groups -- only because of the limited amount of laboratory time we have for this subject. Students who continue to study chemistry will have the opportunity to make full use of modern spectroscopic methods in a course which is focused on the identification of organic compounds. In this course, we will utilize infrared spectroscopy as well as some classical chemical methods to identify the functional groups in an unknown. If you have access to a nuclear magnetic resonance spectrometer and are able to record the nmr spectrum of an unknown, determining its identity will be greatly facilitated. At this time, we assume that nmr spectra will not be available

A brief note should be made concerning the isolation of pure compounds. The separation of complex mixtures into their constituents may present problems as great or greater than identifying the constituents once they are separated. At this time, we will not face the problem of separating mixtures and isolating individual compounds, but instead we will concentrate on identifying compounds which are reasonably pure.

In this experiment, you will be given two or more unknowns from the following groups: aldehydes, ketones, alcohols, phenols, amines, or carboxylic acids. The first task is to determine appropriate physical properties such as melting point, micro-boiling point, refractive index and the solubility in a number of solvents. The data obtained at this initial stage will greatly restrict the number of compounds which need to be considered as possible structures for your unknown.

If an unknown is **water-soluble,** first determine whether the water solution is acidic, neutral or basic. The solubility in ether should also be determined. Most compounds we encounter are soluble in ether, but some highly-polar compounds such as salts and some polyfunctional compounds (e.g., sugars) are insoluble in ether.

For **water-insoluble** compounds, the solubility in dilute aqueous HCl, NaOH and (possibly) $NaHCO_3$ should be determined. These tests will help you decide if the compound is acidic, basic or neutral.

Then you should obtain the infrared spectrum of the unknown. A liquid should be run as a **thin film.** Many solids are sufficiently soluble in CCl_4 to permit its usage as a solvent for running an IR spectrum; in some cases $CHCl_3$ may be required. If solution cells are available and the compound is soluble in carbon tetrachloride or chloroform, the spectrum can be recorded on a solution. Alternatively, one of the methods described earlier for preparing a "mull" can be used for determining the IR spectrum of any solid.

Next several simple chemical tests are performed in order to identify the type of functional groups present in the unknown. Control experiments on known compounds should be run along with the tests on unknowns so that you will see what happens with compounds which definitely give positive or negative tests.

Combining the physical data with the results of chemical tests will narrow the number of possibilities for each unknown to a few compounds. Then by comparing the melting points of **two crystalline derivatives** of each unknown and with the **known melting points** of the corresponding derivatives of the possible compounds, the unknown can be identified.

Because we are working with a restricted selection of unknowns, arriving at identifications should be relatively straightforward. In more general cases, additional derivatives might have to be prepared in order to distinguish between two possibilities. However, the principles remain the same in solving more complex cases even though the amount of data required may be greater.

This approach assumes that the "unknowns" are included in the tables of possible compounds provided. Thus, the unknowns will be restricted to the compounds listed in the tables. But even with that restriction, recognize that you must carefully perform and correctly interpret all experiments or you cannot possibly identify any unknown. A successful conclusion to this experiment should give some satisfaction in the knowledge that you can apply the chemical principles learned earlier to solve chemical problems.

B. Preliminary Measurements

1. Physical Constants

The unknowns are intended to be "pure," but in general, appropriate purification techniques might have to be used to obtain pure samples. These methods include, but are not restricted to, recrystallization, distillation, extraction, chromatography, and possible chemical transformations. Obtaining a pure sample of an unknown can present a serious challenge. We restrict the unknowns to sensibly pure substances.

Examine an unknown and note its physical state (solid or liquid), its color, if any, and even its odor (be cautious in "sniffing" any compound).

If the substance is a solid, determine its melting point, being sure to include the mp-range, because it serves as criterion of purity. Note any unusual behavior upon melting. (Is a gas evolved? Does the substance decompose? Is there a color change?)

If a solid does not melt sharply, it may have to be recrystallized. In that event, try several common solvents. Ethanol is the most commonly used solvent, but others we have used during the previous experiments may be suitable. Discuss the problem with your instructor.

If the unknown is a liquid, determine the micro-boiling point (unless your instructor tells you that the compound has a boiling point which is so high at atmospheric pressure that the compound decomposes). Refer to the procedure on micro-boiling points given in the chapter on distillation. Also determine the refractive index. This physical constant is especially useful because, like a melting point, it will eliminate a large number of compounds from consideration.

2. Solubility

Water: Using 2 microdrops of a liquid or 20 mg of a solid, determine the solubility in water. Add water in 0.1 mL increments up to a total volume of *ca* 0.5 mL. After each addition, stir well with a micro stirring rod. A sample which does not dissolve in 0.5 mL (i.e., the solubility is less than 40 mg/mL) is considered to be insoluble and the molecular structure probably has five or more carbon atoms per molecule.

Low molecular weight alcohols, aldehydes, ketones, amines, and carboxylic acids are water-soluble. But unless the compound is polyfunctional, at about the five-carbon stage (e.g., 1-pentanol), water solubility decreases to the point that the compound is considered to be "insoluble." Polyfunctional compounds which have several polar groups are usually much more soluble than monofunctional compounds and are typically classified as "water-soluble."

If an unknown dissolves in water, determine whether the solution is neutral, acidic, or basic (use pH paper). First wet the strip with distilled water (to ascertain the initial pH of the water, usually *ca* 6) and then wet it with the aqueous solution of the unknown.

Ether: Determine the solubility in ether in the same way. Since many compounds are readily soluble in ether, ether could be used as an "extraction solvent" in a separation scheme.

"Reaction" solvents; *ca* 5% HCl, NaOH, $NaHCO_3$: Determine the solubility in each of these solvents. If a compound is insoluble in NaOH, there is no need to test it in $NaHCO_3$ as $NaHCO_3$ is a weaker base.

Most amines will dissolve in aq. HCl because they form water-soluble salts.

$$RNH_2 + HCl\,(aq) \longrightarrow RNH_3^+\ Cl^-$$

Carboxylic acids and phenols will dissolve in dilute aqueous NaOH by forming water-soluble salts.

$$RCO_2H + NaOH\,(aq) \longrightarrow RCO_2^- + Na^+ + H_2O$$

$$ArOH + NaOH\,(aq) \longrightarrow ArO^- + Na^+ + H_2O$$

Carboxylic acids will dissolve in aq $NaHCO_3$ with liberation of CO_2. Most phenols are not sufficiently acidic to cause this type of reaction.

$$RCO_2H + NaHCO_3\,(aq) \longrightarrow RCO_2^- + Na^+ + CO_2 + H_2O$$

$$ArOH + NaHCO_3\,(aq) \longrightarrow \text{No Reaction for most phenols}$$

Suspend 2 microdrops of a liquid, or *ca* 20 mg of a solid, in *ca* 0.2 mL of water. Then add the reagent dropwise, with shaking after each addition, to see if the unknown dissolves. If the compound does not dissolve, this is a clear indication that the unknown has not reacted with the reagent. This procedure is used (suspending the sample in water and then adding the reagent dropwise) because occasionally one encounters a compound which forms a sparingly soluble salt which, due to the common ion effect, will not dissolve in excess reagent.

If an unknown is water-soluble, the solubilities in aqueous HCl and NaOH need not be run. However, if the compound is acidic, you should determine whether or not it liberates CO_2 from aqueous $NaHCO_3$.

C. Classification Tests

The chemical tests described in this section are all qualitative. Additional tests are available and an impressive number of reactions have been used to define the functional groups in a molecule. We will use only those needed to aid in identifying unknowns available to us. Perform the tests both with known compounds and with unknowns to be sure that you know what clear-cut positive and negative results look like. However, do only those tests which are needed; do not run these tests randomly. For example, if you know that a compound is an alcohol based on the infrared spectrum, the alcohol test makes sense. If the compound is not an alcohol, this test should not be run. Similarly, aldehyde/ketone tests make sense only if the compound has a carbonyl band in the infrared spectrum. To summarize, use the information you have already obtained, then narrow the possibilities with appropriate classification tests.

1. The Jones Oxidation of Alcohols: CrO_3 in H_2SO_4

Primary and secondary alcohols are rapidly oxidized by Cr(VI) in acidic solution, resulting in a change from the orange color of Cr(VI) to the blue-green color of Cr(III). Tertiary alcohols give a negative test.

Place 1 microdrop of liquid or 10 mg of a solid alcohol on a white spot plate or in a small test tube over a white paper. Add 5–10 drops of acetone to dissolve the compound. Then add 1 drop of the test reagent. Immediately stir the mixture. A primary or secondary alcohol will give a positive test within 1 to 3 seconds.

Run a "blank" test on the acetone to ensure that it does not contain impurities that give a false positive test. Note that although tertiary alcohols do not react immediately, they may give a much slower color change as the tertiary alcohol is dehydrated to an alkene and the latter slowly reacts.

2. 2,4-Dinitrophenylhydrazine Reagent (DNP): A Test for Aldehydes and Ketones

Dissolve 2 microdrops of a liquid or 20 mg of a solid unknown in 5 to 10 drops of ethanol. Then add 10-15 drops of the DNP reagent. Mix well. Formation of a yellow-red precipitate should occur with aldehydes and ketones. If no precipitate occurs, warm for a few minutes, but do not evaporate the ethanol (the reagent may precipitate).

Aldehydes and ketones react with the DNP reagent to form 2,4-dinitrophenylhydrazones which are yellow to orange to brick-red. Alcohols do not give this test.

3. The Tollens Reagent: The Silver Mirror Test for Aldehydes

NOTE: This reagent must be freshly made for each test. Do not store it or allow it to become dry, since it can lead to an explosive compound (silver fulminate, AgN_3). Dispose of the test solution after use, washing it down the drain with adequate water.

Aldehydes are oxidized by $Ag(NH_3)_2^+$ in alkaline solution resulting in the formation of a silver mirror inside the test tube. However, the test tube must be absolutely clean to get a mirror. If it is not clean, a black precipitate of Ag forms. (A new test tube works well.)

Add 1 mL of 5% $AgNO_3$ to a **clean** test tube. Then add 2 drops of 10% NaOH solution (do not use more than 2 drops) and shake or stir the mixture to suspend the dark Ag_2O precipitate which forms. Then add 15% aq. NH_3 solution dropwise with swirling until the precipitate just barely dissolves. This solution is the active Tollens reagent and it must be used within 15 minutes or discarded.

Add 1–2 drops of a liquid unknown, or 10–20 mg of a solid, dissolved in *ca* 0.1 mL of 95% ethanol. Mix well and allow to stand. An aldehyde should give rise to a silver mirror within a few minutes. Do not heat the solution.

Remember — dispose of the reagent promptly after the test is concluded.

4. The Iodoform Test for Methyl Ketones

Methyl ketones react with iodine in sodium hydroxide to form a yellow precipitate of CHI_3.

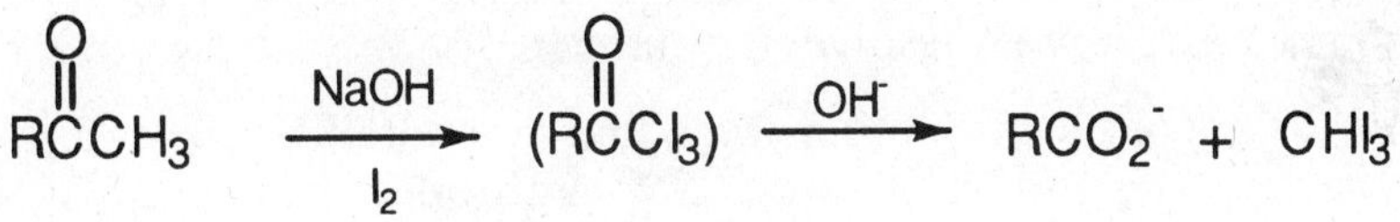

This test can be used to distinguish methyl ketones from those which do not contain a $-COCH_3$ group.

Take 2 microdrops of a liquid or 20 mg of a solid ketone and add 1 mL of 10% NaOH. If the ketone is very insoluble, add 0.2 mL of 1,2-dimethoxyethane as a co-solvent. Then add the iodine/KI reagent dropwise until a definite dark color of iodine persists. If no iodoform separates, as a yellow solid, warm the mixture in a hot-water bath (60°C) for a few minutes. Add more of the iodine reagent if the color fades.

The formation of a yellow solid, CHI_3, indicates that the ketone has a $-COCH_3$ group. Since this reagent oxidizes primary and secondary alcohols, compounds with a $-CH(OH)CH_3$ groups also give positive tests.

5. $FeCl_3$ Test for Phenols

Most phenols (and some enolic compounds) give colored complexes with $FeCl_3$ solution. The color varies with the structure of the phenol and may be yellow, red, purple, or green. Alcohols (and a few phenols) fail to give a positive test. The colors soon fade, so the observation should be made upon mixing.

Dissolve *ca* 10 mg of the unknown in 0.1–0.2 mL of methanol. Add one microdrop of water and then add one drop of the ferric chloride solution (sometimes a second drop is needed). Mix thoroughly. A color change from that of the reagent is a positive test.

6. Hinsberg Test for Amines

Primary, secondary and tertiary amines can usually be differentiated by their reaction with benzenesulfonyl chloride (or p-toluenesulfonyl chloride) in aq. NaOH.

$$RNH_2 + C_6H_5SO_2Cl \xrightarrow{NaOH} [RNSO_2C_6H_5]^- \; Na^+$$

$$[RNSO_2C_6H_5]^- \; Na^+ \underset{NaOH}{\overset{HCl}{\rightleftharpoons}} RNHSO_2C_6H_5$$

$$R_2NH + C_6H_5SO_2Cl \xrightarrow{NaOH} R_2NSO_2C_6H_5$$

$$R_3N + C_6H_5SO_2Cl \xrightarrow{NaOH} \text{No Reaction}$$

Primary amines give a base-soluble sulfonamide; upon acidification, the insoluble sulfonamide is formed. Secondary amines give insoluble sulfonamides (i.e., not soluble in NaOH). Tertiary amines undergo no change; a tertiary amine may remain insoluble in the NaOH solution, but it will dissolve in dilute HCl.

Place *ca* 20 mg of the amine and twice that amount of benzenesulfonyl chloride in a test tube or vial. Add *ca* 1 mL of 10% aq. NaOH. Stopper or cap and shake well periodically for three or four min. Then warm the mixture on the steam bath for a minute or two. Cool and observe. Check with pH paper to ensure that the solution is basic; if not, add more NaOH. If there is no insoluble material, acidify the solution with aq. HCl (a precipitate should form). If the initial reaction mixture has a precipitate, dilute with an equal volume of water, then acidify with HCl. The precipitate from a secondary amine should be insoluble, but a tertiary amine should dissolve.

7. Nitrous Acid Test for Amines

Primary, secondary, and tertiary amines can be differentiated by how they react with nitrous acid.

Aliphatic Amines

$$RNH_2 + HNO_2 \longrightarrow RN_2^+ \longrightarrow R^+ + N_2$$

$R^+ \longrightarrow$ Alkenes or alcohols

$$R_2NH + HNO_2 \longrightarrow R_2N\text{-}N{=}O \quad \text{Insoluble oil}$$

$$R_3N + HNO_2 \longrightarrow \text{No Change}$$

Aromatic Amines

$$ArNH_2 + HNO_2 \longrightarrow ArN_2^+ \xrightarrow{\beta\text{-Naphthol}} \text{Azo Dye}$$

$$ArNHR + HNO_2 \longrightarrow Ar(R)N{-}N{=}O \quad \text{Insoluble solid or oil}$$

$$ArNR_2 + HNO_2 \longrightarrow \text{Colored } {-}C{-}N{=}O \text{ compound}$$

(*p*-Nitroso-N,N-dimethylaniline is green)

Dissolve *ca* 20 mg of the amine in 0.5 mL of 5% aq. HCl. Cool the solution in an ice bath. Dissolve *ca* 20 mg of $NaNO_2$ in 0.5 mL of water and chill the solution in an ice bath. Add this solution dropwise with mixing to the amine solution until there is a slight excess of HNO_2 as indicated by starch-iodide paper. Look for bubbles of N_2 gas (colorless). If no gas is evolved and no solid or oil separates, add a precooled solution of *ca* 20 mg of β-naphthol in 0.5 mL of 5% NaOH dropwise with mixing. A deep color indicates that azo dye has been formed.

Dispose of waste chemicals in labeled waste containers provided in the laboratory.

37. Qualitative Organic Analysis. Part 2: Preparation of Derivatives and Final Identification

When the nature of the functional groups in an unknown has been established, you are ready to prepare derivatives. It is imperative that each derivative be purified so that it has the correct melting point. Refer to the "reported" melting points in the tables of derivatives following this chapter. A melting point of a prepared derivative may be a little different from the reported (literature) melting point, usually slightly lower. But if the difference is more than two or three degrees, you will have trouble identifying the unknown. It may be necessary to recrystallize a derivative several times in order to obtain a sample with a sharp (and probably higher) melting point.

If you have trouble preparing a derivative, or obtaining a sharp melting point, or you find that only one derivative is listed (remember, you are to prepare two crystalline derivatives for each unknown), check with your instructor. It may be possible to prepare some other derivative.

The general procedures which follow will work for most compounds. But there are exceptions. Be observant and ready to modify the procedures when necessary.

You may collect the crystals on the Hirsch funnel, or you may use Craig tube techniques, but this will be your choice. While the general procedures are given, the details are up to you.

A. Derivatives of Aldehydes and Ketones

1. 2,4-Dinitrophenylhydrazones (DNP)

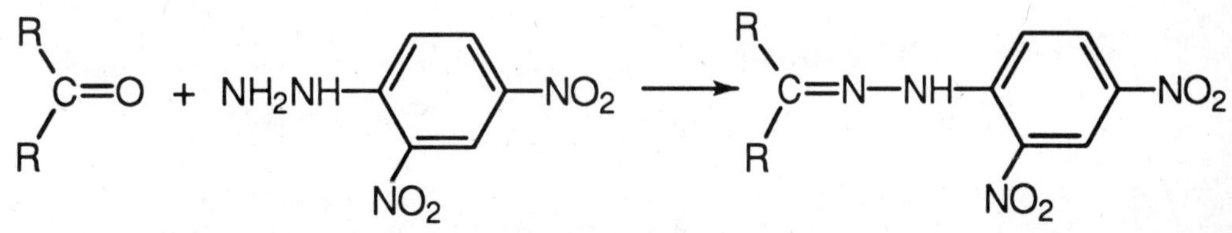

Because it is usually easier to remove an excess of carbonyl compound from the product than it is to remove the DNP reagent, it is a good practice to use an excess of the carbonyl compound so that all of the DNP reagent is used up. Since you do not know the molecular weight of the unknown, you cannot make an exact calculation. But you can make an estimate based on what data you have. Note that a six-carbon ketone (hexanone) has a molecular weight of 100. You can adjust up or down as necessary.

Be sure that you do not contaminate your equipment with acetone, which will give a DNP. If you have used acetone in cleaning equipment, dry it completely and flush out all acetone vapors.

Take 200 mg (1.0 mmol) of 2,4-dinitrophenylhydrazine in a small flask and add 4 mL of ethanol and 0.4 mL of concd. HCl. Warm the mixture to dissolve the reagent. Then add an amount of unknown calculated to be at least 1.2 mmol (120 mg for mol. wt. = 100). Heat the mixture to near boiling for one or two minutes. Then add water dropwise until the product nearly "clouds out." Allow the solution to cool to room temperature. Collect the solid, wash it with a little 1:1 ethanol:water, and recrystallize it from ethanol, ethanol-water, or ethyl acetate.

2. Semicarbazones

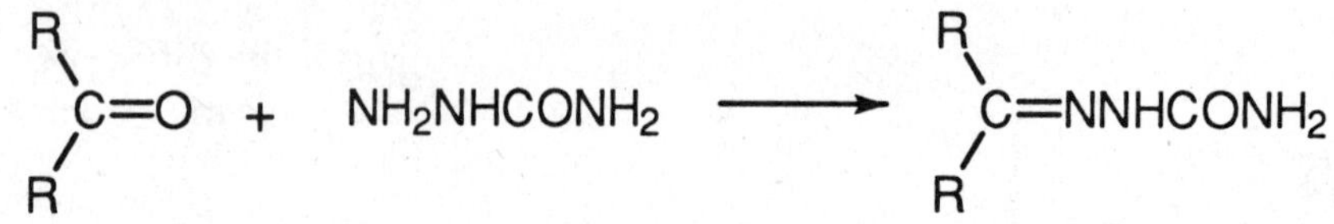

Take 150 mg (1.3 mmol) of semicarbazide hydrochloride ($NH_2CONHNH_2 \cdot HCl$) and 220 mg (1.6 mmol) of sodium acetate trihydrate (or 130 mg of anhydrous sodium acetate).

For water soluble compounds add 1 mL of water and 1.0 mmol of the aldehyde or ketone.

For water insoluble compounds add 0.5 mL of ethanol, 0.5 mL of water, and 1.0 mmol of the aldehyde or ketone.

Heat the solution below the boiling point for at least 10 min. Then cool it, eventually in an ice bath. Collect the crystals, wash with a little ice-cold water, then recrystallize the semicarbazone from ethanol, ethanol-water, or water.

B. Derivatives of Alcohols

1. 3,5-Dinitrobenzoate Esters

The 3,5-dinitrobenzoate (DNB) esters are normally formed by heating the alcohol with 3,5-dinitrobenzoyl chloride in the presence of a base such as pyridine. However, not only is the DNB-chloride difficult to obtain commercially, but the DNB-acid chloride reagent reacts readily with water on storage and therefore soon becomes unusable. For these reasons the DNB-ester will be prepared by first converting the DNB-acid to the acid chloride through reaction with thionyl chloride/pyridine reagent (see step 1 below) and then by treating the acid chloride with the alcohol in the presence of pyridine (step 2).

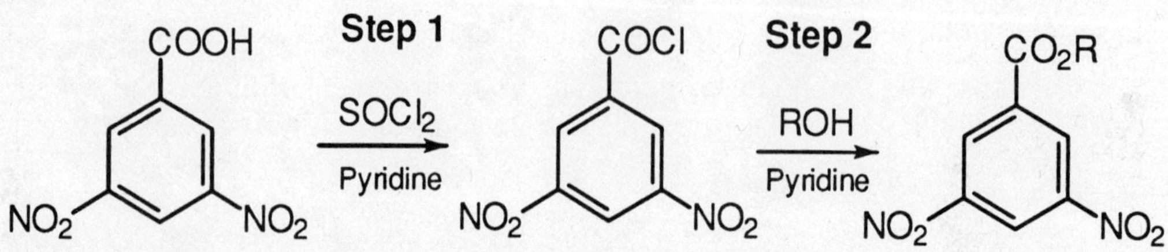

Primary alcohols react in a few minutes, secondary alcohols take 10–20 minutes, and tertiary alcohols take one-half to one hour. Pyridine not only neutralizes the HCl formed, but it actually catalyzes the reactions because it forms a reactive complex with the thionyl chloride reagent and it forms a similar "acyl complex" with the acid chloride.

Some time may be saved by preheating the aluminum block on the hot plate to 130–140°C. **Make sure that all equipment is dry before use.**

Step 1: Weigh 106 mg (0.50 mmol) of 3,5-dinitrobenzoic acid into a dry 5-mL conical vial. Then add 5 microdrops (*ca* 0.6 mmol) of pyridine and 7 microdrops (*ca* 0.8 mmol) of thionyl chloride. Attach a dry air condenser (tighten locknut) and heat the vial on the hot-plate at 130–140°C. Periodically remove the vial, shake it to effect solution, then replace the vial in the heatwell. Gas evolution should occur (frothing). Within two or three minutes the solid should disappear and be replaced by a clear yellow liquid. Continue heating for another 10 minutes.

Step 2: Remove the vial from the heatwell and allow it to cool briefly until the vial can be handled safely. Then remove the air condenser and add *ca* 1.0 mmol of the "unknown" alcohol. (If the unknown is a liquid, assume that one microdrop weighs *ca* 9 mg.) Re-attach the air condenser, shake the vial briefly, then heat it again for 5–10 minutes. A clear solution usually results.

Work-up: Cool the vial until it can be handled safely. In some cases the contents may solidify. Remove the air condenser and add cold water until the vial is *ca* 3/4 full. The product should solidify. Using a microspatula, break up the solid thoroughly and then collect the product using vacuum filtration and a Hirsch funnel. Wash the solid first with a little cold water. Then wash it well with a few mL of 5% potassium carbonate solution to dissolve and remove any of the starting DNB-acid that may be present. To do this, turn off the vacuum, then add the potassium carbonate solution and allow it to stand for one or two minutes before reapplying the vacuum. Then wash the solid on the funnel several times with water and collect the solid. The 3,5-dinitrobenzoate esters can generally be recrystallized from ethanol or ethanol-water mixtures.

2. 1-Naphthylurethanes and Phenylurethanes

Alcohols react with aryl isocyanates to give urethanes, which are usually crystalline solids.

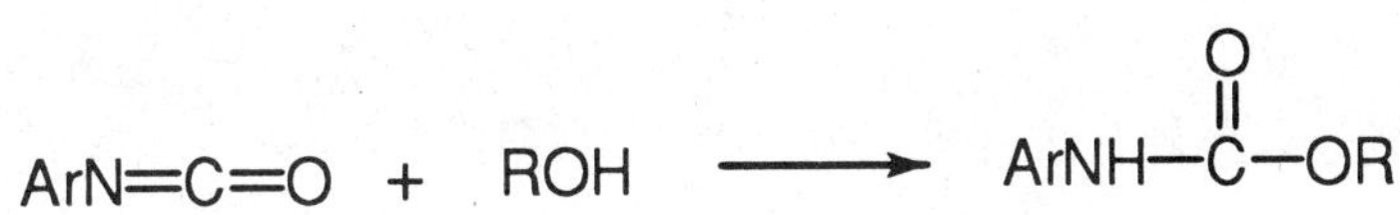

All reagents and glassware must be absolutely free of water. Anhydrous conditions are required because water reacts with the isocyanate to give a primary amine which rapidly reacts with more isocyanate to give *sym*-diaryl urea.

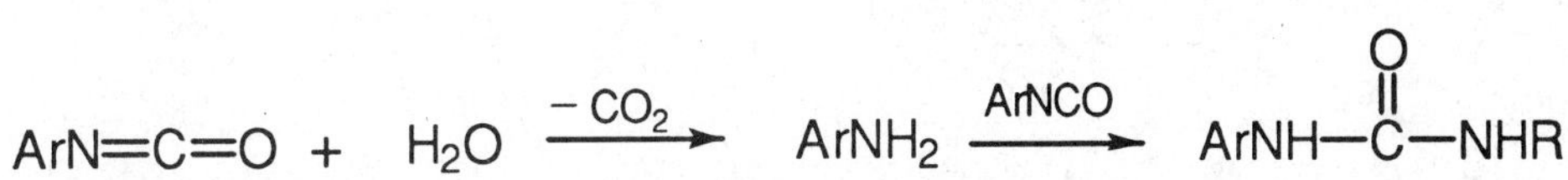

This urea derivative can be difficult to remove from the desired urethane.

Primary alcohols react readily with isocyanates. Secondary alcohols react more slowly and may need a Lewis acid catalyst. Tertiary alcohols may dehydrate (giving water which reacts with the isocyanate). Using sodium acetate as a catalyst and prolonged heating will usually effect reaction while avoiding dehydration.

Primary and secondary alcohols: Mix 1.0 mmol of the isocyanate (170 mg of 1-naphthylisocyanate, or 120 mg of phenylisocyanate) and 1.2 mmol of the alcohol unknown in a dry vial. Allow the mixture to stand for a few minutes. Note if a reaction (heating) occurs.

If no reaction appears to occur, warm the mixture on a hot plate (not over 100°C) for 10 or 15 min.

Work-up: Cool the mixture and add *ca* 2 mL of heptane. Scratch the vial with a stirring rod to induce crystallization. Collect the solid and recrystallize it from heptane. If the solid does not all dissolve in hot heptane, filter the solution while hot through a microfilter to remove the insoluble *sym*-diarylurea. Then allow the solution to cool. Collect the crystals of urethane which should form.

Note: If the reaction fails (with a secondary alcohol), repeat the procedure but add *ca* 20 mg of anhydrous $AlCl_3$ before heating. Then proceed as above.

Tertiary alcohols: Follow the first procedure for primary and secondary alcohols but add *ca* 20 mg of **anhydrous** sodium acetate before heating. Then heat the mixture for 1–2 hr. Add heptane and heat on the hot plate, then filter while hot. Cool the filtrate and induce crystallization by scratching. Collect the product and recrystallize it from heptane. (Filter while hot if necessary).

C. Derivatives of Phenols

1. 3,5-Dinitrobenzoates

Phenols react with acid chlorides in the same way as alcohols, only more slowly. The procedure for tertiary alcohols should be followed, heating for 0.5 to 1 hour. Avoid allowing water to enter in the system.

2. Bromo-Derivatives

Phenols with any unsubstituted "active" positions (i.e., ortho, para positions with H) readily undergo ring bromination with bromine water, as we learned in the bromination of phenol. That procedure is followed here.

Dissolve *ca* 50 mg of the phenol in 1 to 2 mL of water. If the phenol is not soluble in water, dissolve it in 1 mL of methanol or 1,2-dimethoxyethane; then add 0.5 to 1 mL of water. To the solution of the phenol, add a 20% solution of bromine in acetic acid dropwise with thorough mixing until a permanent bromine color persists. Then add a few drops of $NaHSO_3$ solution to reduce the excess bromine. Collect the product, wash it well with water and recrystallize it from an appropriate solvent (usually ethanol).

3. 1-Naphthyl and Phenylurethanes

Mix 1.0 mmol of the isocyanate (170 mg of 1-naphthylisocyanate, or 120 mg of phenylisocyanate) with *ca* 1.2 mmol of the phenol unknown and 2 microdrops of pyridine in a dry vial. Attach an air condenser and heat the vial on the hot plate (set at *ca* 130°C) for 10–15 min with occasional swirling. Then follow the work-up procedure outlined above for alcohols.

D. Derivatives of Amines

1. Benzenesulfonamides

Benzenesulfonamides of primary and secondary amines can be prepared by the procedures used in the classification test with benzenesulfonyl chloride. Scale-up the reaction to about 100 mg of the amine. Collect the benzenesulfonamide and recrystallize after washing well with water. Ethanol is usually suitable as a solvent.

$$RNH_2 + C_6H_5SO_2Cl \xrightarrow[\text{2. HCl}]{\text{1. NaOH}} C_6H_5SO_2NHR$$

2. Benzamides

Primary and secondary amines react with benzoyl chloride to give benzamides which usually are readily crystallized.

$$RNH_2 + C_6H_5\overset{\overset{\displaystyle O}{\|}}{C}Cl \xrightarrow{\text{aq NaOH}} C_6H_5\overset{\overset{\displaystyle O}{\|}}{C}NHR$$

To *ca* 100 mg of the amine, add 1 mL of 10% NaOH and 0.4 g of benzoyl chloride. A small vial is convenient. Seal the vial and shake vigorously for several minutes (5–10 min). Then dilute with 2 or 3 mL of water and collect the product. Wash it with a little dilute NaOH (dilute the 10% reagent to ca 1% or less) and then wash it with a little dilute acetic acid followed by water. Recrystallize from ethanol or ethanol-water.

3. Picrates

Most amines form crystalline salts with picric acid (2,4,6-trinitrophenol) which are called "picrates." These salts usually have useful melting points especially for tertiary amines.

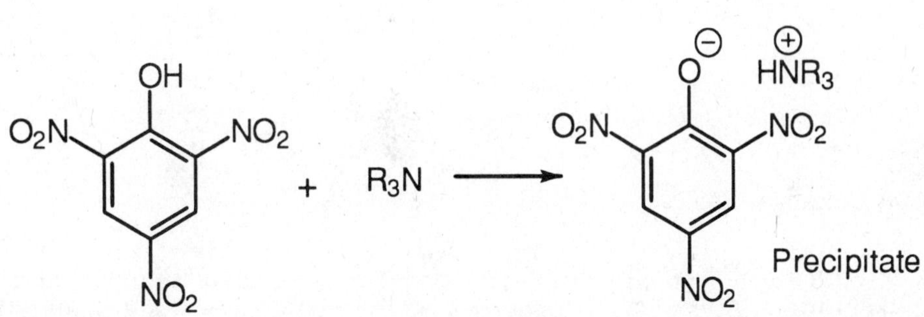

Dissolve *ca* 100 mg of the amine in *ca* 2 mL of ethanol and then add *ca* 2 mL of ethanol (or methanol) saturated with picric acid. Heat the mixture to boiling and then let it cool slowly. Chill in an ice bath, collect the crystals, and wash with a little cold ethanol. The picrates, as formed, are generally sufficiently pure for characterization. Occasionally a picrate may come out as an oil. Extended cooling, scratching the walls of the vial, and much patience may be needed before a crystalline solid forms. Some picrates can be recrystallized from ethanol; others may be too soluble and require mixed solvent systems.

4. Methiodides

Tertiary amines react with methyl iodide to form crystalline quaternary ammonium salts (methiodides) which are useful derivatives. For present purposes, methyl iodide (bp 42°C) is best stored and used as a solution in tetrahydrofuran (bp 67°C).

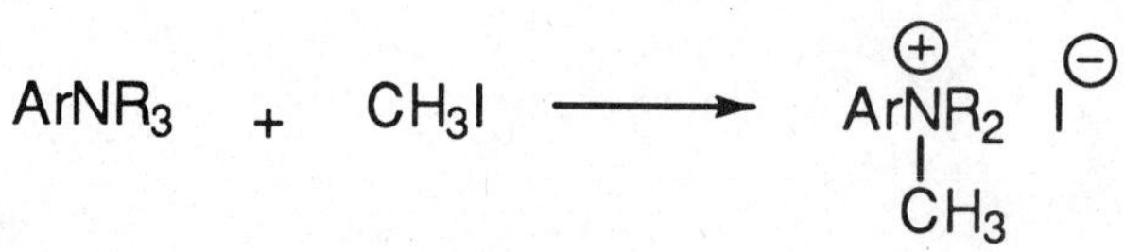

Dissolve *ca* 100 mg of the tertiary amine in 1 mL of 2 **M** methyl iodide in tetrahydrofuran. Seal the vial and allow it to stand at room temperature. Crystals should appear in a few minutes. When the reaction appears to be over, chill the mixture in an ice bath, collect the crystals, and wash them with ice-cold tetrahydrofuran (or diethyl ether). In spite of the fact that they are usually high-melting solids, some methiodides are slow to crystallize. Continued cooling and scratching should induce crystallization. Recrystallization is usually not necessary, but methiodides can be recrystallized from ethanol-tetrahydrofuran.

E. Derivatives of Carboxylic Acids

Carboxylic acids can be converted to acid chlorides by reaction with thionyl chloride, $SOCl_2$, (bp 78°C). The acid chloride can be used to make simple amides or anilides. Alternatively, the acid can be converted to its potassium salt and this can be used to make a crystalline ester by an S_N2 reaction with *p*-nitrobenzyl chloride.

1. Amides

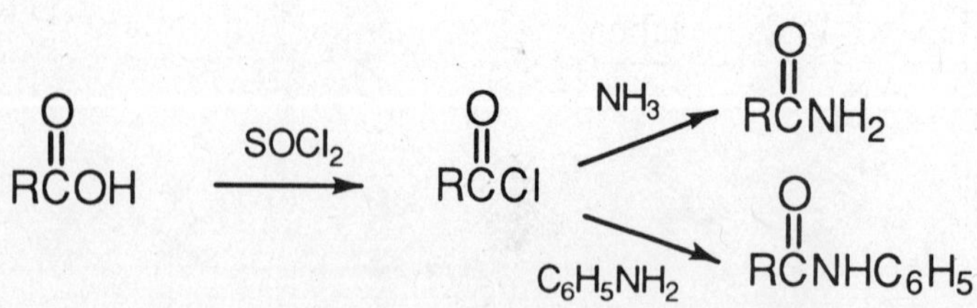

Place *ca* 100 mg of the carboxylic acid in a conical reaction vial and add 0.5 mL of thionyl chloride. Attach an air condenser and heat the vial on a hot-plate (80-90°C) for *ca* 20 min. Then remove the air condenser and evaporate the excess thionyl chloride.

Chill 3 or 4 mL of **concd.** aq. ammonia ("ammonium hydroxide") in a small beaker, then add the acid chloride with a micropipet, and stir well. When no further reaction occurs, collect the product and wash it with cold water. Recrystallize it from ethanol, ethanol-water, or water.

2. Anilides

Note: It is important to remove all excess thionyl chloride before treating the acid chloride with aniline (as the aniline reacts with thionyl chloride to give dark by-products).

Prepare the acid chloride as above. Then remove any remaining thionyl chloride by heating under reduced pressure for a few minutes. Alternatively add *ca* 1 mL of cyclohexane (bp 81°C) and then boil off the cyclohexane. Any remaining thionyl chloride distills with the cyclohexane. Then add *ca* 1 mL of pyridine and *ca* 200 mg of aniline. Heat the mixture on the steam bath for *ca* 5 min. Then pour the mixture into *ca* 5–10 mL of cold water. Stir well. The product should separate, perhaps as an oil which solidifies on cooling in ice. Collect the product and wash it with a little cold dilute acetic acid, then cold water. Recrystallize the product from ethanol or ethanol-water.

3. *p*-Nitrobenzyl Esters

The S_N2 reaction of the salt of an acid is facilitated by keeping the reactants as concentrated as possible. It is desirable to use equivalent amounts of the reactants or a slight excess of the carboxylic acid salt to avoid contamination of the product by the p-nitrobenzyl bromide. The carboxylic acid is neutralized with KOH, rather than with NaOH, because potassium salts tend to be more soluble than sodium salts in ethanol.

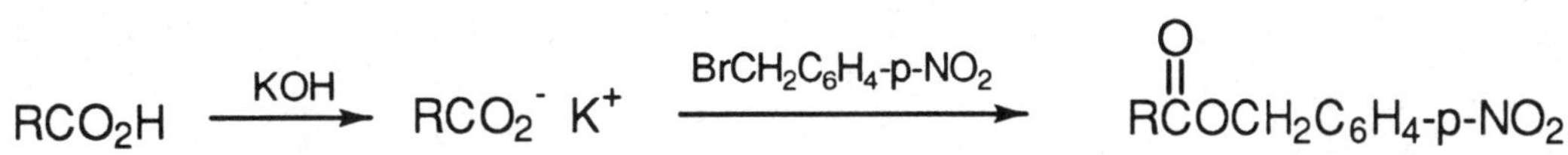

Weigh 1 mmol of the carboxylic acid into a conical reaction vial. Add a few drops of ethanol and one drop of phenolphthalein indicator. Then, using a micropipet, titrate with dilute KOH in ethanol (1 to 2 M) until the solution just turns faintly pink. Then add 1 mmol of *p*-nitrobenzylbromide (mol. wt. 216). Attach an air condenser and heat the mixture at reflux for at least one-half hour. Cool the mixture in an ice bath after adding *ca* 0.5 mL of water (needed to dissolve the KBr which forms). The product should separate as a solid. Collect the product and recrystallize it from ethanol or ethanol-water.

Dispose of waste chemicals in labeled waste containers provided in the laboratory.

38. Tables of Physical Properties of Compounds and their Derivatives

ALCOHOLS	bp °C	n_D^{20} *25°C	mp °C 1-NU	mp °C 3,5-DNB	mp °C PU
2-Methyl-2-butanol	102	1.405	72	117	42
3-Methyl-2-butanol	114	1.397	109	76	68
2-Pentanol	120	1.406	76	62	--
3,3-Dimethyl-2-butanol	120	1.415	101	107	78
3-Methyl-3-pentanol	123	1.411	84	96	44
2-Methyl-1-butanol	129	1.411	82	70	31
4-Methyl-2-pentanol	132	1.401	88	65	143
2-Hexanol	138	1.413	61	38	--
1-Pentanol	138	1.410	68	46	46
Cyclopentanol	141	1.453	118	115	132
2-Methyl-1-pentanol	148	1.419	76	51	--
1-Hexanol	158	1.418	62	61	42
2-Heptanol	159	1.421	54	49	--
Cyclohexanol	161	1.465*	129	113	82
1-Heptanol	177	1.424	62	47	65
2-Octanol	179	1.426	64	32	114
1-Octanol	195	1.427*	67	62	74
1-Phenylethanol	202	1.528	106	94	92
Benzyl Alcohol	206	1.540	134	113	76
2-Phenylethanol	220	1.524	119	108	80

PHENOLS	mp °C	bp °C	mp °C 1-NU	mp °C 3,5-DNB	mp °C Bromo Derivative	mp °C PU
o-Chlorophenol	7	176	120	143	76 (di)	121
Phenol	42	182	132	146	95 (tri)	126
o-Cresol	31	192	142	138	56 (di)	143
o-Bromophenol	5	195	129		95	--
p-Cresol	36	202	146	189	49 (di)	115
m-Cresol	12	203	128	165	84 (tri)	125
2,4-Dimethylphenol	27	212	135	164	179 (tri)	112
o-Methoxyphenol	28	205	118	142	116 (tri)	136
m-Methoxyphenol		243	129		104 (tri)	--
o-Nitrophenol	45	216	113	155	117 (di)	--
p-Chlorophenol	45	217	166	186	90 (di)	148
p-Ethylphenol	47	219	128	133		120
1-Naphthol	94	280	152	217	105 (di)	178
p-tert-Butylphenol	100	237	110		67 (di)	148
2-Naphthol	123	286	157	210	84	156

1-NU = 1-Naphthylurethane; 3,5-DNB = 3,5-Dinitrobenzoate; PU = Phenylurethane

ALDEHYDES	bp °C	n_D^{20}	mp °C SC	mp °C 2,4-DNP
Butanal	75	1.384	106	123
2,2-Dimethylpropanal	75	1.379	190	209
3-Methylbutanal	92	1.394	107	123
Pentanal	103	1.395		107
Hexanal	131	1.407	106	107
Heptanal	155	1.412	109	108
Octanal	170	1.422	101	106
Benzaldehyde	179	1.545	224	237
p-Methylbenzaldehyde	204	1.545	234	233
p-Methoxybenzaldehyde	248	1.573	209	254
trans-Cinnamaldehyde	252	1.619	215	255
	mp °C			
Phenylacetaldehyde	34		156	121
p-Chlorobenzaldehyde	47		232	265
2-Naphthaldehyde	60		245	270
p-Nitrobenzaldehyde	106		221	320

SC = Semicarbazone; DNP = 2,4-Dinitrophenylhydrazone

KETONES	bp °C	n_D^{20}	mp °C SC	mp °C 2,4-DNP
Butanone	80	1.379	115	146
3-Methyl-2-butanone	94	1.388	114	120
2-Pentanone	102	1.390	112	144
3-Pentanone	102	1.392	139	156
3,3-Dimethyl-2-butanone	106	1.396	158	125
4-Methyl-2-pentanone	117	1.396	132	95
2-Hexanone	128	1.401	125	107
Cyclopentanone	131	1.437	210	146
2-Heptanone	151	1.401	127	89
Cyclohexanone	156	1.451	167	162
2,6-Dimethyl-4-heptanone	168	1.417	122	92
Cycloheptanone	180	1.461	162	148
Acetophenone	202	1.532	199	250
Ethyl phenyl ketone	218	1.527	174	191
Propyl phenyl ketone	230	1.520	188	190
	mp °C			
Benzyl methyl ketone	27		198	156
Benzalacetone	42		186	227
Benzophenone	49		165	238
Benzyl phenyl ketone	60		148	206
Fluorenone	83			284

PRIMARY AND SECONDARY AMINES	bp °C	n_D^{20}	mp °C BSA	mp °C BA	mp °C P
Diisopropylamine	84	1.392	94		140
Piperidine	106	1.453	94	48	152
Morpholine	130	1.455	119	75	148
Cyclohexylamine	134	1.459	89	149	
Aniline	183	1.586	112	163	180
Benzylamine	185	1.540	88	105	199
1-Phenylethylamine	185	1.526		120	
N-Methylaniline	196	1.573	79	63	145
2-Phenylethylamine	198	1.533	69	116	167
o-Toluidine	200	1.569	124	144	213
N-Ethylaniline	205	1.556		60	138
Dibenzylamine	300	1.575	68	112	
	mp °C				
p-Toluidine	45		120	158	181
4-Methoxyaniline	58		96	154	

BSA = Benzenesulfonamide; BA = Benzamide; P = Picrate

TERTIARY AMINES	bp °C	n_D^{20}	mp °C P	mp °C M
Tripropylamine	156	1.418	117	208
N,N-Dimethylbenzylamine	184	1.501	93	179
Tributylamine	212	1.428	106	186
N,N-Dimethylaniline	193	1.588	164	228
N-Ethyl-N-methylaniline	201	1.547	134	125
N,N-Dimethyl-*p*-toluidine	211	1.536	130	220
N,N-Diethylaniline	218	1.542	142	102
	mp °C			
N,N-Dibenzylaniline	70		132	135

P = Picrate; M = Methiodide

CARBOXYLIC ACIDS	bp °C	n_D^{20}	mp °C Amide	mp °C Anilide	mp °C PNBE
Propanoic	141	1.387	79	106	31
2-Methylpropanoic	154	1.392	129	105	
Butanoic	163	1.398	115	96	35
Pentanoic	186	1.409	106	63	
Hexanoic	205	1.416	100	95	
	mp °C				
Cyclohexanecarboxylic	31		186	144	
2,2-Dimethylpropanoic	35		154	133	
Phenylacetic	76		157	118	65
Phenoxyacetic	99		101	101	
Glutaric	98		175	224	69
Oxalic·$2H_2O$	101		419d	246	204
o-Toluic	105		143	125	91
m-Toluic	111		95	126	87
Benzoic	121		129	162	89
Cinnamic	133		147	153	117
Decanedioic (sebacic)	133		209	202	73
Maleic	135		181	187	89
Malonic	135d		170	225	86
Octanedioic (suberic)	142		217	187	85
Diphenylacetic	148		168	180	
Hexanedioic (adipic)	152		220	239	106
1-Naphthoic	162		202	163	
p-Toluic	178		159	146	104
p-Methoxybenzoic	184		162	169	132
2-Naphthoic	185		192	170	
Butanedioic (succinic)	185		260	230	88

PNBE = *p*-Nitrobenzyl Ester

39. Appendix

REPORT FORMS

3. Report: Melting Points

Name ______________________________ Desk No.________

mp Range of Benzoic Acid:	
mp Range of 2-Naphthol:	
mp of Mixtures of Benzoic Acid and 2-Naphthol: 1:1 Benzoic Acid 2-Naphthol 1:10 Benzoic Acid 2-Naphthol 10:1 Benzoic Acid 2-Naphthol	 __________ __________ __________
mp Range of Unknown Sample:	
Possible compounds and their melting points:	
Mixed mps of Unknown Sample with Possible Compounds:	
Identity of Unknown and Evidence for this Conclusion:	

4. Results: Crystallization

Name ______________________________ Desk No.________

One Gram Scale — Acetanilide
Weight of Acetanilide Used
Weight of Acetanilide Recovered________ mp__________°C
Recovery (%): Show Calculations.

One–tenth Gram Scale — Naphthalene
Weight of Naphthalene Used
Weight of Naphthalene Recovered__________ mp__________°C
Recovery (%): Show Calculations.

Turn in to your Instructor.

5. Results: Distillation

Name ________________________________ Desk No.________

Record of Temperature *vs*. Volume During Distillation

Volume Distilled mL	Temperature °C	Volume Distilled mL	Temperature °C	Volume Distilled mL	Temperature °C
1st drops		8		16	
1		9		17	
2		10		18	
3		11		19	
4		12		20	
Sample 2		13		21	
5		Sample 3		22	
6		14		23	
7		15		Sample 4	Stop

Type of Distillation, Simple_______, Fractional ________.
Boiling points
Fraction 1, bp_________°C
Fraction 2, bp_________°C
Fraction 3, bp_________°C
Fraction 4, bp_________°C
Fraction 5, bp_________°C
Micro bp, Fraction Number________, bp___________°C

Graph of the Distillation

Plot Temperature *vs*. Volume Distilled

Type of Distilllation, Simple __________, Fractional __________.

TEMPERATURE

Volume Distilled

Plot the distillation curve on graph paper and turn in to your instructor.

6. Results: Gas Chromatography

Name ______________________________ Desk No.________

Sample No.	bp, °C	Retention times (mm)	Peak Areas	Composition %	R.I. n_D
1.		MA			
		EA			
		PA			
2.		MA			
		EA			
		PA			
3.		MA			
		EA			
		PA			
4.		MA			
		EA			
		PA			
5.		MA			
		EA			
		PA			

On the basis of the refractive index of sample #3, does this sample seem to be a pure, or a nearly pure, component, or a complex mixture? Does the GC analysis support this conclusion? Using the composition of #3 as measured by the GC, and the refractive indices of the pure components, calculate the expected refractive index of sample #3 by means of the equation, $\boldsymbol{n} = \Sigma \boldsymbol{n}_i \mathbf{W}_i$, that was described above.

Refractive index literature values at 20°C: Methyl Acetate 1.3614; Ethyl acetate 1.3720; Propyl acetate 1.3842.

Calculations: (Use reverse side of sheet.)

Turn in to your instructor together with the gas chromatograms.

7. Results: Thin-Layer Chromatography

Name ______________________________ Desk No.________

Analysis of Drug Store Items

Knowns	R_f values
Aspirin	
Acetaminophen	
Ascorbic Acid	
Unknown	
Identity of unknown Mixture	

Analysis of Paprika pigments

Spot Number	R_f	Color

Turn in report to your instructor.

8. Results: Extraction, Separation of an Unknown Mixture

Name ______________________________ Desk No.________

Assume that unknown mixtures are 50:50 by weight.

Macroscale			
Solvent Used:			
Starting Weights:			
Compound	Weight recovered	Percent recovered	mp °C
Organic Acid			
Neutral Compound			

Results: Extraction, Separation of an Unknown Mixture

Microscale			
Solvent Used:			
Starting Weights:			
Compound	Weight recovered	Percent recovered	mp °C
Organic Acid			
Neutral Compound			

Mixed Melting Point Data

Name of Neutral Component of the Mixture________________

Name of Acid Component of the Mixture________________

13. Results, Preparative Scale GLC. Separation of Carvone and Limonene from Caroway Seed Oil

Name ______________________________ Desk No.________

Volume of Caroway Seed Oil Injected __________µL
Wt. of Fraction #1 ________mg
Wt. of Fraction #2 ________mg
Identity of Fraction #1 Based on IR Analysis ____________________
Identity of Fraction #2 Based on IR Analysis ____________________
Vol. of Carvone (Density 0.959 g/mL) ________µL
Vol. of Limonene (Density 0.840 g/mL) ________µL
Percent of Total Recovery ________%
Infrared Analysis: Fraction #1
Infrared Analysis: Fraction #2

14. Results: S_N1 and S_N2 Reactions

Name ________________________________ Desk No.________

Part 1. Synthesis of *p*-Nitrobenzylthiocyanate
Mol. Wts.: *p*-Nitrobenzylbromide _________; *p*-Nitrobenzylthiocyanate_______
Weight of *p*-Nitrobenzylbromide: Weight_______mg; mmol________
p-Nitrobenzylthiocyanate: Theoretical _______mmol; ________mg
Weight of *p*-Nitrobenzylthiocyanate obtained ___________mg
% Yield__________%
mp of Product:

Part 2. Relativity reactivities by S_N1 and S_N2 processes.			
Tube No.	Organic Halide	Formation of Precipitate **List Both Time and Temperature**	
		S_N2 18 % NaI	S_N1 1% $AgNO_3$
1			
2			
3			
4			
5			
6			
7			
8			
9			
10			
11	Unknown		

Turn in to your instructor.

15. Results, E2 Elimination Reactions, Dehydrobromination of the Butyl Bromides, C_4H_9Br

Name ______________________________ Desk No.________

Starting Alkyl Bromide	Retention times and Percent Composition of Evolved Alkenes			
	1-Butene	*cis*-2-Butene	*trans*-2-Butene	2-Methyl-propene
1-Bromobutane				
2-Bromobutane				
2-Bromo-2-methylpropane				
1-Bromo-2-methylpropane				

Compare the observed products with the products expected based on the mechanism of E2 elimination.

1-Bromobutane

2-Bromobutane

2-Bromo-2-methylpropane

1-Bromo-2-methylpropane

16. Results, E1 Elimination Reactions, Dehydration of Butyl Alcohols

Name ______________________________ Desk No.________

Alcohol	Retention Times; Percent Composition			
	1-Butene	*cis*- 2-Butene	*trans*- 2-Butene	Isobutene
tert-Butyl Alcohol				
sec-Butyl Alcohol				
Isobutyl Alcohol				
n-Butyl Alcohol				

Compare the observed products with the products expected based on the mechanism of the reaction (see Figures 3, 4, 5 and 6).

tert-Butyl Alcohol

sec-Butyl Alcohol

Isobutyl Alcohol

n-Butyl Alcohol

17. Results, Hydroboration-Oxidation of Alkenes

Name ______________________________ Desk No.________

<table>
<tr><td colspan="3">Molecular Weights: Octene__________; Octanol__________</td></tr>
<tr><td colspan="3">Molecular Weights: Styrene__________; Phenylethanol__________</td></tr>
<tr><td colspan="3">Starting Amounts: mg___________, mmol________</td></tr>
<tr><td colspan="3">Theoretical Yield: mg________, mmol______</td></tr>
<tr><td colspan="3">Actual Yield: mg_______, mmol_______</td></tr>
<tr><td colspan="3">Percent Yield:_________%</td></tr>
<tr><td>GLC Retention times</td><td>Alcohol________________
T_R__________</td><td>Alcohol________________
T_R__________</td></tr>
<tr><td colspan="3">Ratio of Primary to Secondary Alcohol</td></tr>
<tr><td colspan="3">Infrared Peaks and Assignments</td></tr>
</table>

Turn in to your instructor together with the GLC and IR spectrum.

Calculations:

18. Results, Reduction of Ketones to Alcohols with Sodium Borohydride

Name ______________________________ Desk No.________

Reduction of 4-*tert*-Butylcyclohexanone		
Molecular Weights:	4-*tert*-Butylcyclohexanone, ____________ 4-*tert*-Butylcyclohexanol, ____________	
Starting Material: mg_________, mmol_________		
Theoretical Yield: mg_________, mmol_________		
Actual Yield: mg_________, mmol____________		
Percent Yield_________%		
Melting Point Range:________________°C		
GLC Retention times for 4-*tert*-Butylcyclohexanone Isomers	*cis*-	*trans*-
GLC Percentages	Percent *cis*-	Percent *trans*-
Infrared Peaks and Assignments		

Reduction of Benzoin
Molecular Weights: Benzoin_________; Hydrobenzoin_________
Starting Material (Benzoin): mg_________, mmol_________
Theoretical Yield of Hydrobenzoin: mg_________, mmol_________
Actual Yield: mg_________, mmol____________
Percent Yield__________%
Melting Point Range:________________°C

Hand in to your instructor together with the IR spectrum.

19. Results, Stereospecific Bromination of *cis*- and *trans*-Stilbene

Name ______________________________ Desk No.________

Part 1. *trans*-Stilbene Dibromide
Molecular Weights: Stilbene________, Stilbene dibromide________
Starting Material: mg________, mmol________
Theoretical Yield: mg________, mmol________
Actual Yield: mg_________, mmol___________
Percent Yield_________%
Melting Point Range:_______________°C

Part 2. *cis*-Stilbene Dibromide
Starting Material: mg________, mmol________
Theoretical Yield: mg________, mmol________
Actual Yield: mg_________, mmol___________
Percent Yield_________%
Melting Point Range:_______________°C

Turn in to your instructor.

20. Results, Diphenylacetylene and (*E*)-1,2-Dibromostilbene

Name ________________________________ Desk No.________

Part 1. Diphenylacetylene
Mol. Wts.: Stilbene Dibromide________, Diphenylacetylene__________
Starting Material: mg________, mmol________
Theoretical Yield: mg________, mmol_______
Actual Yield: mg_________, mmol___________
Percent Yield_________%
Melting Point Range:_______________°C

Part 2. Debromination of *meso*-Stilbene Dibromide
Results and Explanations:

Part 3. (*E*)-1,2-Dibromostilbene
Mol. Wts.: Diphenylacetylene________, 1,2-Dibromo-1,2-diphenylethene________
Starting Material: mg________, mmol________
Theoretical Yield: mg________, mmol_______
Actual Yield: mg_________, mmol___________
Percent Yield_________%
Melting Point Range:_______________°C

21. Results, Addition Polymerization

Name ______________________________ Desk No.________

Melting Point Range of Poly(methyl methacrylate):______________°C
Infrared peaks (Only those between 4000 and 1500 cm^{-1}) and assignments.
Polystyrene:
Poly(methyl methacrylate):
Changes in the IR spectra between monomer and polymer:
Polystyrene:
Poly(methyl methacrylate):
Comparison of the IR spectra of your polymers with the IR of commercial polymers:
Polystyrene:
Poly(methyl methacrylate):

22. Results, Diels-Alder Reaction

Name ______________________________ Desk No.________

Molecular Weights: Cyclopentadiene______, Maleic Anhydride_______
Cyclopentadiene: __________mg, ________mmol
Maleic Anhydride: __________mg, ________mmol
Theoretical Yield: __________mg, ________mmol
Actual Yield: __________mg, ________mmol
Percent Yield: ________%
Melting Point Range:__________°C

Molecular Weights: Sulfolene______, Maleic Anhydride_______.
Sulfolene: __________mg, ________mmol
Maleic Anhydride: __________mg, ________mmol
Theoretical Yield: __________mg, ________mmol
Actual Yield: __________mg, ________mmol
Percent Yield: ________%
Melting Point Range: _________°C

Molecular Weights: Furan______, Maleic Anhydride_______.
Furan: __________mg, ________mmol
Maleic Anhydride: __________mg, ________mmol
Theoretical Yield: __________mg, ________mmol
Actual Yield: __________mg, ________mmol
Percent Yield: ________%
Melting Point Range:__________°C

Report IR peaks and assignments on the back.

23. Results, Aromatic Substitution I, Bromination of Aniline, Acetanilide, and Phenol

Name ______________________________ Desk No.________

Bromination of Aniline
Molecular Weight: Aniline________; Tribromoaniline_________
Starting Material: _________mg, __________mmol
Theoretical Yield: ________ mg, ________mmol
Actual Yield: ________mg, ________mmol
Percent Yield: _________%
Melting Point Range:________________°C

Bromination of Acetanilide
Molecular Weight: Acetanilide_________ *p*-Bromoacetanilide_________
Starting Material: _________mg, __________mmol
Theoretical Yield: _________mg, __________mmol
Actual Yield: _________mg, __________mmol
Percent Yield: _________%
Melting Point Range:________________°C

Bromination of Phenol
Molecular Weight: Phenol__________ Tribromophenol_________
Starting Material: _________mg, __________mmol
Theoretical Yield: _________mg, __________mmol
Actual Yield: _________mg, __________mmol
Percent Yield: _________%
Melting Point Range:________________°C

Turn in results to your instructor.

24. Results, Aromatic Substitution II: Nitration of Methyl Benzoate: Relative Reactivity of Aromatic Compounds: A Competitive Nitration

Name ______________________________ Desk No.________

Nitration of Methyl Benzoate	
Mol. Wts.	Methyl Benzoate______ Methyl *m*-nitrobenzoate________
Starting Material: ________mg, ________mmol	
Theoretical Yield: ________mg, ________mmol	
Actual Yield: ________mg, ________mmol	
Percent Yield: ________%	
Melting Point Range:_______________°C	
Infrared peaks and assignments:	

Competitive Nitration	
Molecular Weights:	Methyl Benzoate_____, Methyl *m*-Nitrobenzoate________ Acetanilide______, *p*-Nitroacetanilide ________ Nitric Acid ________
Limiting Reagent: _______mg 70% Nitric Acid, _________mmol	
Melting Point Range:________________°C	
Product of the Competitive Reaction:	
Starting Material with Greater Reactivity:	
Actual Yield: __________mg	
Percent Yield (based on nitric acid as the limiting reagent): _________%	

25. Results, The Grignard Reaction: The Synthesis of Triphenylmethanol

Name ______________________________ Desk No.________

Molecular Weights:	Bromobenzene______, Methyl Benzoate______ Triphenylmethanol______, Magnesium______
Magnesium (the limiting reagent*): ________mg, ________mmol * Corrected for the amount that reacts with 1,2 dibromoethane	
Bromobenzene: ________mg, ________mmol	
Methyl Benzoate: ________mg, ________mmol	
Theoretical Yield of Product: ________mg, ________mmol	
Actual Yield: ________mg, ________mmol	
Percent Yield: _________%	
Melting Point Range:________________°C	
Infrared Peaks and Assignments:	

Turn in results to your instructor.

26. Results, Chromic Acid Oxidation of *para*-Nitrotoluene to *para*-Nitrobenzoic Acid

Name ______________________________ Desk No.________

Molecular Weights:	*para*-Nitrotoluene____________ *para*-Nitrobenzoic Acid__________
Starting Material: _________mg, ________mmol	
Theoretical Yield: _________mg, ________mmol	
Actual Yield: _________mg, ________mmol	
Percent Yield: _________%	
Melting Point Range:_______________°C	

Turn in results to your instructor.

27. Results: Oxidation of Alcohols Using a Chromate-Impregnated Resin: Following a Reaction by TLC

Name ______________________________ Desk No.________

Oxidation of 9-Hydroxyfluorene
Molecular Weights: 9-Hydroxyfluorene ________, 9-Fluorenone________
9-Hydroxyfluorene: ________mg, ________mmol
Theoretical Yield of 9-Fluorenone: __________mmol, __________mg
Actual Yield of 9-Fluorenone: _________mg, ________mmol
Percent Yield: _________%
Melting Point Range: __________
Average R_f values: 9-Hydroxyfluorene_________, Fluorenone ________

Oxidation of Benzoin
Molecular Weights: Benzoin ________, Benzil________
Benzoin: ________mg, _________mmol
Theoretical Yield of Benzil: __________mmol, __________mg
Actual Yield of Benzil: _________mg, ________mmol
Percent Yield: _________%
Melting Point Range: ____________
Average R_f values: Benzoin_________, Benzil ________

Evidence for the confirmation of structure of the product:

28. Results, The Haloform Reaction. Oxidation of Aryl Methyl Ketones to Aromatic Carboxylic Acids

Name ______________________________ Desk No.________

Oxidation of an Aryl Methyl Ketone
Starting Material: ________mg
Theoretical yield: ________mg, Assume MW same as starting material.
Actual Yield: ________mg
Percent Yield: ________%
Melting Point Range: ________________°C
Infrared Peaks and Assignments:

Iodoform Test for Methyl Ketones
Results:
Melting Point Range:________________°C

Turn in results to your instructor.

29. Results, Synthesis of Isopentyl Acetate by Fischer Esterification

Name ______________________________ Desk No.________

Molecular Weights:	Acetic Acid Isopentyl Alcohol__________ Isopentyl Acetate__________
Limiting Reagent: __________________________	
Limiting Reagent: ________mg, ________mmol	
Theoretical Yield: ________mg, ________mmol	
Actual Yield: ________mg, ________mmol	
Percent Yield: ________%	
Boiling Point: _______°C Refractive Index__________	
Infrared Peaks and Assignments:	
GLC Results:	

30. Results, Acetylation: Aspirin and N-Acetylanthranilic Acid

Name ________________________________ Desk No.________

Actylsalicylic Acid
Molecular Weights: Salicylic Acid_________, Acetylsalicylic Acid_________
Salicylic Acid: _______mg, _______mmol
Theoretical Yield of Acetylsalicylic Acid: _______mg, _______mmol
Actual Yield of Acetylsalicylic Acid: ________mg
Percent Yield of Acetylsalicylic Acid: ________%
Melting Point Range: _________

N-Acetylanthranilic Acid
Molecular Weights: Anthranilic Acid _______, N-Acetylanthranilic Acid ______
Anthranilic Acid: ________mg, ________mmol
Theoretical Yield of N-Acetylanthranilic Acid: _______mg, _______mmol
Actual Yield of N-Acetylanthranilic Acid: ________mg, ________mmol
Percent Yield of N-Acetylanthranilic Acid: ________%
Melting Point Range:________

31. Results, Preparation of a Liquid Crystal: the Enol Acetate of 4-Cholestene-3-one

Name ______________________________ Desk No.________

Synthesis of Enol Acetate of 4-Cholestene-3-one
Molecular Weights: 4-Cholestene-3-one ______, Enol Acetate: ________
Quantity of 4-Cholestene-3-one: __________mg, ________mmol
Theoretical Yield of Enol Acetate: ________mg, ________mmol
Actual Yield: ________mg, ________mmol
Percent Yield: ________%
mp Range: ______________°C, Colors Observed:

Observations of Liquid Crystal Behavior:
1. Colors Observed on Melting between Glass Slides:
a. Reflected Light:
b. Transmitted Light:
2. Color changes observed when one slide is moved over the other one:
3. Observation of sample between crossed polarizers. What is the observed effect?
4. Observation when the sample is heated to higher temperature:
5. Observation when sample is allowed to cool:

32. Results, The Claisen-Schmidt Reaction: Benzalacetone and Dibenzalacetone

Name ______________________________ Desk No.________

Dibenzalacetone	
Molecular Weights:	Benzaldehyde___________ Acetone____________ Dibenzalacetone____________
Limiting Reagent: ______________________________	
Quantity of Limiting Reagent: ________ mg, ________ mmol	
Theoretical Yield: ________ mg, ________ mmol	
Actual Yield: ________ mg, ________ mmol	
Percent Yield: _________%	
Melting Point Range:________________°C	
Infrared Peaks and Assignments:	

Benzalacetone Semicarbazone
Molecular Weight, Benzalacetone Semicarbazone: __________
Limiting Reagent: ________________________
Quantity of limiting reagent: mg_________, mmol_________
Theoretical Yield: ________ mg, ________ mmol
Actual Yield: ________ mg, ________ mmol
Percent Yield: _________%
Melting Point Range:________________°C

33. Results, Aryldiazonium Ion Reactions: Formation, Displacement, Coupling

Name ______________________________ Desk No.________

Mol. Wts.:	Anthranilic Acid______ Dimethylaniline_______	Salicylic Acid______ o-Iodobenzene_______

Diazotization: Amt. Anthranilic Acid: ________mg, ________mmol

Synthesis of Methyl Red
Amt. of Anthranilic Acid used: _______mg, _______mmol (1/3 of above Amt.)
Amt. of N,N-Dimethylaniline (d = 0.956): _______mL, ______mg, ______mmol
Yield of Methyl Red: ________mg, ________mmol
Theoretical Yield: ________mg, ________mmol
Percent Yield: _______%
Melting Point Range:________________°C
Color in Acid: ____________, Color in Base: ______________

Synthesis of Salicylic Acid
Amt. of Anthranilic Acid used: _______mg, _______mmol (1/3 of above Amt.)
Yield of Salicylic Acid: ________mg, ________mmol
Theoretical Yield: ________mg, ________mmol
Percent Yield: _______%
Melting Point Range:________________°C

Synthesis of *o*-Iodobenzoic Acid
Amt. of Anthranilic Acid used: _______mg, _______mmol (1/3 of above Amt).
Theoretical Yield: ________mg, ________mmol
Yield: ________mg, ________mmol
Percent Yield: _______%
Melting Point Range:________________°C

34. Results, Dyes Produced by Diazo Coupling and Diazo Imaging

Name ______________________________ Desk No.________

Dyes by Diazo Coupling

Coupler	First Color	Final Color

Diazo Imaging

Dye Coupler	Height of Light	Exposure Time	Result

35. Results, Imaging Dyes in Color Photography

Name ______________________________ Desk No.________

Testing Dye Couplers

Dye Coupler	Color

Conversion of a Photograph to a Dye-colored Photograph

Coupler
Color produced

Draw below the chemical structures of the three dyes produced.

36-37. Results, Qualitative Organic Analysis Unknown Report #1

Name ________________________________ Desk No.________

1. Physical Properties

Color: ________________	Odor: ____________________
mp: ______ or Micro bp: _______	Refractive Index (°C): __________

2. Infrared Spectrum

Sample-handling, i.e., thin-film or solution (solvent) or mull:
Significant bands present/absent and interpretation:

3. Solubility

Water Soluble Compounds:

pH of Water-soluble Compounds):
Solubility in Ether:

Water Insoluble Compounds. Check solubility in:

5% aq. HCl	5% aq. NaOH	5% aq. $NaHCO_3$

4. Classification Tests

Jones Oxidation Reagent, 2,4-Dinitrophenylhydrazine Reagent, Tollen's Reagent, Iodoform Test, $FeCl_3$ Test, Hinsberg Test, Nitrous Acid Test, and any others you perform.

Do only those tests which are applicable based on your previous observations. **Perform each test which is appropriate on known compounds to see how a positive and a negative test appears.** Then, run the test on your unknown. For each test you perform, report both observations and conclusions. Use additional pages if necessary.

5. Derivatives

Prepare derivatives which are appropriate for the kind of unknown you have. Also note any additional chemical tests (if any) which may be useful in identifying the unknown.

In the summary, list the names and physical constants for your unknown and the melting points of the derivatives you prepared. Also list the names and physical constants for compounds which might be your unknown and the melting points of the derivatives of these compounds.

	Physical Constants of Derivatives Give type of Derivative	
Type:		
Names of Possible Unknown Compound With mp, or bp and Refractive Index	mp	mp
Unknown:		

Additional chemical tests and conclusions. Attach extra page if necessary.

6. The unknown is:

Reasons for conclusion:

7. Submit this report and a sample of each derivative sufficient for a mp determination to your TA. Secure a label to each mp tube (at the top) with transparent tape and tape them to this sheet.

36-37. Results, Qualitative Organic Analysis Unknown Report #2

Name ______________________________ Desk No.________

1. Physical Properties

Color: ________________	Odor: ________________
mp: ______ or Micro bp: ______	Refractive Index (°C):__________

2. Infrared Spectrum

Sample-handling, i.e., thin-film or solution (solvent) or mull:
Significant bands present/absent and interpretation:

3. Solubility

Water Soluble Compounds:

pH of Water-soluble Compounds):
Solubility in Ether:

Water Insoluble Compounds. Check solubility in:

5% aq. HCl	5% aq. NaOH	5% aq. $NaHCO_3$

4. Classification Tests

Jones Oxidation Reagent, 2,4-Dinitrophenylhydrazine Reagent, Tollen's Reagent, Iodoform Test, $FeCl_3$ Test, Hinsberg Test, Nitrous Acid Test, and any others you perform.

Do only those tests which are applicable based on your previous observations. **Perform each test which is appropriate on known compounds to see how a positive and a negative test appears.** Then, run the test on your unknown. For each test you perform, report both observations and conclusions. Use additional pages if necessary.

5. Derivatives

Prepare derivatives which are appropriate for the kind of unknown you have. Also note any additional chemical tests (if any) which may be useful in identifying the unknown.

In the summary, list the names and physical constants for your unknown and the melting points of the derivatives you prepared. Also list the names and physical constants for compounds which might be your unknown and the melting points of the derivatives of these compounds.

	Physical Constants of Derivatives Give type of Derivative	
Type:		
Names of Possible Unknown Compound With mp, or bp and Refractive Index	mp	mp
Unknown:		

Additional chemical tests and conclusions. Attach extra page if necessary.

6. The unknown is:

Reasons for conclusion:

7. Submit this report and a sample of each derivative sufficient for a mp determination to your TA. Secure a label to each mp tube (at the top) with transparent tape and tape them to this sheet.